P9-DNC-254

Phytochemical Diversity
A Source of New Industrial Products

Phytochemical Diversity

A Source of New Industrial Products

Edited by

Stephen Wrigley

Xenova Limited, Slough, Berkshire UK

Martin Hayes

Glaxo Wellcome Research and Development, Stevenage, Hertfordshire, UK

Robert Thomas

University of Sussex, Brighton, Sussex, UK

Ewan Chrystal

Zeneca Agrochemicals, Bracknell, Berkshire, UK

The proceedings of the symposium, organised by the Biotechnology Group (Industrial Division) of The Royal Society of Chemistry, on 'Phytochemical Diversity: A Source of New Industrial Products' held at the University of Sussex, Brighton, UK on 15–17 April 1996.

Special Publication No. 200

ISBN 0-85404-717-4

A catalogue record for this book in available from the British Library

Published by The Royal Society of Chemistry,
Thomas Graham House, Science Park, Milton Road,
Cambridge CB4 4WF, UK

Printed by Bookcraft (Bath) Ltd

Preface

The past century has seen the development of many varied industrial applications of natural and synthetic organic chemicals which have greatly enhanced both human health and the quality of life. Despite the popular aphorism, 'natural products good - synthetic chemicals bad', there is growing public awareness of the intrinsic value of both categories of compounds, as perceived in products ranging from penicillins to plastics.

Such artificial distinctions are increasingly blurred by technological developments in the production of industrially important organic chemicals, both synthetic and naturally derived. In the natural arena recent developments in genetic manipulation offer many exciting prospects, such as the design of functional mix and match combinatorial hybrid enzymes and enzyme products. However, for the foreseeable future, the unique structural features of complex natural products, particularly those with multiple chiral centres, will ensure their continued exploitation by pharmaceutical, agrochemical and other major industries as economic synthons for the preparation of high value-added products.

From the standpoint of useful sources of natural products, throughout the past fifty years the highly successful commercial development of penicillin and other antibiotics has inevitably focused attention on microorganisms. While initially viewed as the source of new products which had therapeutic or other useful applications *per se*, following the development of disease-related enzyme inhibitor, receptor-based and mammalian cell-based high throughput screening protocols, they now primarily attract attention as sources of novel leads.

Despite the enormous variety of known microbial products and the certainty of many more awaiting discovery, the key to successful screening remains maximising access to structural diversity. This can be achieved through utilisation of the metabolites not only of microbes but also of plants, marine organisms and insects. Although these often appear to be superficially related, most exhibit distinguishing structural characteristics which reflect minor but significant variations on common biosynthetic themes.

Such considerations have stimulated the systematic search of different phyla for potentially useful natural products. Furthermore, this is taking place at a time of growing awareness of the importance of conserving our planet's rapidly declining biodiversity and the need to promote sustainable exploitation of its irreplaceable genetic resources.

While the study of the chemicals produced by plants is arguably the oldest of the natural product sciences, the above concerns, together with the availability of methodology for systematic genetic manipulation and plant tissue culture, have re-kindled industrial interest in phytochemical screening.

In response to these trends, the Biotechnology Group of the Industrial Division of the Royal Society of Chemistry decided that it was timely to hold an international conference in April 1996 on '*Phytochemical Diversity: A Source of New Industrial Products*', the proceedings of which form the subject matter of this book.

The contrasting research approaches of industrial, academic, government and developing country laboratories are presented, together with the views of representatives of major international agencies on the implications of the industrial exploitation of phytochemical diversity, as exemplified by the UN Convention on Biological Diversity.

The primary focus of the conference proceedings concerns the discovery of plant constituents with particular reference to potential utility in the pharmaceutical and agrochemical industries and emphasises recent technological developments. These include automated high throughput screening and the requirement for efficient dereplication procedures, which involve the early recognition both of known compounds and the presence of multiple samples containing the same active chemical. While the high capacity of modern screens is optimally suited to processing large numbers of extracts of plants collected at random, the complementary role of ethnobotanical screening based on established traditional medicine usage is described by leading research contributors from developing countries and by a US company specialising in this strategy.

Collaboration with plant source countries requires the assurance of sustainable supplies and the formulation of equitable agreements which elaborate intellectual property and royalty rights, training and technology transfer options and ensure fair sharing of the benefits of bioprocessing. These are key concerns of the UN Convention on Biological Diversity and form the subject of chapters by representatives of the European Commission Institute for Prospective Technological Studies and the Organisation for Economic Cooperation and Development.

In conclusion, it is a pleasure to acknowledge the valuable contributions of the speakers who have kindly provided manuscripts detailing their presentations and also the Publications Section of the Royal Society of Chemistry for their assistance in presenting these papers for publication.

December 1996

Robert Thomas
Stephen Wrigley
Martin Hayes
Ewan Chrystal

Contents*

* The author who presented the paper at the meeting is italicised

Abbreviations/Acronyms

AIDS	Acquired Immune Deficiency Syndrome
BHC	Benzene hexachloride
CAS	Chemical Abstracts Service
CATIE	Agricultural Technology Research and Educational Centre, Costa Rica
CBD	Convention on Biological Diversity
CCNSC	Cancer Chemotherapy National Service Centre (NCI)
CGIAR	Consultative Group for International Agricultural Research
CNRS	Centre National de la Recherche Scientifique
CNS	Central Nervous System
CoA	Co-enzyme A
COBIDEC	Heteronuclear shift correlation NMR experiment
COMPARE	Computerised Pattern-Recognition Algorithm NCI
COSY	Correlated Spectroscopy (homonuclear shift correlation NMR experiment)
CPATU-EMBRAPA	Eastern Amazonian Agroforestry Research Centre
CRADA	Co-operative Research and Development Agreement (NCI)
CRC	Cancer Research Campaign, UK
CTA	Clinical Trial Agreement (NCI)
DCTDC	Division of Cancer Treatment, Diagnosis and Centres (NCI)
DEREP	Natural product database
DDT	Dichlorodiphenyltrichloroethane (1, 1-bis (4-chlorophenyl)-2,2,2-trichloroethane)
DG	Directorate General (EC)
DMDP	2*R*,5*R*-Dihydroxymethyl-3*R*,4*R*-dihydroxypyrrolidine
DMSO	Dimethyl sulphoxide
DNP	Chapman and Hall Dictionary of Natural Products
DOC	Chapman and Hall Dictionary of Organic Compounds
DTP	Developmental Therapeutics Program (NCI)
ECU	European Commercial Unit

EORTC	European Organisation for Research and Treatment of Cancer
EtOAc	Ethyl acetate
FAO	Food and Agriculture Organisation (UN)
FAST	Forecasting and Assessment in Science and Technology (EC)
GABA	Gamma-aminobutyric acid
GATT	General Agreement on Tariffs and Trade
GBE	*Gingko biloba* extract
GC	Gas Chromatography
GC-MS	Gas Chromatography - Mass Spectrometry
HIV	Human Immunodeficiency Virus
HMBC	Heteronuclear Multiple Bond Connectivity (NMR experiment)
HMQC	Heteronuclear Multiple Quantum Coherence (NMR experiment)
HPLC	High Performance Liquid Chromatography
HTS	High Throughput Screening
IARCs	International Agricultural Research Centres
IC_{50}	Concentration at which 50% inhibition is observed
ICBG	International Co-operative Biodiversity Group (ICBG)
IGER	Institute of Grassland and Environmental Research
INDA	Investigational New Drug Application
INPA-CPPN	Natural Products Division of the National Amazonian Research Institute
IPR	Intellectual Property Rights
IPTS	Institute for Prospective Technological Studies (EC)
IR	Infra-red
LOC	Letter Of Collection (NCI)
MAO	Master Agreement Order (NCI)
MDR	Multi-Drug Resistance
MIC	Minimum Inhibitory Concentration
MRR, XTT	Colorimetric Assays for Cell Viability
MS	Mass Spectrometry

MTA	Material Transfer Agreement
NAPRALERT	Natural products database
NCDDG	National Co-operative Drug Discovery Group (NCI)
NCI	National Cancer Institute
NCNPDDG	National Co-operative Natural Product Drug Discovery Group (NCI)
NDA	New Drug Application
NIDDM	Non-Insulin Dependent Diabetes Mellitus
NGO	None Governmental Organisation
NIH	National Institute of Health
NMR	Nuclear Magnetic Resonance
NOESY	Nuclear Overhauser Effect Spectroscopy
NPR	NCI Natural Products Repository, Frederick, Maryland
NSF	National Science Foundation
OD	Optical Density
OECD	Organisation for Economic Co-operation and Development
Pgp	P-glycoprotein
RSM	Respiratory Syncytial Virus
SAR	Structure Activity Relationship
TCA	Tricarboxylic acid
SRB	Sulforhodamine B
TRIP	Trade-Related Intellectual Property
USAID	US Agency for International Development
USDA	United States Department of Agriculture
UV	Ultra-Violet
WTO	World Trade Organisation

Screening of Natural Products of Plant, Microbial and Marine Origin: The NCI Experience

G. M. Cragg[1], M. R. Boyd[1], M. A. Christini[1], R. Kneller[3], T. D. Mays[2], K. D. Mazan[2], D. J. Newman[1] and E. A. Sausville[1]

[1] DEVELOPMENTAL THERAPEUTICS PROGRAM, DIVISION OF CANCER TREATMENT, DIAGNOSIS AND CENTERS AND [2] OFFICE OF TECHNOLOGY DEVELOPMENT, NATIONAL CANCER INSTITUTE, BETHESDA, MD, 20892, USA

1 INTRODUCTION

The United States National Cancer Institute (NCI) was established in 1937, its mission being "to provide for, foster and aid in coordinating research related to cancer." In 1955, NCI set up the Cancer Chemotherapy National Service Center (CCNSC) to coordinate a national voluntary cooperative cancer chemotherapy program, involving the procurement of drugs, screening, preclinical studies, and clinical evaluation of new agents. By 1958 the initial service nature of the organization had evolved into a drug research and development program with input from academic sources and massive participation of the pharmaceutical industry. The responsibility for drug discovery and preclinical development at NCI now rests with the Developmental Therapeutics Program (DTP), a major component of the Division of Cancer Treatment, Diagnosis and Centers (DCTDC). Thus, NCI has, for the past forty years, provided a resource for the preclinical screening of compounds and materials submitted by grantees, contractors, pharmaceutical and chemical companies, and other scientists and institutions, public and private, worldwide, and has played a major role in the discovery and development of many of the available commercial and investigational anticancer agents. [1] During this period, more than 400,000 chemicals, both synthetic and natural, have been screened for antitumor activity.

Initially, most of the materials screened were pure compounds of synthetic origin, but the program also recognized that natural products were an excellent source of complex chemicals with a wide variety of biological activities. During the early years of the CCNSC, the screening of natural products was concerned mainly with the testing of fermentation products, and, prior to 1960, only about 1,500 plant extracts were screened for antitumor activity. Earlier work on the isolation of podophyllotoxin and other lignans exhibiting *in vivo* activity against the murine sarcoma 37 model from *Podophyllum peltatum* L.,[2,3] and the discovery and development of vinblastine and vincristine from *Catharanthus roseus* (L.) G. Don, [4,5] however, provided convincing evidence that plants could be sources of a variety of novel potential chemotherapeutic agents. A decision was made to explore plants more extensively as sources of agents with antitumor activity, and, in 1960, an interagency agreement was established with the United States Department of Agriculture (USDA) for the collection of plants for screening in the CCNSC program. A small number of animal extracts, mainly of marine origin, were also tested beginning in 1960, but by the end of 1968 only 1,000 animal extracts had been screened. The pace of investigation of marine invertebrates accelerated in the 1970s and, by 1982, over 16,000 extracts had been screened.

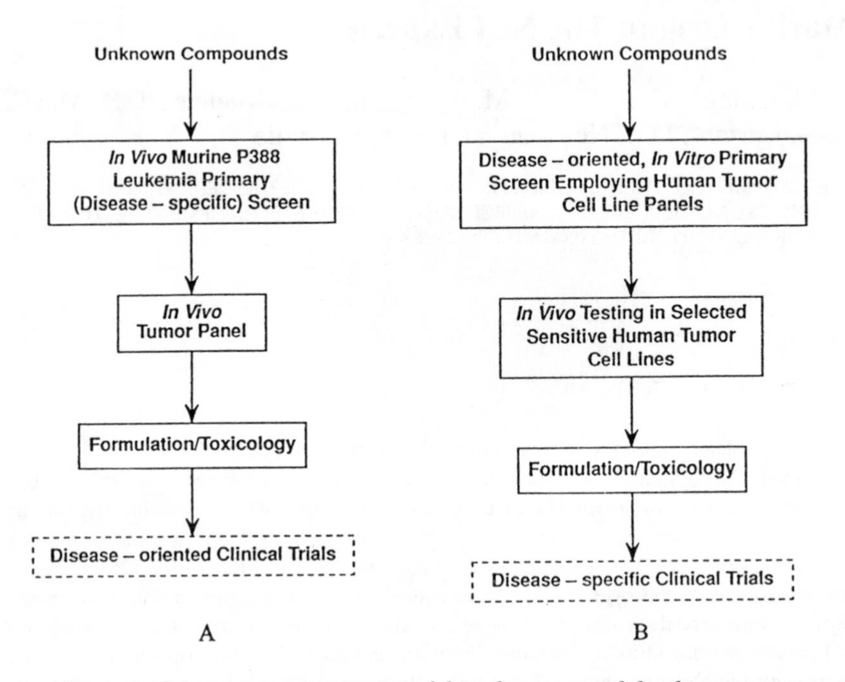

Figure 1 *Schematic representation of drug discovery and development strategy using the P388 primary screen (A) and the* in vitro *human tumor cell line screen (B).*

In contrast, however, from 1960 to 1982, over 180,000 microbial fermentation products and over 114,000 plant-derived extracts were tested for *in vivo* antitumor activity, mainly using the L1210 and P388 mouse leukemia models. Extracts showing significant activity were subjected to bioassay-guided fractionation, and the isolated active agents were submitted for secondary testing against panels comprising four to eight animal tumor models and human tumor xenografts.[6,7] Those agents showing significant activity in the secondary panel were assigned priorities for preclinical and clinical development (Figure 1A). In assessing the performance of these screens it is clear that they have done well in detecting a diverse range of compounds having a variety of mechanisms of action. These compounds include: alkylating agents; purine and pyrimidine antimetabolites; antifolates; mitotic inhibitors; DNA-interactive compounds, including intercalators, alkylators, minor groove binders, single- and double-strand breakers, DNA polymerase inhibitors, and topoisomerase inhibitors, and inhibitors of protein synthesis.[8]

2 COMMERCIAL ANTICANCER DRUGS AND AGENTS IN ADVANCED DEVELOPMENT

Much of the drug discovery effort was carried out through collaborations with research organizations and the pharmaceutical industry, which either submitted compounds on a

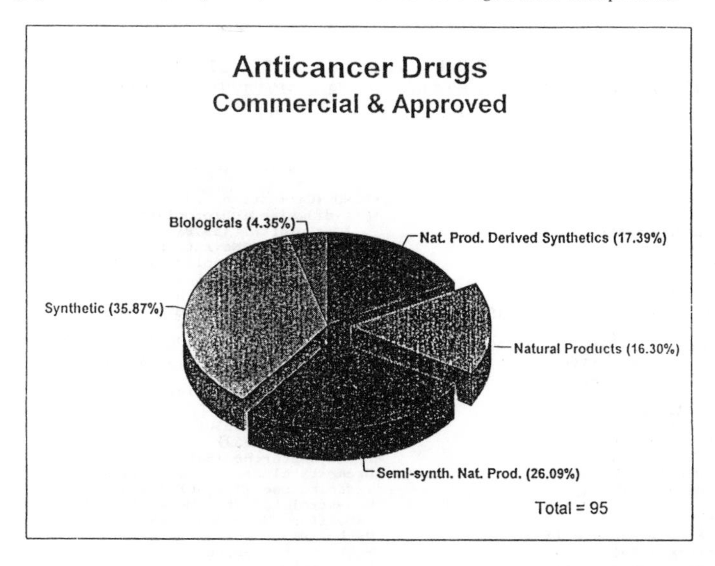

Figure 2

voluntary basis or were supported by NCI through contract or grant funding mechanisms. Of the 95 commercial and approved anticancer drugs currently available, close to 60% may be classified as being of natural origin (Figure 2).[9] The fifteen original natural products (N) include ten microbial and three plant-derived agents, and two steroidal products, while those derived from a natural product (ND), generally through semi-synthesis, include four microbial- and three plant-derived agents, two nucleosides, fourteen steroidal products and a choline derivative. Sixteen have been classified as being synthetics modeled on a natural product parent (S*), and include nucleosides and peptides. Mitoxantrone, which combines the key structural features of anthracyclines and anthracene diones, has also been classified as S*.[10]

While the majority of these drugs were discovered outside the NCI program, the NCI did play a significant role in the development of many of them (Table 1). Two plant-derived agents, paclitaxel (taxol®) and camptothecin, were discovered through the NCI program, and, while camptothecin failed as a clinical candidate in the 1970s, its derivatives, topotecan, 9-amino camptothecin and irenotecan (CPT-11), are currently showing clinical efficacy against a variety of cancer disease types.[11] Both paclitaxel and topotecan are discussed elsewhere in these proceedings. Other plant-derived drugs in clinical trials are homoharringtonine isolated from the small Chinese evergreen tree, *Cephalotaxus harringtonia* var. *Drupacea* (Sieb. & Zucc.) *Koidzumi*, [12] and 4-ipomeanol, a pneumotoxic furan derivative produced by sweet potatoes (*Ipomoea batatas*) infected with the fungus, *Fusarium solani*.[13] Homoharringtonine has shown activity against various leukemias,[14] while ipomeanol is in early clinical trials for treatment of patients with lung cancer[15]. A number of plant-derived agents were entered into clinical trials by the NCI, but the trials were terminated due to lack of efficacy or unacceptable toxicity.[16] Amongst these agents were acronycine, bruceantin, maytansine and thalicarpine, all of which could serve as cytotoxins for linking to monoclonal antibodies targeted to specific tumors (Section 8.3).

ALL ANTICANCER AGENTS THAT HAVE RECEIVED FDA APPROVAL FOR MARKETING IN THE UNITED STATES (As of January 1, 1996)

Alkylating Agents - 12

Mechlorethamine (Mustargen) (1949)
TEM (1953)
Busulfan (Myleran) (1954)
*Chlorambucil (Leukeran) (1957)
*Cyclophosphamide (Cytoxan) (1959)
*Thiotepa (1959)
Uracil Mustard (1962)
*Melphalan (L-PAM, Alkeran) (1964)
Pipobroman (Vercyte) (1966)
*Streptozotocin (Zanosar) (1982)
*Ifosfamide (Ifex) (1988)
*Melphalan (IV) (1993)

Antimetabolites - 9

*Mercaptopurine (6-MP) (1953)
*Methotrexate (1953)
Fluorouracil (5-FU) (1962)
*Thioguanine (6-TG) (1966)
*Cytosine Arabinoside (Ara-C) (1969)
*FUDR (1970)
*Fludarabine Phosphate (1991)
*Pentostatin (1991)
*Chlorodeoxyadenosine (1992)

Plant Alkaloids and Antibiotics - 15

Vinblastine (Velban) (1961)
*Vincristine (Oncovin) (1963)
*Actinomycin D (Cosmegen) (1964)
*Mithramycin (Mithracin) (1970)
*Bleomycin (Blenoxane) (1973)
*Doxorubicin (Adriamycin) (1974)
*Mitomycin C (Mutamycin) (1974)
*L-Asparaginase (Elspar) (1978)
*Daunomycin (Cerubidine) (1979)
*VP-16-213 (Etoposide) (1983)
Idarubicin (Idamycin) (1990)
*VM-26 (Teniposide) (1992)
*Taxol® (Paclitaxel) (1992)
Navelbine (1994)
Doxorubicin Liposome Inj. (Doxil) (1995)

Synthetics - 13

*Hydroxyurea (Hydrea) (1967)
*Procarbazine (Matulane) (1969)
*o-p'-DDD (Lysodren, Mitotane) (1970)
*Dacarbazine (DTIC) (1975)
*CCNU (Lomustine) (1976)

Synthetics (continued)

*BCNU (Carmustine) 1977
*Cis-diamminedichloroplatinum (Cis-platin) (1978)
*Mitoxantrone (Novantrone) (1988)
*Carboplatin (Paraplatin) (1989)
Levamisole (Ergamisol) (1990)
Hexamethylmelamine (Hexalen) (1990)
*All-trans retinoic Acid (Vesanoid) (1995)
*Porfimer sodium (Photofrin) (1995)

Hormones and Steroids - 23

Ethinyl Estradiol (1949)
*DES (1950)
Testosterone (1950)
*Prednisone (1953)
*Fluoxymesterone (Halotestin) (1958)
*Dromostanolone (Drolban) (1961
*Testolactone (Teslac) (1970)
Megestrol Acetate (Megace) (1971)
Tamoxifen (Nolvadex) (1978)
*Methyl Prednisolone
Methyltestosterone
*Prednisolone
Triamcinolene
Chlorotrianisene (TACE)
Hyroxyprogesterone (Delalutin)
Aminoglutethimide (Cytadren)
Estramustine (Emcyt)
Medroxyprogesterone Acetate (Depo-Provera, Provera)
Leuprolide (Lupron) (1985)
Flutamide (Eulexin) (1989)
*Zoladex (1989)
Bicalutamide (Casodex) (1995)
Anastrozole (Arimidex) (1995)

Biologicals - 5

*Alpha Interferon (Intron A, Roferon-A) (1986)
*BCG (TheraCYs, TICE) (1990)
*G-CSF (1991)
*GM-CSF (1991)
*Interleukin-2 (Proleukin) (1992)

Grand Total - 77

*IND sponsored by the National Cancer Institute (total INDs sponsored by NCI = 52)

Table 1

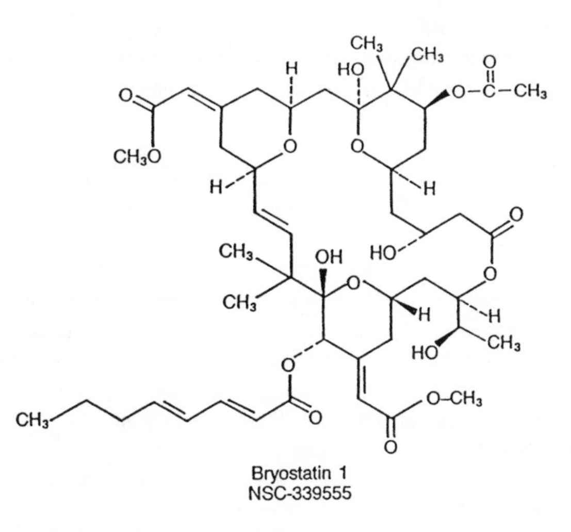

Figure 3

Many of the commercial drugs of microbial origin, such as actinomycin D, doxorubicin (adriamycin) and mitomycin c, were discovered by research groups associated with the pharmaceutical industry, and this trend continues, generally in close collaboration with the NCI in the developmental phases. Much of the drug discovery effort in the marine area, however, has been supported by the NCI through contract or grant mechanisms. While no marine-organism-derived agent has yet been approved for commercial development, several agents, including bryostatin 1 (Figure 3) and didemnin B, are in clinical trials;[17] bryostatin 1 is showing some promising activity in trials against melanoma.[18]

Most of the drugs currently available for cancer therapy are effective predominantly against rapidly proliferating tumors, such as leukemias and lymphomas, but, with some notable exceptions such as paclitaxel, show little useful activity against the slow-growing adult solid tumors, such as lung, colon, prostatic, pancreatic, and brain tumors. In retrospect, these results might be attributed to the use of a single disease-specific model as the primary screen that filtered out those agents with potential specificity against tumors other than mouse leukemia or closely related human diseases.

In an attempt to overcome this deficiency, NCI has developed an alternative, disease-oriented, preclinical anticancer drug discovery strategy aimed at the discovery of new agents for disease-specific clinical trials in relevant cancer patient populations.[19]

3 PRIMARY ANTICANCER SCREENING AT THE NCI: CURRENT STATUS

Cancer comprises an extremely complex and diverse group of diseases that share some common biological characteristics, collectively defined as malignancy. Specific forms of cancer, however, often exhibit distinctive characteristics reflected by differences in heterogeneity, accessibility, size and diffuseness, histologic appearance, growth rates, and other features. From the point of view of chemotherapy, the most critical difference among animal (including human) tumors is that of drug sensitivity. No single drug can be considered truly broad spectrum, and often apparently similar tumors exhibit markedly different drug response patterns. The nature of cancer thus presents serious problems in the development of suitable screening models.

A disease-oriented screening strategy should employ multiple disease-specific (e.g., tumor-type specific) models and should permit the detection of either broad-spectrum or disease-specific activity. The use of multiple *in vivo* animal models for such a screen is not practical, given the scope of requirements for adequate screening capacity and specific tumor-type representation. The availability of a wide variety of human tumor cell lines representing many different forms of human cancer offered, however, a suitable basis for development of a disease-oriented *in vitro* primary screen. Moreover, since many established human tumor cell lines can be propagated *in vivo* in athymic, nude mice, there existed the basis for secondary *in vivo* testing of any agents exhibiting line- or panel-selective cytotoxicity in the primary screen. Such agents would be evaluated *in vivo* in selected cell lines from the primary screen and, if shown to be sufficiently effective *in vivo*, could be advanced through preclinical development (formulation, pharmacology, toxicology) to disease-specific clinical trials (Figure 1B).

During 1985-1990 a new *in vitro* primary screen based upon a diverse panel of human tumor cell lines was developed (Figure 1B).[19] The screen currently comprises sixty cell lines derived from nine cancer types, and organized into subpanels representing leukemia, lung, colon, central nervous system, melanoma, ovarian, renal, prostate and breast. Details of the identities and characteristics of the cell lines, as well as the screening methodology are provided elsewhere; [19,20] suffice it to say that each cell line meets minimal quality assurance criteria (testing for mycoplasma, membrane-associated proteins (MAP), human isoenzyme, karyology, *in vivo* tumorigenicity) and is adaptable to a single growth medium. Mass stocks of each line were prepared and cryopreserved, and these provide reservoirs for replacement of the corresponding lines used for screening after no more than twenty passages in the screening laboratory. After extensive investigation of alternative assays, a protein-staining procedure using sulforhodamine B (SRB) was selected as a suitable method for determination of cellular growth and viability in the screen. [20] Repetitive screening of a set of 175 known compounds, comprising commercially marketed (NDA-approved) anticancer agents, investigational (INDA-approved) anticancer agents, and other candidate antitumor agents in preclinical development, has established the reproducibility of the screening data and permitted development of procedures for quality control monitoring.

In the routine primary screening, each agent is tested over a broad concentration range against every cell line in the current panel. All lines are inoculated onto a series of standard ninety-six-well microtiter plates on day zero, in the majority of cases at 20,000 cells/well, then preincubated in absence of drug for twenty-four hours. Test drugs are then added in five tenfold dilutions, starting from the highest desired concentration, and incubated for a further forty-eight hours. Following this, the cells are fixed *in situ*, washed, and dried. SRB is added, followed by further washing and drying of the stained adherent cell mass. The bound stain is solubilized, and optical densities are measured on automated plate readers, followed by computerized data acquisition, processing, storage and availability for display and analysis.

Each successful test of a compound in the full screen generates 60 dose-response curves, which are printed in the NCI screening data report as a series of composites comprising the tumor-type subpanels, plus a composite comprising the entire panel. Data for any cell lines failing quality control criteria are eliminated from further analysis and are deleted from the

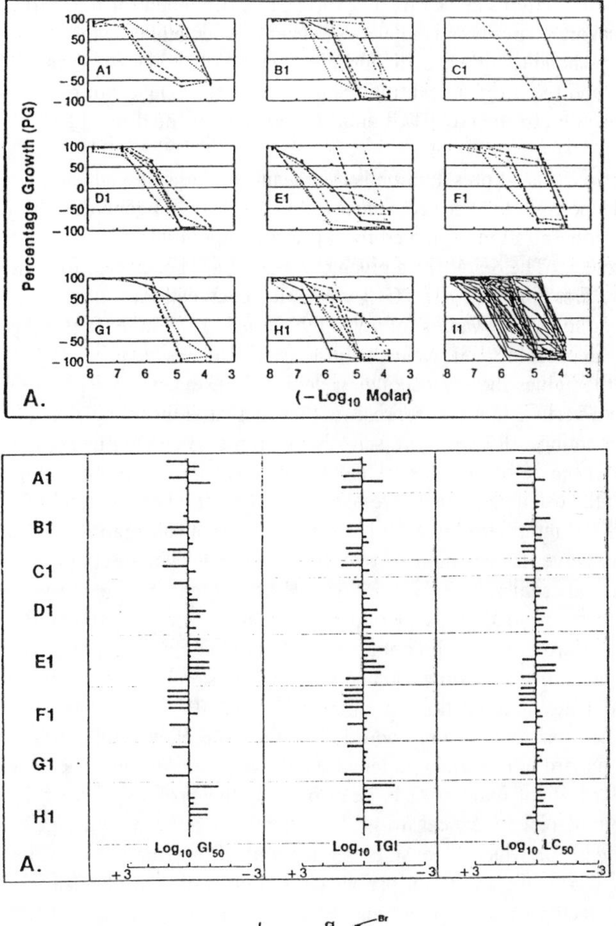

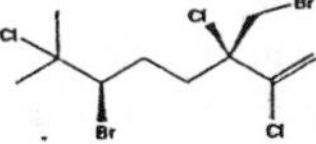

Figure 4 *A composite of nine sets of dose-response curves and the GI_{50}, TGI and LC_{50} mean graphs derived from these curves from the testing of halomon in the NCI in vitro human tumor cell line screen. The subpanel identifiers are: A1, leukemia/ lymphoma; B1, non-small cell lung cancer; C1, small -cell lung cancer; D1, colon cancer; E1, brain tumors; F1, melanoma; G1, ovarian cancer; H1, renal cancer. Graph II is a composite of all the subpanels together. Reprinted with permission of the American Chemical Society from Fuller, et. al., J. Med. Chem., 1992, **35**, 3007.*

screening report. Patterns in the dose response curves obtained for the marine algal-derived compound, halomon, are shown in Figure 4. The "percentage growth" (PG) terms of +50, 0 and -50 correspond to 50% growth inhibition, total growth inhibition, and 50% cell killing, respectively, and the drug concentrations producing these levels of response against each cell line correspond to the GI_{50}, TGI and LC_{50} values for the drug against the relevant cell lines.

An alternative visual presentation is the mean-graph display which is a pattern created by plotting the positive and negative values (termed "deltas") generated from a set of GI_{50}, TGI, or LC_{50} concentrations obtained for a given compound tested against each cell line in the *in vitro* screen. The deltas for a given compound are generated, for example, from the GI_{50} data, by first converting the GI_{50} values for each cell line successfully tested to the corresponding $\log_{10} GI_{50}$ values followed by averaging the individual $\log_{10} GI_{50}$ values to obtain the mean panel $\log_{10} GI_{50}$ value; subtracting each individual $\log_{10} GI_{50}$ value from the panel mean then gives the corresponding deltas. These deltas are plotted horizontally in reference to a vertical line that represents the calculated mean $\log_{10} GI_{50}$ value. Negative deltas, representing cell lines more sensitive than the calculated mean, are plotted to the right of the mean reference line, while the positive deltas, representing cell lines less sensitive than the calculated mean, are plotted to the left. The TGI and LC_{50} mean-graphs are prepared and interpreted in a similar manner. The mean-graphs corresponding to the dose-response curves for halomon are shown in Figure 4. More detailed discussions of the interpretation and usefulness of the mean-graph display, and possible pitfalls in interpreting its significance, are beyond the scope of this presentation, but are given by Boyd and Paull.[19]

Valuable information can be obtained by determining the degree of similarity, or lack thereof, of mean-graph profiles generated on the same or different compounds. A computerized, pattern-recognition algorithm called COMPARE has been developed for this purpose.[21] The consistency and reproducibility of the screen is monitored by comparing the profiles of standard compounds screened repetitively against the same panel of cell lines.

A standard agent database has been built of the profiles of over 170 compounds, including commercial anticancer drugs, investigational anticancer drugs, and other agents in development, for which a considerable amount of information is available about their preclinical and/or clinic antitumor properties and presumed mechanism(s) of action. The profile of a selected standard agent may be used as a seed to probe other available mean-graph databases to see if they contain any closely matching profiles, or, conversely, the profile of a compound from one of these databases may be used to probe the standard agent database to determine if there are any closely matching standard agent profiles. Use of COMPARE has led to the observation that matching mean-graph patterns are often associated with compounds having related chemical structures; furthermore, it has been determined that compounds of either related or unrelated structures exhibiting similar mean-graph patterns frequently share the same or related biochemical mechanisms of action. In this manner, compounds having profiles matching standard agents having known or presumed known mechanisms of action can be identified and selected for further study; compounds of previously unknown mechanisms of action have now been classified into a number of different known mechanistic classes of interest, such as tubulin-interactive antimitotics, alkylating agents, topoisomerase inhibitors, and DNA binders. Such a comparison can also be performed at the natural product extract level permitting the rapid elimination of extracts having mean-graph profiles resembling those of well-known classes of active agents. On the other hand, compounds or extracts which exhibit profiles bearing little or no resemblance to the profile of any of the standard agents, nor to any of the known agents in the entire database of tested compounds, would be given priority for further study.

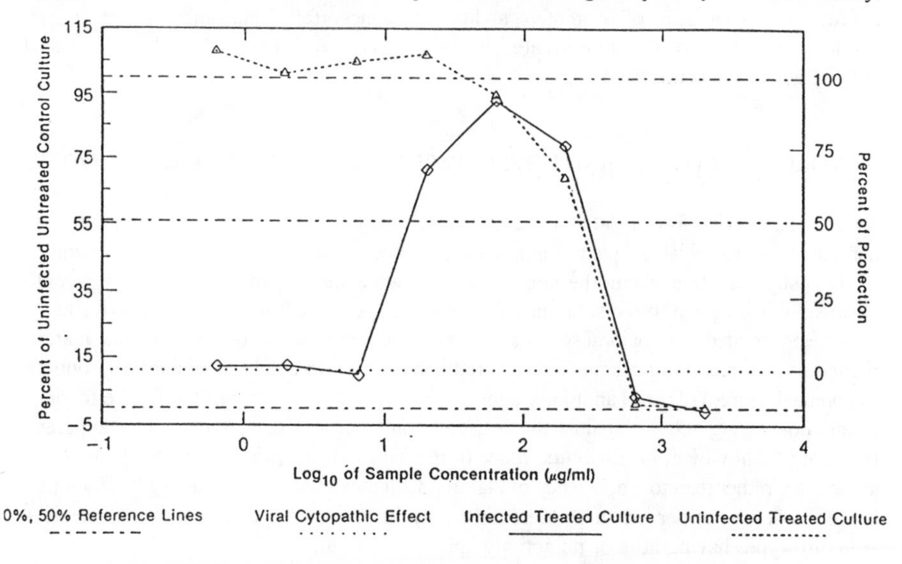

Figure 5 *Anti-HIV activity graph of plant extract*

4 PRECLINICAL ANTI-HIV SCREENING

As part of the response of the National Institutes of Health (NIH) to the AIDS epidemic in 1987, the DTP of the NCI developed a screening program for the large-scale testing of synthetic and natural products for anti-HIV activity.[22] The anti-HIV screening assay[23] uses somewhat similar technology for determination of cellular growth or viability as the *in vitro* human tumor cell line primary screen described above. Human lymphoblastoid cells (CEM-SS cells) are grown in microtiter plate wells in the presence or absence of the human immunodeficiency virus (HIV-1) and in the presence or absence of test material. Anti-HIV activity is indicated by an enhanced growth or survival of the virus-infected cells in the presence of the test material. As an important control, the relative growth or survival of uninfected cells in the presence of the test material gives a measure of the direct cytotoxicity of the test material to the host cells. The degree of cell survival is measured quantitatively by a colorimetric procedure using a soluble tetrazolium reagent (XTT), which is metabolically reduced by viable cells to yield a soluble, colored formazan product. Uninfected, viable cells, or cells that are protected from the cytopathic effects of the virus by the test material and have continued to proliferate, produce the soluble orange formazan, and these cultures give high optical densities (OD). Cells not protected by the test material are killed by the virus and do not proliferate, resulting in less formazan production and lower OD. Data are expressed as percent of formazan produced by untreated control cells.

Results may be expressed graphically as exemplified in Figure 5 for a typical *in vitro* active crude extract. A plot of the percentage untreated control formazan for treated uninfected cells against the log concentration (upper curve) gives a measure of cytotoxicity; the concentration that inhibits formazan production to 50% of that in untreated, uninfected control cells is called the IC_{50} value (IC=inhibitory concentration). A plot of the percentage untreated control formazan for treated infected cells against the log concentration (lower curve) gives a measure of protective ability; the concentration that increases formazan production to 50% of that in untreated, uninfected control cells is called the EC_{50} value (EC=effective concentration).

5 DRUG DISCOVERY AND DEVELOPMENT AT THE NCI: CURRENT STATUS

As discussed earlier, from 1960 to 1982 a considerable number of natural product extracts, particularly of microbial and plant origin, were screened for antitumor activity, and a number of clinically effective chemotherapeutic agents were developed. In the early 1980s, however, this program was discontinued since it was perceived that few novel active leads were being isolated from natural sources. This conclusion applied equally to plants, marine organisms and micro-organisms, and resulted in a general de-emphasis of natural products as potential sources of novel antitumor agents. Of particular concern was the failure to yield agents possessing activity against the resistant solid tumor disease-types. This apparent failure might however, be attributed more to the nature of the primary screens being used at the time, rather than to a deficiency of Nature. Continued use of the primary P388 mouse leukemia screen appeared to be detecting only known active compounds or chemical structure-types having little or no activity against solid tumors.

The revision of the antitumor screening strategy in 1985, and the development of the new *in vitro* human cancer cell line screen led to the implementation of a new NCI natural products program involving new procurement, extraction and isolation components. The initiation, in 1987, of a major new program within the NCI for the discovery and development of anti-HIV agents provided yet further impetus and resources for the revitalization of the NCI's focus upon natural products. Contracts for the cultivation and extraction of fungi and cyanobacteria, and for the collection of marine invertebrates and plants, were initiated in 1986 and, with the exception of fungi and cyanobacteria, these programs are continuing to operate. Marine organism collections have focused in the Southwestern Pacific Ocean, but are now expanding to the Indian Ocean off East Africa through a contract with the Coral Reef Research Foundation. Terrestrial plant collections have been carried out in over 25 countries in tropical and subtropical regions worldwide through contracts with Missouri Botanical Garden (Africa and Madagascar), New York Botanical Garden (Central and South America), and the University of Illinois at Chicago (southeast Asia).

In carrying out these collections, the NCI contractors work closely with qualified organizations in each of the source countries. To date, botanists and marine biologists from source country organizations have collaborated in field collection activities and taxonomic identifications, and their knowledge of local species and conditions has been indispensable to the success of the NCI collection operations. Source country organizations provide facilities for the preparation, packaging, and shipment of the samples to the NCI natural products repository in Frederick, Maryland. The collaboration between the source country organizations and the NCI collection contractors has, in turn, provided support for expanded research activities by source country biologists, and the deposition of a voucher specimen of each species collected in the national herbarium or repository is expanding source country holdings of their biota. When requested, NCI contractors also provide training opportunities for local personnel through conducting workshops and presentation of lectures. In addition, through its Letter of Collection (LOC) and agreements based upon it, the NCI invites scientists nominated by Source Country Organizations to visit its facilities, or equivalent facilities in other approved U.S. organizations for 3-12 months to participate in collaborative natural products research, while representatives of most of the source countries have visited the NCI and contractor facilities for shorter periods to discuss collaboration.

Dried plant samples (0.3-1 kg dry weight) and frozen marine organism samples (~ 1 kg wet weight) are shipped to the NCI Natural Products Repository (NPR) in Frederick where they are stored at -20°C prior to extraction with a 1:1 mixture of methanol: dichloromethane and water to give organic solvent and aqueous extracts. All the extracts are assigned discrete NCI extract numbers and returned to the NPR for storage at -20°C until requested for screening or further investigation. After testing in the *in vitro* human cancer cell line and the anti-HIV screens, active extracts are subjected to bioassay-guided fractionation to isolate and characterize the pure, active constituents. Agents showing significant activity in the primary *in vitro* screens are selected for secondary testing. The *in vivo* activity of potential anticancer agents is first assessed using a hollow fiber encapsulation/ implantation methodology, whereby sterilized polyvinylidene fluoride hollow fibers are filled with cell suspensions of twelve tumors representing six of the subpanels in the *in vitro* screen, heat-sealed, and implanted in anesthetized athymic mice in either the peritoneal cavity or subcutaneously.[24] The agent may be administered intraperitoneally or at a distant site, and after six days, the hollow fibers are retrieved and the viable cell masses contained within the fibers determined colorimetrically using the MTT assay. Up to six hollow fibers (3 ip and 3 sc) may be implanted in a single mouse, providing a relatively inexpensive preliminary method to assess the capacity of a potential drug to reach tumor cells growing in two distinct physiologic compartments at pharmacologically active concentrations. Those agents exhibiting significant activity are then assessed against athymic mouse xenografts which reflect the complex phenomena which occur when the tumor cells are growing in and interacting with the host's normal tissues (e.g. angiogenesis effects, metastatic potential). In the case of potential anti-AIDS drugs, the hollow fiber methodology using HIV-infected cells is used to assess efficacy.[25] Those agents showing significant *in vivo* activity are presented to the NCI Division of Cancer Treatment, Diagnosis and Centers (DCTDC) Decision Network Committee (DNC) and, if approved by the DNC, the agent is entered into preclinical and clinical development. The Decision Network Process divides the drug development process into stages designated as DNIIA, DNIIB and DNIII (Figure 6). All testing prior to entry into the DNIIA process is considered routine and non-proprietary (Section 7.1 and Appendix 1, Stage 1). Approval by the DNC for entry into stage DNIIA involves a commitment of substantial NCI resources to the development of the drug and use of non-routine/proprietary operations (Section 7.1 and Appendix 1, Stage 2).

Of the 23 anticancer agents currently in active development in the DN process (excluding biologics), 17 are either natural products or derived from natural products, including 14 of microbial and 2 of marine origin, and one nucleoside (Figure 7). Of the 10 anti-HIV agents in active development in the DN process (excluding biologics), 5 are natural product derived, including 3 of plant and 1 of microbial origin, and one nucleoside (Figure 8).

6 NATURAL PRODUCT DRUG DEVELOPMENT: THE SUPPLY ISSUE

The critical first step in the development of any natural product drug is the procurement of an adequate supply to meet the requirements for preclinical and clinical investigation. While total synthesis may be considered as a potential route for bulk production of the active agent, it is worth noting that the structures of most bioactive natural products are extremely

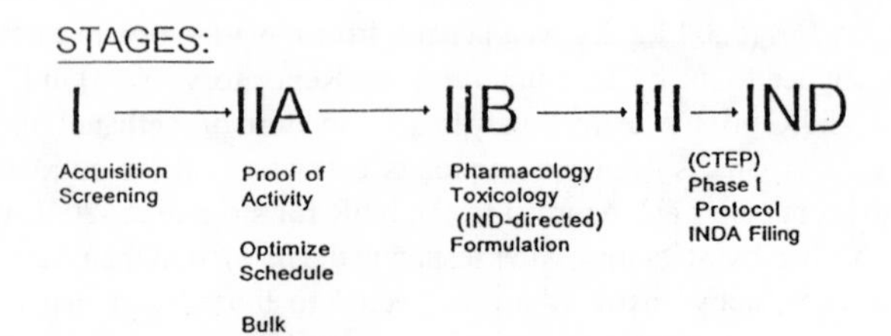

Figure 6

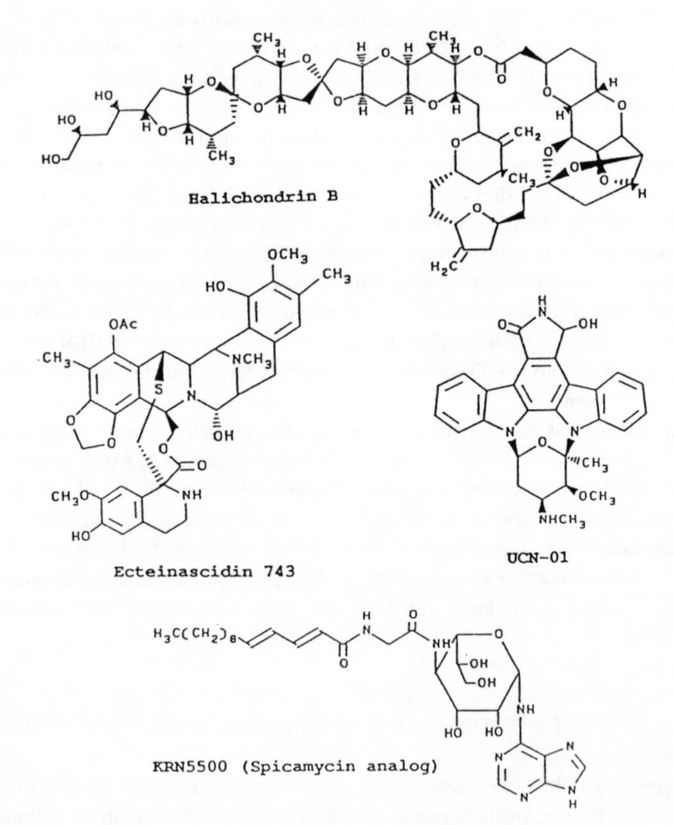

Figure 7 *Some anticancer agents in development in the NCI Decision Network process.*

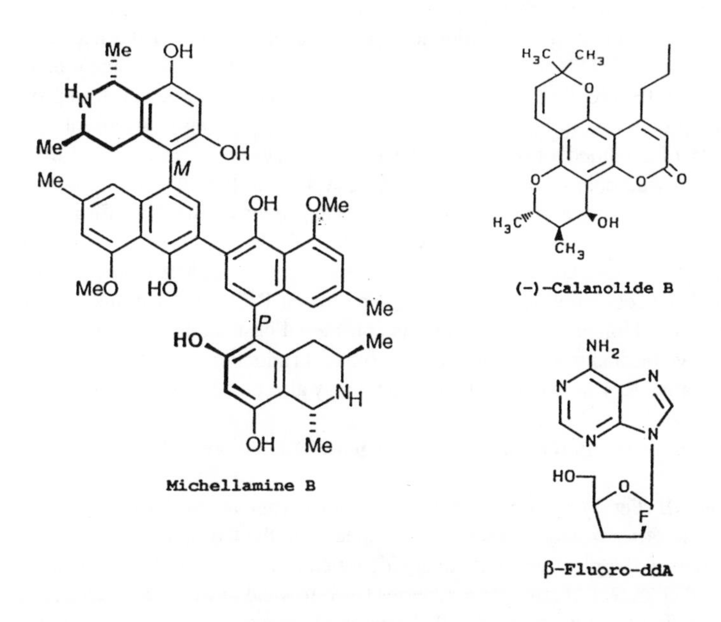

Figure 8 *Some anti-HIV agents in development in the NCI Decision Network process.*

complex, and bench-scale syntheses often are not readily adapted to large-scale economic production. Isolation from the natural source, therefore, often provides the most economically viable method of production. It should be noted that, of the established plant-derived commercial anticancer drugs, vinblastine and vincristine are still produced by isolation from *Catharanthus roseus* grown in various regions worldwide, while etoposide and teniposide are semisynthetically produced from natural precursors isolated from *Podophyllum emodii* harvested in India and Pakistan. The problems associated with the large-scale production of paclitaxel have also been resolved through semisynthesis from natural precursors, such as baccatin III and 10-desacetylbaccatin III, isolated from the needles of various *Taxus* species.[26]

The initial raw material collection sample (0.3-1.0 kg) will generally yield enough extract (10-40 g) to permit isolation of the pure, active constituent in sufficient milligram quantity for complete structural elucidation. Subsequent secondary testing and preclinical development, however, might require gram or even kilogram quantities, depending on the degree of activity and toxicity of the active agent.

In order to obtain sufficient quantities of an active agent for early preclinical development, recollections of 5-50 kg of the raw material, preferably from the original collection location, might be necessary. Should the preclinical studies justify development of the agent towards clinical trials, considerably larger amounts of material would be required. The performance of large recollections necessitates surveys of the distribution and abundance of source organism as well as determination of the variation of drug content in the various parts in the case of plants, and the fluctuation of content with the time and season of harvesting. In addition, the potential for mass cultivation or aquaculture of the source organism would need to be assessed. If problems are encountered due to scarcity of the wild source organism or inability to adapt it to cultivation, a search for alternative sources would be necessary. Other species of the same genus, or closely related genera, can be analyzed for drug content, and techniques, such as tissue culture, can be investigated.

The experience gained in the production of paclitaxel highlighted the necessity for studying various methods of biomass production at an early stage of development of a new agent. To this end, the NCI has implemented a Master Agreement mechanism to promote the large-scale production of biomass for the isolation of promising new natural product agents. Pools of qualified organizations have been established with expertise in the cultivation and tissue culture of source plants, as well as in the aquaculture and tissue culture of source marine organisms. When an agent is approved by the DNC for preclinical development, Master Agreement Order (MAO) Requests for Proposals (RFPs) for projects in one or more of the above areas are issued to the relevant pools of MA Holders who then submit technical and cost proposals addressing the particular RFP specifications. An award is made to the MA Holder whose proposal is considered best suited to the Government needs. Alternative mechanisms for supporting biomass production projects, such as Small Business Innovative Research contracts or grants, may also be considered.

6.1 Development of the Potential Anti-HIV Agent, Michellamine B

Michellamine B (Fig. 8) was isolated as the main *in vitro* active anti-HIV agent from the leaves of the liana, *Ancistrocladus korupensis*, collected in the Korup region of southwest Cameroon. Initially the plant was tentatively identified as *A. abbreviatus*, but collections of this and all other known *Ancistrocladus* species failed to yield any michellamines or show any anti-HIV activity. Subsequent detailed taxonomic investigation of the source plant compared to authentic specimens of *A. abbreviatus* revealed subtle but distinctive morphological differences, and the species was determined to be new to science, and officially named *Ancistrocladus korupensis*.[27] Michellamine B shows *in vitro* activity against a broad range of strains of both HIV-1 and HIV-2, including several resistant strains of HIV-1.[28] The species appears to be mainly distributed within the Korup National Park, and vine densities are of the order of one large vine per hectare. Fallen leaves collected from the forest floor do contain michellamine B, and current collections of these leaves should provide sufficient biomass for the isolation of enough drug for completion of preclinical development and possible preliminary clinical evaluation. It is clear, however, that extensive collections of fresh leaves could pose a possible threat to the wild source. Thus far, no other *Ancistrocladus* species has been found to contain michellamine B, and investigation of the feasibility of cultivation of the plant as a reliable biomass source was initiated in 1993 through a MAO contract with the Center for New Crops and Plant Products of Purdue University working in close collaboration with the University of Yaounde 1, the World Wide Fund for Nature Korup Project, Missouri Botanical Garden, Oregon State University and the NCI contractor, SAIC. An extensive botanical survey has been undertaken, and the range and distribution of the species has been mapped out. Dried leaf samples from representative vines were shipped to NCI for analysis of michellamine B content. Plants indicating high concentrations were re-sampled for confirmatory analysis, and those showing repeated high concentrations were targeted for cloning via vegetative propagation. A medicinal plant nursery has been established to hold and maintain the *A. korupensis* collection at the Korup Park Headquarters in Mundemba. In keeping with the NCI policies of collaboration with source countries, all the cultivation studies are being performed in Cameroon, and involve the local population, particularly those in the regions adjacent to the Korup National Park. In performing this project the cooperation and support of the scientific community and the Government of the Republic of Cameroon has been indispensable, and is greatly appreciated by the NCI and its collaborating organizations.

Based on the observed activity, the NCI has committed michellamine B to INDA-directed preclinical development. Unlike many natural products, formulation presents no problem since the drug is readily water-soluble as its diacetate salt. Continuous infusion studies in dogs indicate that *in vivo* effective anti-HIV concentrations can be achieved at nontoxic dose levels. However, despite these observations and the *in vitro* activity against an impressive range of HIV-1 and HIV-2 strains, there are some serious disadvantages which could preclude advancement of michellamine B to clinical trials. The difference between the toxic dose level and the anticipated level required for effective antiviral activity is small, indicative of a very narrow therapeutic index. In view of some of the toxicities observed in the toxicology studies this narrow therapeutic index is a concern to clinicians considering the drug as a candidate for preliminary clinical trials. In addition, administration by continuous infusion over a period of several weeks is a decided disadvantage compared to oral administration but, unfortunately, michellamine B is not orally bioavailable.

Even if michellamine B does show some activity in a preliminary clinical trial (assuming it advances that far), it is clear that extensive research will be necessary to determine if the pharmacological and toxicological profiles can be improved through analogue synthesis. Such studies could require substantial quantities of the natural product, or the successful synthetic studies of Bringmann and his group could provide a satisfactory solution.[29] The isolation of the novel antimalarial compounds, the korupensamines, from *A. korupensis*, provides another class of potential medicinal agents from this plant.[30] The korupensamines, which are equivalent to the "monomeric" units of the michellamines, are essentially inactive against HIV, whereas the michellamines exhibit only very weak antimalarial activity.

6.2 Development of Bryostatin 1

In 1969, an aqueous isopropanol extract of the colonial bryozoan, *Bugula neritina*, collected off the west coast of Florida in 1966, was found by the Pettit group at Arizona State University to exhibit moderate activity against the *in vivo* P388 mouse leukemia system. A recollection in early 1970 was found to be inactive, but modest activity was again observed in a 1971 recollection, though fractions of this extract were found to be inactive. In March 1976, a series of collections in the Gulf of California by the Pettit group once more gave active extracts, and recollections from the same region in March, 1980, yielded significantly active material. Bioassay-guided fractionation using both the *in vivo* and *in vitro* P388 systems led to the isolation of bryostatin 1 (Figure 3) and 2 in January, 1981, thirteen years after the original collection off the Florida coast. These studies, and subsequent isolation of more than 17 related compounds, have been reported.[31] Secondary testing against other animal tumor models and human tumor xenografts demonstrated only mediocre activity, and the compounds were dropped by NCI as candidates for clinical development.

Earlier correlation of antitumor test results of extracts of *Bugula neritina* with results observed for phorbols prompted further biochemical studies which demonstrated that the bryostatins competitively bind to the phorbol receptor site on protein kinase C. Several large-scale collections of *B. neritina* off the coast of southern California provided sufficient bryostatin 1 to permit a limited Phase I clinical study by the Cancer Research Campaign (CRC) in the United Kingdom in 1987. At the same time, potent selective activity was observed against the panel of leukemia cell lines in the NCI human tumor cell line screen (Figure 1B), and biochemical studies demonstrated that bryostatin 1, unlike most antileukemic agents, stimulated bone marrow cell production. After approval for preclinical and clinical development by the NCI Decision Network Committee, a large-scale collection of 13,000 kg (wet weight) of *B. neritina* was performed off the coast of southern California in 1988, and the isolation of 18 grams of bryostatin 1 under GMP (Good Manufacturing Practices) conditions was performed in NCI facilities through a collaborative agreement between the NCI contractor, Program Resources, Inc., the NCI, and Bristol-Myers Squibb.[33]

Bryostatin 1 is currently in Phase II clinical trials in the United States. Currently there is sufficient drug available for completion of clinical trials, but if bryostatin 1 advances to a marketable status, a reliable source of biomass will be required. While *Bugula neritina* is an ubiquitous fouling organism, the collection history outlined above has demonstrated considerable variation in bryostatin content, and only three populations are known to produce these compounds. Studies on the aquacultural production of bryostatin 1 by *Bugula neritina* were funded by the NCI through SBIR contract with the California company, CalBioMarine Technologies during the period 1991 to 1994. The company designed and built a pilot system using a proprietary seawater recycling system and, using animals collected off the southern California coast, determined the optimal feeding systems necessary to grow and maintain in culture animals producing bryostatin 1 at levels comparable to those found in freshly collected, wild material. Use of proper controls demonstrated that the bryostatin 1 was produced through *de novo* synthesis, and was not simply the result of dilution of the original broodstock's content. In the spring of 1995, laboratory-settled colonies of the bryozoan were transferred to PVC-plastic growout panels, mounted on a zinc-plated steel structure anchored at a mean depth of 12 meters, 1 km offshore of the Scripps Institution of Oceanography pier in La Jolla, California. After 5 months, high densities of healthy colonies of *B. neritina* were attained, averaging 2.88 kg/m^2 of panel area, and yielding 7-8 ug/g of bryostatin 1, comparable to yields from wild material.[34] Working in collaboration with the chemical engineering company, Aphios, an efficient, patented supercritical fluid extraction technique has been developed which permits the extraction and partial purification of bryostatins from both wild and aquacultured materials. The methodology for the growth and processing of *Bugula neritina* now exists for the production of quantities of purified bryostatins if needed for advanced trials and commercial development.

7 COLLABORATION IN DRUG DISCOVERY AND DEVELOPMENT

As noted earlier, much of the NCI drug discovery and development effort has been and continues to be, carried out through collaborations with research organizations and the pharmaceutical industry worldwide (Table 1).

Many of the naturally derived anticancer agents were developed through such collaborative efforts. Thus, the discovery and preclinical development of etoposide and teniposide, semisynthetic derivatives of the natural product epipodophyllotoxin, were performed by Sandoz investigators, and the NCI played a substantial role in the clinical development. Though paclitaxel (taxol®) was discovered by Wall and Wani with NCI contract support, the key to solving the supply problem was the semisynthetic conversion of baccatin III derivatives to taxol (and taxol analogs) pioneered by the French group led by Poitier, followed by the development of improved conversion methods by the Holton group, supported by the NCI, and Bristol-Myers Squibb.[26] The semisynthetic analog, taxotere, produced through a collaborative agreement between the Centre National de la Recherche Scientifique (CNRS) and Rhone-Poulenc Rorer, is undergoing extensive clinical evaluation in Europe and North America under auspices of organizations, such as the European Organization for Research and Treatment of Cancer (EORTC), and the Canadian and U.S. national cancer institutes[35]. Indeed, there is close collaboration between the

EORTC, the United Kingdom Cancer Research Campaign (CRC) and the NCI in the preclinical and clinical development of many anticancer agents, including agents such as bryostatin 1, dolastatin 10, aphidicolin glycinate, rhizoxin, pancratistatin and phyllanthoside.

Drugs, such as bleomycin, aclacinomycin and deoxyspergualin, were discovered by the Umezawa group at the Institute of Microbial Chemistry in Japan, and developed in collaboration with the NCI; a number of the agents currently in advanced preclinical development at the NCI, such as quinocarmycin, UCN-01 and a spicamycin analog, are the result of collaboration between Japanese companies, such as Kyowa Hakko and Kirin Breweries, and the NCI.

The Developmental Therapeutics Program (DTP) of the NCI thus complements the efforts of the pharmaceutical industry and other research organizations through taking good leads which industry might consider too uncertain to sponsor, and conducting the "high risk" research necessary to determine their potential utility. In promoting drug discovery and development, the DTP/NCI has formulated various mechanisms for establishing collaborations with research groups worldwide.

7.1 Screening Agreement Between Compound Providers and the NCI Division of Cancer Treatment, Diagnosis and Centers

In the case of organizations wishing to have pure compounds tested in the NCI drug screening program, such as pharmaceutical and chemical companies or university research groups, the DTP/NCI has formulated a screening agreement which includes terms stipulating confidentiality, patent rights, routine and non-proprietary screening and testing versus non-routine and proprietary screening and testing, and levels of collaboration in the drug development process. Individual scientists and research organizations wishing to submit pure compounds for testing generally consider entering into this agreement with the NCI Division of Cancer Treatment, Diagnosis and Centers (DCTDC).

The issue of routine/non-proprietary screening and testing versus non-routine/proprietary screening referred to above designates those NCI operations which could result in intellectual contributions in the development of a compound by NCI scientists, and which may rise to the level of inventorship as determined under United States patent law. Routine/nonproprietary screening and testing (Stage 1) generally refers to the standard NCI *in vitro* and *in vivo* anticancer and anti-HIV screens, and preliminary formulation and toxicology studies necessary to perform the animal *in vivo* screens. Non-routine/proprietary screening and testing (Stage 2) encompasses more advanced formulation, pharmacological and toxicological studies aimed at determining optimal parameters for clinical trials, as well as analogue development and mechanism of action studies. These latter operations (Stage 2) require significant research involving intellectual input which could result in sole NCI scientist inventorship. These two stages of screening and testing are presented in more detail in Appendix 1. In signing a screening agreement, NCI agrees that it will not proceed with Stage 2 operations without the prior written consent of the supplier organization. In addition, the supplier organization may designate which of the Stage 2 operations it wishes to delete from the scope of the screening agreement prior to execution, and DTP/NCI could consider amending the agreement to incorporate them at a later stage. If no limitation in the scope of testing is requested by the supplier organization, it is assumed that DTP/NCI may proceed directly from Stage 1 to Stage 2 testing. Should a compound show promising anticancer or anti-AIDS activity through Stage 1 and designated State 2 testing, the NCI will propose the establishment of a more formal collaboration, such as a Cooperative Research and Development Agreement (CRADA) or a Clinical Trial Agreement (CTA).

7.2 National Cooperative Drug Discovery Group (NCDDG) Program

In the late 1970s and early 1980s, many significant discoveries were being made in such fields as biochemistry, molecular biology, embryology and carcinogenesis, that had the potential for the development of new strategies and agents for cancer treatment; most investigators, however, were working only in their own areas of expertise without the benefit of close liaison with experts in the many disciplines required to discover and develop new therapies and strategies. In response to the need to coordinate these research efforts, the NCI initiated the NCDDG Program in the early 1980s with the goal of bringing together scientists from academia, industry and government, in the form of consortia, in a focused effort aimed at the discovery of new drugs for cancer treatment.[36] The inclusion of an industrial component in each consortium has had strong positive effects in helping to orient the academic component(s) towards drug development, and maintaining a focus on the final outcomes of drug discovery in terms of clinical trials and marketable products, as well as contributing high quality scientists and resources to the Program. Involvement of NCI Staff has enabled the NCI to contribute its considerable resources and expertise in cancer drug development, including extensive computerized databases and repositories of compounds tested over more than 35 years, primary and secondary screening systems, and all the resources necessary for preclinical development of agents meeting DNC selection criteria. The consortia, headed by a Principal Investigator, submit proposals based on independent ideas, rather than in response to specific topics proposed by the NCI, thereby permitting the widest scope and the greatest degree of innovative science, and encouraging diversity in the discovery of new drugs and therapeutic approaches.

The National Cooperative Natural Product Drug Discovery Group (NCNPDDG) Program is one of four such programs, the other three being directed at studies of Mechanisms of Action, Specific Diseases (e.g. lung and colon cancer), and Preclinical Model Development. Since 1989, twelve NCNPDDGs have been awarded encompassing the study of all natural sources, including plants, marine bacteria and invertebrates, microalgae, cyanophytes and dinoflagellates, and using a variety of assays, such as molecular targets, mechanism-based assays, cell lines and *in vivo* systems.

It is, of course, unrealistic to expect that every NCDDG will produce a lead for clinical development, since many of the targets of screening are speculative, in that no drugs working by the target mechanisms may have been established as being clinically useful in the treatment of cancer (e.g. inhibitors of kinases or phosphatases). The likelihood of finding an effective drug acting through modulation of such cell signalling mechanisms, therefore, is unknown, and this research entails a much higher risk than searches for novel mitotic or topoisomerase inhibitors. The inclusion of projects associated with promising new science, but possible high risk, is an important part of the design of the NCDDG Program which emphasizes the most current technologies and the possibility of the discovery of a breakthrough drug identified with a mechanistically novel class of anticancer agents. Some of the drugs to emerge as clinical candidates from the NCDDG Program are topotecan, polyamines (e.g. N^1, N^{14}-diethylhomospermine), hydroxyurea (for treatment of ovarian cancer), and pentosan polysulfate. Though these drugs might not have been discovered in the NCDDG Program, program support was critical to their development.

7.3 International Cooperative Biodiversity Group (ICBG) Program

The ICBG Program resembles the NCDDG Program in structure, in that consortia are formed comprising academic, industrial and U.S. government organizations, but organizations from developing countries are also required components. This Program is jointly sponsored by the National Science Foundation (NSF), the U.S. Agency for International Development (USAID) and components of the National Institutes of Health (NIH), including the Fogarty International Center (FIC), the NCI, the National Institute of Allergy and Infectious Diseases (NIAID), the National Heart, Lung and Blood Institute (NHLBI), and the National Institute of Mental Health (NIMH). The goals of the Program are research into drug discovery from natural sources, linked to the identification, inventory and conservation of biodiversity, a primary concern of the NSF, and economic development in developing countries, part of the mission of USAID[36]. All these goals are linked to the provision of suitable training and infrastructure building.

Five awards, four involving countries in Central and South America and one involving the West African countries of Cameroon and Nigeria, were awarded in 1993 and 1994, and are administered through the NIH Fogarty International Center. A significant challenge in the development of the ICBGs was the establishment of principles related to intellectual property rights and the protection of the rights of the participating source (developing) countries, including communities and indigenous peoples. While it was possible to develop guidelines for use in negotiating contracts and agreements, no single set of contractual terms could apply to all participants, and awardees have developed unique mechanisms and agreements to suit the particular circumstances of organizations and countries involved[37] .As integrated conservation and development projects, the long term evaluation of this Program will depend on how successful the projects are in demonstrating the economic value of biodiversity in providing new pharmaceuticals and sustainable natural products-based industries for the participating developing countries.

7.4 Source Country Collaboration and Compensation

The recognition of the value of the natural resources (plant, marine and microbial) being investigated by the NCI, and the significant contributions being made by source country scientists and traditional healers in aiding the performance of the NCI collection programs, have led the NCI to formulate policies aimed at facilitating collaboration with, and compensation of, countries participating in the NCI drug discovery program. Many of these countries are developing nations which have a real sensitivity to the possibility of exploitation of their natural resources by developed country organizations involved in drug discovery or other programs searching for novel bioactive agents.

The letter of collection formulated by the NCI contains both short term and long term measures aimed at assuring countries participating in NCI-funded collections of its intentions to deal with them in fair and equitable manner. In the short term, the NCI periodically invites scientists from source country organizations to visit the drug discovery facilities at the Frederick Cancer Research and Development Center (FCRDC) to discuss the goals of the program, and to explore the scope of collaboration in the drug discovery effort. When laboratory space and resources permit, suitably qualified scientists are invited to spend periods of up to 12 months working with scientists in NCI or other suitable facilities on projects related to natural products drug discovery, such as the testing of extracts and the bioassay-guided isolation and structure determination of active agents, preferably from organisms collected in their home countries.

As test data become available from the anticancer and anti-HIV screens, these are provided to the NCI collection contractors for dissemination to interested scientists in countries participating in the NCI collection programs. Each country receives only data obtained from extracts of organisms collected within its borders, and scientists are requested to keep data on active organisms confidential until the NCI has had sufficient time to assess the potential for the development of novel drugs from such organisms. Confidentiality is an important issue, since the NCI will apply for patents on agents showing particular promise; such patents may be licensed to pharmaceutical companies for development and eventual marketing of the drugs. As part of the licensing agreement, the NCI requires the successful licensee to negotiate and enter into agreement(s) with an appropriate organization(s) in the country of origin of the organism yielding the drug. Such an agreement will address the concern on the part of the source country that pertinent agencies, institutions and/or persons or communities receive royalties and other forms of compensation, as appropriate. Such agreements apply equally to instances where the drug is structurally based on the isolated natural product, though the percentage of royalties or compensation may vary depending on the relationship of the marketed drug to the originally isolated product. Such compensation is regarded as a potential long term benefit, since development of a drug to the stage of marketing can take from 10 to 20 years from its time of discovery.

Another potential benefit to the country of origin is the development of a large-scale cultivation program to supply sufficient raw material for bulk production of the drug. In licensing a patent on a new drug to a pharmaceutical company, the NCI will require the licensee to seek, as its first source of supply, the raw material produced in the country of origin. The policies concerning compensation and raw material supplies will also apply to inventions made by other organizations screening extracts from the NCI Natural Products Repository for activities against diseases related to the NCI mission (Section 7.5). In addition, Master Agreement Holders investigating the large-scale production of biomass for the isolation of drugs of interest to NCI will be required to explore collaboration with the relevant source countries in the performance of their Master Agreement Order tasks.

The discovery and development of michellamine B (Section 6.1) illustrates the potential for international collaboration resulting from the contract collection program supported by the NCI. A further example of a productive collaboration is the development of the calanolides, isolated from the Sarawak (Malaysia) plants, *Calophyllum lanigerum* and *C. teysmanii*.[38] In this instance, the Sarawak State Government, through its Department of Forests, has collaborated in and cost-shared the collection of large quantities of the latex of *C. teysmanii* for the production of (-)-calanolide B, and is supporting an investigation of the cultivation of both *Calophyllum* species.

In the development of the marine organism-derived, potential anticancer agent, halichondrin B, the NCI has collaborated closely with New Zealand organizations in exploring the production of this agent from one of the source sponges, a *Lissodendoryx* species. This sponge was collected in 1993 during a deep-water dredging expedition in New Zealand territorial waters and shown to contain halichondrin B and several congeners by the University of Canterbury marine chemistry group led by Blunt and Munro. Prior to considering a large-scale recollection, the NCI funded an ecological survey of the ocean bed performed by the New Zealand Oceanographic Institution (NZOI), in collaboration with the Royal New Zealand Navy, which demonstrated that significant amounts of the sponge could be removed by dredging without deleterious environmental effects; subsequent studies indicated that the sponge reseeded rapidly. The University of Canterbury group and NZOI, with NCI support, have also demonstrated the feasibility of transporting sponge samples collected from depths of 70 to 100 meters to water at 10 to 30 meters while maintaining production of halichondrins; this has led to successful "in sea" aquacultural experiments at depths of about 30 meters, jointly supported by the NCI and the New Zealand Government through its Foundation for Research, Science and Technology. Similar joint funding has enabled performance of an ecologically-sensitive large-scale collection of the sponge, and the isolation and purification of sufficient halichondrin B for preclinical studies, through the establishment of a "Halichondrin Joint Venture" involving the National Institute of Water and Atmospheric Research (NIWA), New Zealand Pharmaceuticals and the University of Canterbury. Results of this successful joint venture are being prepared for publication.

In addition to the contract acquisition programs, direct collaborations have been established between the NCI and research organizations in countries not covered by the present collection contracts, or organizations studying organisms not included in the NCI program. Medicinal plants from Yunnan Province of the Peoples' Republic of China, Korea, India and Russia are being studied in collaboration with the Kunming Institute of Botany, the Korean Research Institute of Chemical Technology in Seoul, the Central Drug Research Institute in Lucknow, and the Cancer Research Center in Moscow, respectively. Collaborations have also been established with Instituto Nacional de Biodiversidad (INBio) in Costa Rica, the South American Organization for Anticancer Drug Development in Brazil, the Institute of Chemistry at the National University of Mexico, the HEJ Research Institute in Karachi, Pakistan, and the Zimbabwe National Traditional Healers Association (ZINATHA) and the University of Zimbabwe. In establishing these collaborations, NCI undertakes to abide by the same policies of collaboration and compensation as apply to source countries participating in the contract collection programs. The terms of collaboration are generally stated in a Memorandum of Understanding (MOU) signed by the organization and the NCI. The terms, based on the policies of the NCI Letter of Collection, are summarized in a schematic diagram (Figure 9). The NCI Letter of Collection (formerly known as the Letter of Intent) was initiated in 1988, pre-dating the U.N. Convention of Biological Diversity by some four years. A number of other organizations and companies have implemented similar policies[39].

7.5 Distribution of Extracts from the NCI Natural Products Repository

In carrying out the collection and extraction of thousands of plant and marine organisms samples worldwide, the NCI has established a Natural Products Repository (NPR) which is a unique and valuable resource for the discovery of potential new drugs and other bioactive agents. In recognition of this potential, the NCI has developed policies for the distribution of these extracts to qualified organizations, subject to the signing of a legally-binding Material Transfer Agreement (MTA). While the current policies only apply to the testing of extracts in screens pertaining to activity against cancer, AIDS, and related opportunistic infections, as well as diseases of concern to developing countries (e.g. malaria, parasitic diseases), the extension to testing against all human diseases is currently being considered.

To be considered for access to the NIH, organizations have to submit short proposals outlining the nature of their screening systems, and demonstrating the capability to process active extracts and develop any active agents isolated towards clinical trials and commercial production. Proposals are treated as confidential, and are reviewed for approval by a committee comprising of senior staff of the NCI Division of Cancer Treatment, Diagnosis, and Centers (DCTDC). Approved organizations have to enter into a MTA with DCTDC, with one of the key terms being the requirement for the recipient organization to negotiate suitable terms of collaboration and compensation with the source country(ies) of any extract(s) shown to exhibit significant activity in the organization's screens. While the NCI will provide sufficient milligram quantities of extracts for completion of primary and secondary testing, procurement of larger quantities to enable fractionation, isolation and development of active agents requires interaction with the relevant source countries; in such instances, the NCI will refer the organization to the NCI contractor who originally collected the organism(s) of interest for assistance in the procurement of additional material, or the organization may deal directly with the source country(ies).

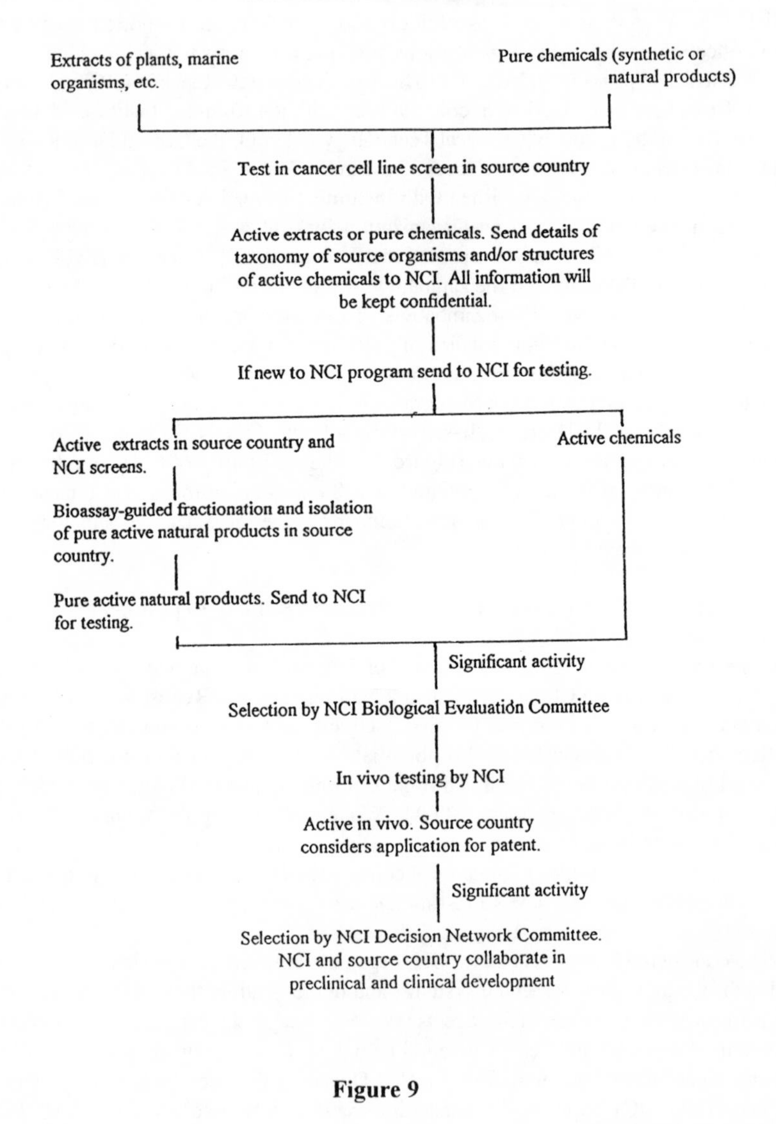
POSSIBLE COLLABORATION WITH NCI
Extracts of plants, marine organisms, etc.
Pure chemicals (synthetic or natural products)
Test in cancer cell line screen in source country
Active extracts or pure chemicals. Send details of taxonomy of source organisms and/or structures of active chemicals to NCI. All information will be kept confidential.
If new to NCI program send to NCI for testing.
Active extracts in source country and NCI screens.
Active chemicals
Bioassay-guided fractionation and isolation of pure active natural products in source country.
Pure active natural products. Send to NCI for testing.
Significant activity
Selection by NCI Biological Evaluation Committee
In vivo testing by NCI
Active in vivo. Source country considers application for patent.
Significant activity
Selection by NCI Decision Network Committee. NCI and source country collaborate in preclinical and clinical development

Figure 9

7.6 DTP WWW Homepage

The NCI Development Therapeutics Programe (DTP) offers access to a considerable body of data and background information through its WWW Homepage, http://epnws1.ncifcrf.gov:2345/dis3d/dtp.html.

Publicly available data include results from the human tumor cell line screen and AIDS antiviral drug screen, the expression of molecular targets in cell lines, and 2D and 3D structural information. Background information is available on the drug screen and the behavior of "standard agents" (Section 3), NCI investigational drugs, analysis of screening data by COMPARE, the AIDS antiviral drug screen, and the 3D database. It must be noted that data and information is only available on so-called "open compounds" which are not subject to the terms of confidential submission.

In providing screening data on extracts, the extracts are identified by code numbers only; details of the origin of the extracts, such as source organism taxonomy and location of collection, may only be obtained by individuals or organizations prepared to sign agreements binding them to terms of confidentiality and requirements regarding collaboration with, and compensation of, source countries. Such requirements are in line with the NCI commitments to the source countries through its Letter of Collection and the Material Transfer Agreement.

8 NEW DIRECTIONS IN NATURAL PRODUCT DRUG DISCOVERY AND DEVELOPMENT

8.1 Source Country Collaboration

With the increased awareness of genetically-rich source countries to the great value of their natural resources, and the confirmation of source country sovereign rights over these resources by the U.N. Convention of Biological Diversity, organizations involved in drug discovery and development are increasingly adopting policies of equitable collaboration and compensation in interacting with these countries[39]. Particularly in the area of plant-related studies, source country scientists and governments are committed to performing more of the operations in-country, as opposed to the export of raw materials. The NCI has recognized this fact for several years and has negotiated Memoranda of Understanding with a number of source country organizations suitably qualified to perform in-country processing (Section 7.4). In considering the continuation of its plant-derived drug discovery program, the NCI has de-emphasized its contract collection projects in favor of expanding closer collaboration with qualified source country scientists and organizations.

A logical pathway for source countries to follow is to build up well-documented collections of plant extracts, maintained under conditions of careful quality control in low temperature repositories. While some source countries are suitably equipped to undertake such a task, many still lack the necessary equipment and infrastructure to perform collections and extractions on a large scale. While the NCI is not permitted to support infrastructure development, it has proposed mechanisms to assist interested source countries launch extract repositories; one such mechanism would involve the extraction of plant samples sent by a source country to the NCI facility in Frederick, Maryland, and the return of 60% of each extract to the source country for storage and use by the local scientific community or sale to screening organizations. The costs of shipment of the plant materials by the source country, and extraction and return shipment of the extracts, would be borne by the NCI, with 40% of each extract being retained by the NCI for testing in its human cancer line and anti-HIV screens. Any active extracts would be investigated in collaboration with the source country. Such a mechanism would permit the source country to establish a repository, and eventually generate sufficient revenue to become self-supporting in the extraction and processing of its natural materials.

8.2 Applications of the Human Cancer Cell Line Screen

Analysis of the 60 cell lines used in the NCI *in vitro* screen for their content of molecular targets, such as p-glycoprotein, p53, Ras and BCL2, permits their use as seeds in COMPARE analyses of the entire NCI database, and the identification of compounds that can selectively attack cells high (or, alternatively low) in these targets[21]. Analysis of 58 of the 60 cell lines for their ability to efflux the fluorescent dye, rhodamine 123, provided an assay for content of P-glycoprotein (Pgp) which is associated with multidrug resistance[40] Significant rhodamine efflux, associated with high Pgp content, was observed for 12 cell lines, nine of which were colon or kidney lines. COMPARE analysis of the NCI database of compounds, using the rhodamine efflux profile as seed, identified hundreds of compounds with high correlation coefficents, many of them natural products. Some of these compounds were tested and shown to be Pgp substrates by reversal of resistance with the Pgp antagonist, verapamil, and in some instances the cytotoxicity could be modulated several thousandfold. Such results could have important implications in the clinical development of new drugs, in that single agent trials of compounds determined to be Pgp substrates are likely to result in their failure to show activity in the clinical setting. It is possible, however, that addition of a Pgp antagonist could prevent such resistance.

8.3 Targeting Natural Products

A recurring liability of natural products, at least in the area of cancer chemotherapy, is that, although many are generally very potent, they have limited solubility in aqueous solvents and exhibit narrow therapeutic indices. These factors have resulted in the demise of a number of promising leads, such as bruceantin and maytansine (Section 2).

An alternative approach to utilizing such agents is to investigate their potential as "warheads" attached to monoclonal antibodies specifically targeted to epitopes on tumors of interest. While this is not a new area of research to the NCI, the DTP is well placed to refine and expand this approach to cancer therapy. The DTP has a wide range of potent, natural product chemotypes to explore as potential "warheads", and also has the capability to produce clinical grade monoclonal antibodies (Mabs) through its Biological Resources Branch. One of the approaches being investigated is the selection of a less potent member of a chemical class (e.g. ansamycin antibiotics) as the "warhead" in order to avoid undue toxic effects in the event of cleavage of the "warhead-Mab" bond prior to delivery of the agent to the desired target tumor cells; the potential for severe toxicity in such instances when using "warheads" of the potency of ricin or calicheamicin is substantial.

Another strategy of interest is the use of antibodies as vectors for enzymes capable of activating a nontoxic drug precursor (prodrug) to a potent cytotoxic moiety.[41] After injection, and localization of an antibody-enzyme conjugate at the tumor, a nontoxic prodrug is administered, and while remaining innocuous to the normal tissues, it is converted to the cytotoxin by the enzyme localized at the tumor. This approach, called "antibody-directed enzyme prodrug therapy" (ADEPT) provides further potential for the application of potent natural products to cancer treatment.

8.4 *In Vivo* Combinatorial Biochemistry

Advances in the understanding of bacterial aromatic polyketide biosynthesis have lead to the identification of multifunctional polyketide synthase enzymes (PKSs) responsible for the construction of polyketide backbones of defined chain lengths, the degree and regiospecificity of ketoreduction, and the regiospecificity of cyclizations and aromatizations, together with the genes encoding for the enzymes [42]. A set of rules for manipulating the early steps of aromatic polyketide biosynthesis through genetic engineering has been developed, permitting the biosynthesis of polyketides not generated naturally (unnatural natural products). Since polyketides constitute a large number of structurally diverse natural products exhibiting a broad range of biological activities (e.g. tetracyclines, doxorubicin, avermectin), the potential for generating novel molecules with enhanced known bioactivities, or even novel bioactivities, appears to be high.

9 CONCLUSIONS

The NCI has the full capability to advance compounds showing promising preliminary anticancer or anti-AIDS activity through all the phases of preclinical and clinical development. The NCI welcomes the opportunity to collaborate with scientists and organizations in the discovery and/or development of anticancer and anti-AIDS agents, and has formulated agreements (Memorandum of Understanding or Screening Agreement) outlining terms of collaboration covering issues of confidentiality, inventorship and patent rights, and compensation should a commercial product be developed. The NCI, as a U.S. taxpayer supported, non-profit institution, is firmly committed to protecting the rights of suppliers and Source Countries during all phases of drug discovery and development.

References

1. M. R. Boyd, In: Current Therapy in Oncology. Section 1. Introduction to Cancer Therapy, J. Niederhuber (ed.), B. C. Decker, Philadelphia, 1993, p. 11.
2. J. L. Hartwell, and A. W. Schrecker, In: 'Progress in the Chemistry of Organic Natural Products', L. Zechmeister (ed.), Springer-Verlag, Vienna, Austria, 1958, **15**, 83.
3. I. Jardine, In: 'Anticancer Agents Based on Natural Product Models,' J. M. Cassady and J. D. Douros (eds.), Academic Press, New York, 1979, p. 319.
4. I. S. Johnson, H. F. Wright and G. H. Svoboda, *Proc. Am. Assoc. Cancer Res.*, 1960, **3**, 122.
5. S. K. Carter and R. B. Livingston, *Cancer Treat. Rep.*, 1976, **60**, 1141.
6. J. S. Driscoll, *Cancer Treat. Rep.*, 1984, **68**, 63.
7. J. M. Venditti, R. A. Wesley and J. Plowman, *Adv. Pharmacol. Chemotherapeutics*, 1984, **20**, 1.
8. B. A. Chabner and J. M. Collin, 'Cancer Chemotherapy: Principles and Practice', Lippincott, Philadelphia, 1990.
9. G. M. Cragg, D. J. Newman and K. M. Snader, *J. Nat. Prod.*, submitted.
10. R. Johnson, *Ann. Rep. Med. Chem.*, 1992, **28**, 168.
11. M. Potmesil and H. Pinedo, 'Camptothecins: New Anticancer Agents,' CRC Press, Inc., Boca Raton, Florida, 1995.
12. M. Suffness and G. A. Cordell, In: 'The Alkaloids. Antitumor Alkaloids', A. Brossi (ed.), Academic Press, New York, 1985, **25**, p. 57.

13. M. R. Boyd, L. T. Burka, T. M. Harris, and B. J. Wilson, *Biochem. Biophys. Acta*, 1974, **337**, 184.
14. P. O'Dwyer, S. A. King, D. F. Hoth, M. Suffness and B. Leyland-Jones, *J. Clin. Oncol.*, 1986, **4**, 1563.
15. M. C. Christian, R. E. Wittes, B. Leyland-Jones, T. L. McLemore, A. C. Smith, C. K. Grieshaber, B. A. Chabner and M. R. Boyd, *J. Natl. Cancer Inst.*, 1990, **82**, 1420.
16. G. M. Cragg, M. R. Boyd, J. H. Cardellina II, M. R. Grever, S. A. Schepartz, K. M. Snader and M. Suffness, In: 'Human Medicinal Agents from Plants', A. D. Kinghorn and M. F. Balandrin (eds.), ACS Symposium Series 534, Amer. Chem. Soc., Washington, D.C., 1993, p. 80.
17. M. Suffness, D. J. Newman and K. M. Snader, In: 'Bioorganic Marine Chemistry', P. Scheuer (ed.), Springer-Verlag, Berlin, 1989, **3**, p 131.
18. P. A. Philip, D. Rea, P. Thavasu, J. Carmichael, N. S. A. Stuart, H. Rockett, D. C. Talbot, T. Ganesan, G. R. Pettit, F. Balkwill and A. L. Harris. *J. Natl. Cancer Inst.*, 1993, **85**, 1812.
19. M. R. Boyd and K. D. Paull, *Drug Development Research*, 1995, **34,** 91.
20. A. Monks, D. Scudiero, P. Skehan, R. Shoemaker, K. Paull, D. Vistica, C. Hose, J. Langley, P. Cronise, A. Vaigro-Wolff, M. Gray-Goodrich, H. Campbell and M. Boyd, *J. Natl. Cancer Inst.*, 1991, **83**, 757.
21. K. D. Paull, R. H. Shoemaker, L. Hodes, A. Monks, D. A. Scudiero, L. Rubinstein, J. Plowman and M. R. Boyd, *J. Natl. Cancer Inst.*, 1989, **81**, 1088.
22. M. R. Boyd, In: 'AIDS: Etiology, Diagnosis, Treatment and Prevention', V. T. DeVita, S. Hellman and S.A. Rosenberg (eds.), J.B. Lippincott, Philadelphia, 1988, p. 305.
23. O. S. Weislow, R. Kiser, D. L. Fine, J. Bader, R. H. Shoemaker and M. R. Boyd, *J. Natl. Cancer Inst.*, 1989, **81**, 577.
24. M. G. Hollingshead, M. C. Alley, R. F. Camalier, B. J. Abbott, J. G. Mayo, L. Malspeis and M. R. Grever, *Life Sciences*, 1995, **57,** 131.
25. M. G. Hollingshead, J. Roberson, W. Decker, R. Buckheit, Jr., C. Elder, L. Malspeis, J. G. Mayo and M. R. Grever, *Antiviral Research*, 1995, **28**, 265.
26. G. M. Cragg, S. A. Schepartz, M. Suffness and M. R. Grever, *J. Nat. Prod.*, 1993, ***56***, 1657.
27. D. W. Thomas and R. E. Gereau, *Novon*, 1993, **3**, 494.
28. M. R. Boyd, Y. F. Hallock, J. H. Cardellina II, K. P. Manfredi, J. W. Blunt, J. B. McMahon, R. W. Buckheit, Jr., G. Bringmann, M. Schaffer, G. M. Cragg, D. W. Thomas and J. G. Jato, *J. Med. Chem.*, 1994, **37**, 1740.
29. G. Bringmann, S. Harmsen, J. Holenz, T. Geuder, R. Gotz, P.A. Keller , R. Walter, Y. F. Hallock, J. H. Cardellina II and M. R. Boyd, *Tetrahedron*, 1994, **50**, 9643.
30. Y. F. Hallock, K. P. Manfredi, J. W. Blunt, J. H. Cardellina II, M. Schaffer, K-P. Gulden, G Bringmann, A. Y. Lee, J. Clardy, G. Francois and M. R. Boyd, *J. Org. Chem.*, 1994, **59**, 6349.
31. G. R. Pettit, D. Sengupta, P. M. Blumberg, N. E. Lewin, J. M. Schmidt and A. S. Kraft, *Anti-cancer Drug Design*, 1992, **7**, 101.
32. W. S. May, S. J. Sharkis, A. H. Esa, V. Gebbia, A. S. Kraft, G. R. Pettit and . Sensenbrenner, *Proc. Natl. Acad. Sci.USA*, 1987, **84**, 8483.
33. D. E. Schaufelberger, M. P. Koleck, J. A. Beutler, A. M. Vatakis, A. B. Alvarado, P. Andrews, L. V. Marzo, G. M. Muschik, J. Roach, J. T. Ross, W. B. Lebherz, M. P. Reeves, R. M. Eberwein, L. L. Rodgers, R. T. Testerman, K. M. Snader and S. Forenza, *J. Nat. Prod.*, 1991, **54**, 1265.

34. D. Mendola, C. J. Sheehan and B. J. Javor, Poster Presentation, Marine Natural Products Gordon Research Conference, Ventura, California, Feb. 25, 1996.
35. J. E. Cortes and R. Pazdur, *J. Clin. Oncol.*, 1995, **13**, 2643.
36. M. Sufffness, G. M. Cragg, M. R. Grever, F. J. Grifo, G. Johnson, J. A. R. Mead, S. A. Schepartz, J. M. Venditti and M. Wolpert, *Internat. J. Pharmacognosy*, 1995, **33**, (Supplement 5).
37. J. R. Rosenthal, Proceedings of the Organization for Economic Cooperation and Commercial Development International Conference on Biodiversity Incentive Measures, Cairns, Australia, March, 1996, in press.
38. Y. Kashman, K. R. Gustafson, R. W. Fuller, J. H. Cardellina, II, J. B. McMahon, M. J. Currens, R. W. Buckheit, S. H. Hughes, G. M. Cragg and M. R. Boyd, *J. Med. Chem.*, 1992, **35**, 2735.
39. J. T. Baker, R. P. Borris, B. Carte, G. A. Cordell, D. D. Soejarto, G. M. Cragg, M. P. Gupta, M. M. Iwu, D. R. Madulid and V. E. Tyler, *J. Nat. Prod.*, 1995, **58**, 1325.
40. J.-S. Lee, K. D. Paull, M. Alvarez, C. Hose, A. Monks, M. R., Grever, A. T. Fojo and S. E. Bates, *Mol. Pharmacol.*, 1994, **46**, 627.
41. R. G. Melton and R. F. Sherwood, *J. Natl. Cancer Inst.*, 1996, **88**, 153.
42. R. McDaniel, S. Ebert-Khosla, D. A. Hopwood and C. Khosla, *Nature*, 1995, **375**, 549.

Appendix 1

Step 1: Routine or Non-Proprietary Screening and Testing

1. Entry of the structure of a pure synthetic compound or natural product into NCI's confidential data base of structures.

2. Testing of the pure synthetic compound or natural product in NCI's *in vitro* Cancer Drug Screen in a two-day format and a six-day format AND/OR testing of the pure synthetic compound or natural product in NCI's *in vitro* AIDS Drugs Screen against wild type and mutant strains of HIV.

3. Comparison of the pattern of activity in the Cancer or AIDS Screen of the pure synthetic compound or natural product against the pattern of activity of other compounds in NCI's data base, and with the expression of molecular targets in NCI's Cancer Screen in the case of compounds tested in the Cancer Screen.

4. For compounds with evidence of antiproliferative activity in the primary screen, initiation of

   a. Preliminary toxicology with determination of Maximal Tolerated Dose (MTD) according to routine iv and ip protocols.

   b. initiation of formulation studies to determine appropriate vehicle to allow *in vivo* testing.

c. assessment of *in vivo* activity against athymic mouse xenografts of human tumors dosed according to standard protocols for the tumor types chosen. The tumors will represent those cell types suggested to be sensitive by the *in vitro* screening data. In the case of anti-AIDS drugs, study in *in vivo* mouse models bearing HIV-infected cells to observe evidence of anti-retroviral effect.

5. For compounds with evidence of anti-proliferative activity in the primary screen, study in selected routine subscreens to allow assessment of activity in *in vitro* and *in vivo* models of

   a. AIDS related lymphoma
   b. prostate carcinoma (primary)
   c. breast carcinoma (metastatic)
   d growth of tumor vasculature (as they may be available to DCTDC)
   e. differentiation

Step 2: Proprietary or Non-Routine Screening and Testing

1. Determine the optimal schedule for routine demonstration of anti-tumor or anti-retro viral activity in selected *in vivo* models.

2. Produce structural analogs of the initial compound that would optimize pharmacologic and or pharmaceutic features.

3. Determine or develop the optimal formulation for the agent to allow clinical use, including verification of

   a. stability of drug as bulk substances
   b. dilution stability in common diluents
   c. stability as aliquoted in clinical dose forms

4. Develop a detailed assay for the bulk drug as well as drug dissolved in vehicles and body fluids.

5. Determine the range of doses that would cause toxic effect in two species according to a dosing schedule that would mimic the proposed clinical use. Also:

   a. determine effect on marrow progenitor growth from at least two species
   b. determine initial clinical pathology parameter changes in response to drug

6. Perform detailed toxicologic evaluation in two species at three dose levels to allow estimation of a "low toxic" dose from which clinical starting doses for a Phase I protocol might be estimated. These studies would include.

   a. detailed gross anatomic and histopathologic studies
   b. correlation of plasma drug concentration with pathology
   c. correlation of clinical pathology with 6a and 6b.

7. If the drug and formulation appears practicable, to file an Investigational New Drug Application (INDA) with the U.S. Food and Drug Administration with NCI as holder of the INDA.

8. Verification that the compound or its analog possess a mechanism of action suggested by comparison of its pattern of activity with known agents, or determine its mechanism of anti-neoplastic or anti-retroviral effect by detailed evaluation in non-routine assays.

Plants and Microbes as Complementary Sources of Chemical Diversity for Drug Discovery

M. Inês Chicarelli-Robinson*, Simon Gibbons, Carole McNicholas, Neil Robinson, Michael Moore, Ursula Fauth and Stephen K. Wrigley

XENOVA LIMITED, 240 BATH ROAD, SLOUGH, BERKSHIRE SL1 4EF, UK

1 INTRODUCTION

Plants and microorganisms are of great importance to the pharmaceutical industry, both as a source of many drugs and as a vast reservoir of chemical diversity for screening programmes aimed at new drug discovery. High throughput screening of microbial samples began with the "Golden Era" of antibiotic discovery[1] in the 1940's and 1950's and has continued to yield important and extremely successful drugs and agrochemicals with a wide range of biological activities such as the serum cholesterol lowering agent mevinolin,[2] the immunosuppressant cyclosporin,[3] the β-lactamase inhibitor clavulanic acid[4] and the antiparasitic avermectins.[5] Screening programmes based on plant samples have been particularly successful in identifying clinically useful anticancer agents such as vincristine/vinblastine[6] and the semi-synthetic epi-podophyllotoxin derivatives etoposide[7] and teniposide. Two of the more recent natural product-derived drugs to be launched are also anticancer agents from plants where the initial discoveries were made by screening: the microtubule assembly-promoter taxol[8,9] and the topoisomerase I inhibitor topotecan, which is a semi-synthetic derivative of camptothecin.[8, 10]

Xenova initially based its biologically-active natural product discovery operations on the screening of microbial fermentation extracts and subsequently decided to use plants as a source of additional chemical diversity. Two examples of screening targets which illustrate how microbial and plant extracts can fruitfully be used as complementary sources of chemical diversity for drug discovery are provided by Xenova's CNS and multi-drug resistance (MDR) programmes.

2 MICROBIAL AND PLANT RESOURCES: COMPARISON OF PRODUCTIVITY

2.1 Microbial Resource

Xenova's microbial culture collection consists of over 24,000 isolates, approximately two-thirds of which are fungal strains and one third filamentous bacteria. The taxonomic breakdown of this collection into different genera is illustrated in Figure 1.

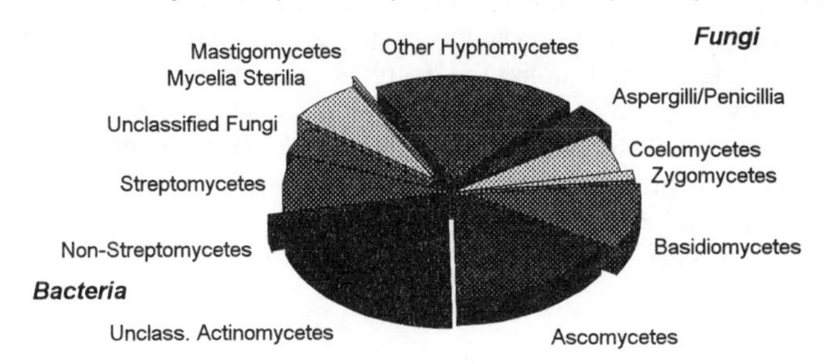

Figure 1 *Taxonomic breakdown of microbial culture collection*

The rationale behind using this unique collection of fungi and actinomycetes as a source of chemical diversity for drug discovery is based upon interactions between organisms with the focus on rare organisms from unusual ecological habitats. These considerations assume that different physiological activities which allow the existence and survival in distinct ecological niches are the reflection of specific metabolic capabilities of those organisms. Taxonomic classification and source ecology are the criteria initially used for collecting organisms. Classifications of the fungi by source ecology and of the actinomycetes by climatic zones and source ecology are shown in Figure 2.

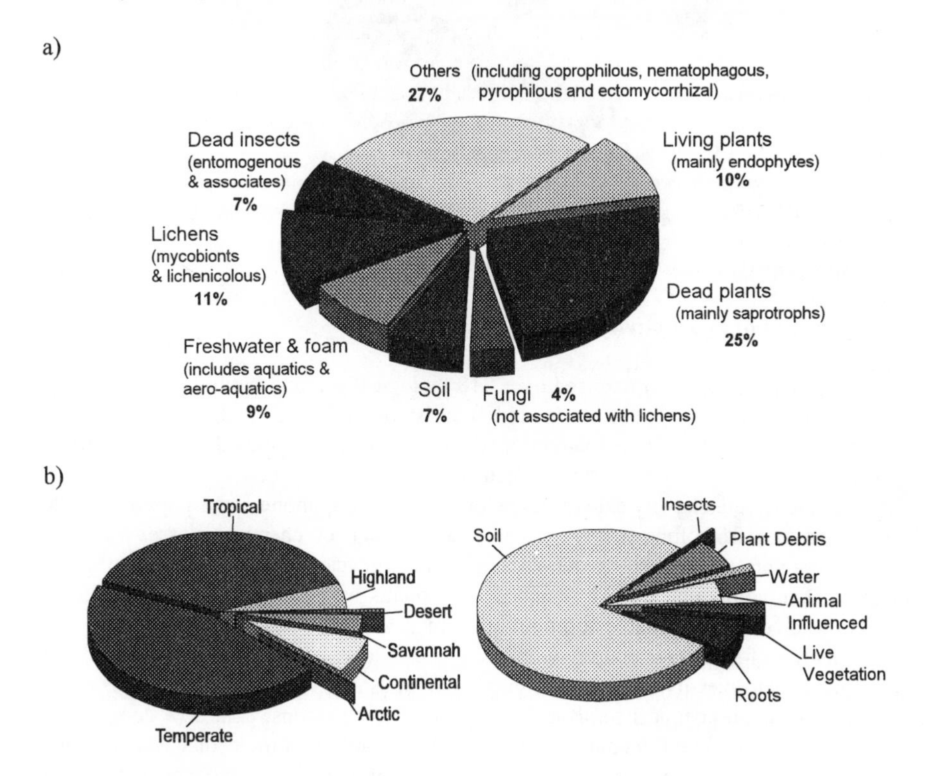

Figure 2 a) *Classification of fungi by source ecology,* ***b)*** *Classification of actinomycetes by climatic zones and source ecology*

Following the isolation of pure cultures, analysis of metabolic capacities through chemical profiling and nutrient utilisation patterns is performed to ensure maximum biosynthetic diversity within the collection. Conditions which promote growth and more importantly stimulate the production of small molecules are investigated experimentally for different ecological and taxonomic groups. The results of these experiments are then compiled to design a variety of fermentation conditions and systems with the aim of capturing the full biosynthetic potential of the organisms.

2.2 Plant Resource

Xenova has access to over 7,000 plant species, collected principally by collaborators in Asia and Africa and targeting both plants with known medicinal properties and species for which there is little phytochemical or ethnobotanical data. The major sub-classes of plants are well represented as illustrated by the taxonomic breakdown shown in Figure 3. All extracts are treated using proprietary technologies to remove interfering substances prior to screening.

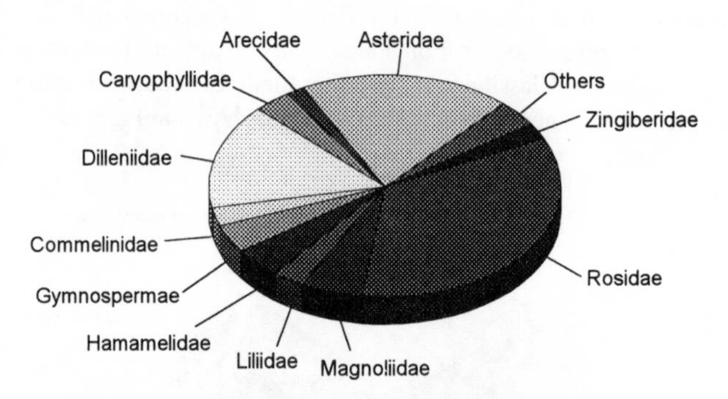

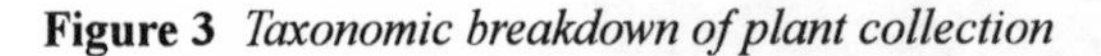

Figure 3 *Taxonomic breakdown of plant collection*

2.3 Comparison of Productivity

To date Xenova has characterised over 318 biologically active natural products from high-throughput screens. These compounds consist of molecules with novel biological activities and structurally related analogues, which are widely distributed across the major chemical classes of naturally occurring molecules, as shown in Figure 4.

The actinomycetes have yielded a range of macrolides, quinones, and peptides, while active compounds from fungi have covered a wide range of chemical classes with the largest representation amongst oxygen-containing heterocycles, terpenoids, quinones and aromatics. The compound classes obtained from plants have been similar to the fungal output but also contain a valuable contribution from alkaloids.

The degree of structural novelty within this portfolio is high (37%) indicative of the inherent molecular diversity within the originating organisms. The novel compounds (117) comprise 53 different chemical series and 17 new carbon skeletons. The 318 compounds comprise over 120 different chemical series and fall primarily in a molecular weight range of 200-600 daltons, which compares well with that of the top selling small molecule pharmaceutical products.

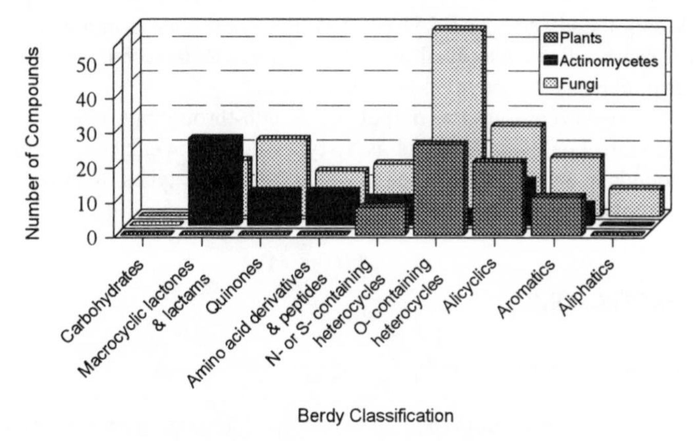

Figure 4 *Analysis of natural products characterised at Xenova by chemical class according to the 'Berdy Classification'*[1]

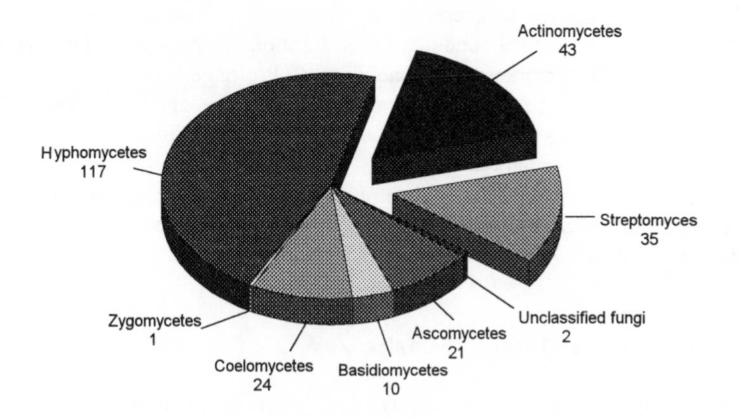

Figure 5 *Distribution of microbial metabolites characterised at Xenova by taxonomy*

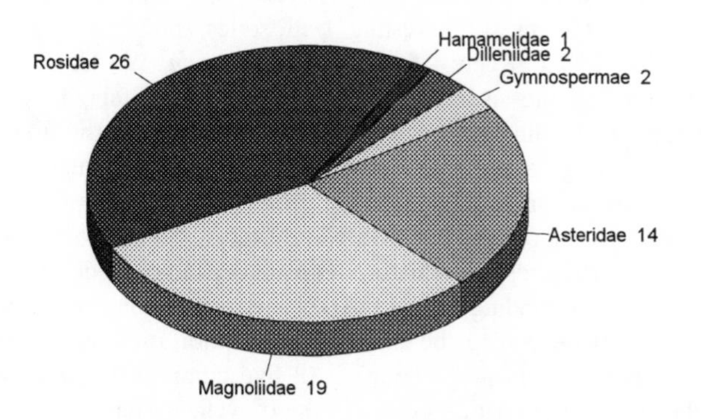

Figure 6 *Distribution of plant metabolites characterised at Xenova by higher plant subclass. The sub-division Gymnospermae is included for convenience.*

The 318 compounds comprise 253 (95 novel) of microbial origin and 65 (22 novel) derived from plants. Both microbial and plant derived compounds are derived from a broad biological diversity, as shown in Figures 5 and 6.

These compounds represent the output of 51 high-throughput assays, each of which can be classified into one of three broad assay types, namely: enzyme, receptor and cell-based. Compounds have been discovered in each of the three types, with a slightly greater average productivity observed to date for cell-based assays.

3 CNS PROGRAMME

3.1 Objective and Screening Assay

Benzodiazepines with tranquillising and anticonvulsant activity act by modulating the major inhibitory neurotransmitter, gamma-aminobutyric acid (GABA). Activation of the $GABA_A$ receptor by GABA agonists causes the chloride channel to open, and the resulting influx of chloride anions inhibits neuronal firing by generating hyperpolarisation. The effect of benzodiazepines is to increase the frequency of channel opening without significantly changing the channel conductance or duration of opening. A large number of benzodiazepines have been marketed as tranquillisers. Prolonged benzodiazepine therapy, although effective, can lead to the development of dependence and has a number of side effects including increasing tolerance and sedation. Acute withdrawal can result in rebound anxiety. Xenova embarked on a CNS programme to find selective, orally active agonists of the $GABA_A$-benzodazepine receptor with potential for the treatment of anxiety and epilepsy. The screening assay was based on the binding of flunitrazepam to $GABA_A$-benzodiazepine receptors in a synaptosome membrane preparation from ox cerebral cortex.[11]

3.2 Results of Screening Microbial Samples

Inhibitors of flunitrazepam binding to $GABA_A$-benzodiazepine receptor preparations were detected in a number of microbial fermentation samples and several of these were refermented and their active components purified and characterised, including the fungal macrocyclic lactones curvularin from a *Cordyceps* sp. and dehydrocurvularin from *Aspergillus carneus*. The most interesting lead series, however, was obtained from fermentations of the hyphomycete, *Acremonium strictum* W. Gams. Assay-guided purification resulted in the isolation of five related metabolites with $GABA_A$-benzodiazepine receptor binding activity. Structure elucidation of the first member of the series was achieved through a combination of spectroscopic techniques which indicated a novel carbon skeleton of mixed biosynthetic origin, consisting of a sesquiterpenoid humulene ring fused to an oxygenated tricyclic system.[11] This compound was named Xenovulene A®. The structures of the naturally-produced xenovulenes and their IC_{50}'s for inhibition of flunitrazepam binding to the $GABA_A$-benzodiazepine receptor are shown in Figure 7 (a manuscript describing the structure elucidation of xenovulenes B, C, D and xenovulene A-7,8-epoxide is in preparation). All five compounds possess the humulene ring. Most of the structural variation occurs in the tricyclic system fused to the humulene ring which contains a cyclopentenone moiety in Xenovulene A®, tropolone moieties in xenovulenes B and C and a diphenolic moiety in xenovulene D. The tricyclic portions of these molecules are probably polyketide-derived. Studies to confirm the biosynthetic origin of the xenovulenes are currently in progress.

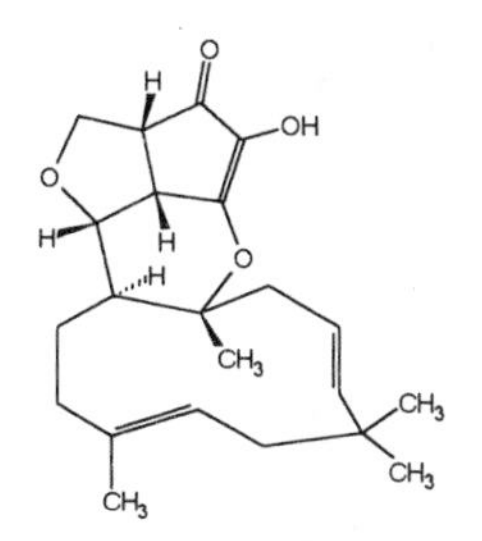

Xenovulene A® (40nM)

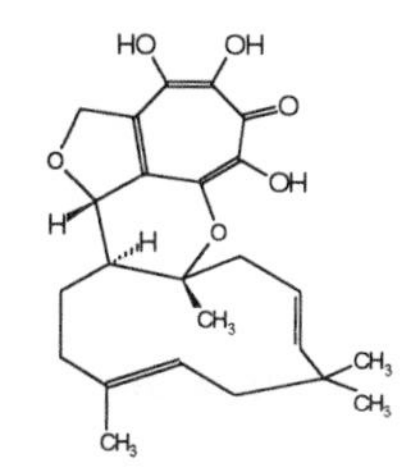

Xenovulene B (5μM)

Xenovulene C (20μM)

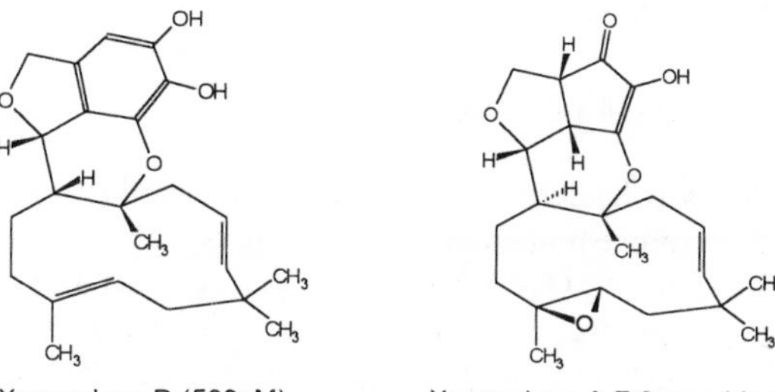

Xenovulene D (500nM)

Xenovulene A-7,8-epoxide (40nM)

Figure 7 *Structures and* IC_{50}*'s for* $GABA_A$*-benzodiazepine receptor binding of the xenovulenes*

The most potent and selective member of the series is Xenovulene A®. Its low initial fermentation titre was increased by more than 100-fold to enable detailed chemical and biological evaluation.[12] Secondary biological evaluation showed that Xenovulene A® is an extremely selective $GABA_A$-benzodiazepine receptor agonist, acting only on a subset of receptors (unpublished). Efficacy has been established in *in vivo* models, with little sign of induction of sedation. Profile optimisation has been explored through synthesis of a number of semi-synthetic analogues of Xenovulene A®.

3.3 Active Plant Metabolites

Screening of plant extracts resulted in the detection of a number of samples which inhibited binding of flunitrazepam to the $GABA_A$-benzodiazepine receptor. The two most active compounds obtained on bioassay-guided purification of these samples were the flavone oroxylin A from *Oroxylum indicum* (Bignoniaceae) and the furocoumarin pimpinellin from *Heracleum nepalense* (Umbelliferae). The structures and IC_{50}'s of these compounds are shown in Figure 8. Both compounds had been isolated previously: oroxylin A from *Gomphrena martiana* (Amaranthaceae)[13] and pimpinellin from *Pimpinella saxifraga* (Umbelliferae).[14]

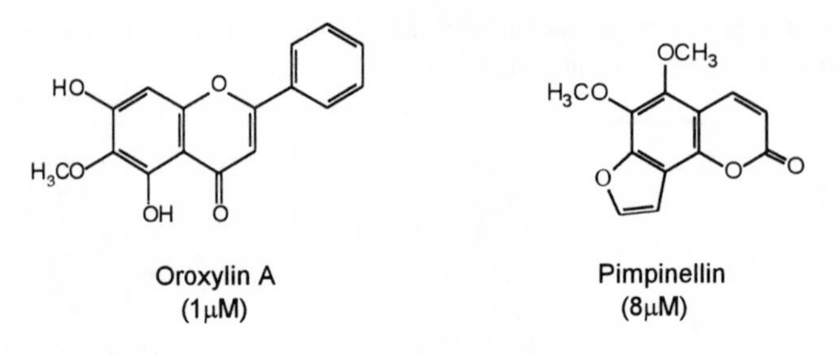

Figure 8 *Structures and $GABA_A$-benzodiazepine-binding IC_{50}'s of oroxylin A and pimpinellin*

4 MDR PROGRAMME

4.1 Objective and Screening Assay

Cancer treatment is hindered by the failure of chemotherapy. Some cancer types are intrinsically resistant to chemotherapy whereas others, although initially sensitive, develop resistance during treatment. One MDR phenotype is characterised by cross resistance to a broad range of cytotoxic drugs differing in both their structures and their mechanisms of action. These drugs include natural products such as the *Vinca* alkaloids, anthracyclines, podophyllotoxins and taxol. This "classical" MDR has been linked clinically to the overexpression of the *mdr1* gene product, P-glycoprotein (P-gp).[15, 16] P-gp is a 170 kD transmembrane ATP-dependent efflux pump which is expressed normally in a wide range of human tissues and shows homology to bacterial transport proteins. Its physiological role is presumed to be excretion of xenobiotics. Compounds which reverse the MDR phenotype have been identified, such as the cardiovascular drug verapamil and cyclosporin A, but have mostly proven ineffective in man to date because of adverse side effects. Xenova's MDR programme was established to discover agents which reverse P-gp mediated drug efflux with potential use in the treatment of chemoresistant solid tumours and leukemias. The screening assay was a cell-based assay monitoring accumulation of tritiated daunomycin.

4.2 Active Microbial Metabolites

The most interesting microbial metabolites with MDR-modulating activity were discovered at an early stage of the programme. They were produced in fermentations of a *Streptomyces* sp. and identified as a series of diketopiperazine derivatives, the structure of one member of which is shown in Figure 9. These compounds were capable of restoring the sensitivity of the MDR cancer cells with moderate potency, potentiating the cytotoxicity of doxorubicin, vinblastine and taxol. Although less potent than some of the plant metabolites discovered later (see Section 4.3) these compounds were highly amenable to synthetic modification. Analogues with enhanced potency were synthesised and shown to cause true inhibition of the MDR pump with a clean pharmacological profile. A drug development candidate has been selected and has been approved for phase 1 clinical trials.

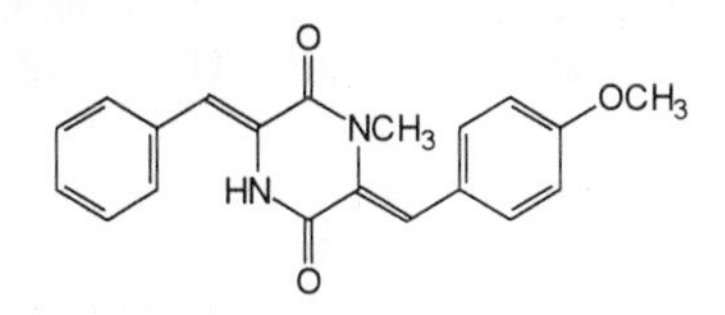

Figure 9 *Structure of a member of the diketopiperazine series of MDR modulators*

Other MDR-modulating activities produced by microorganisms were found to be caused by mycotoxins, such as verruculogen and fumitremorgin B from *Aspergillus fumigatus*.

4.3 Results of Screening Plant Extracts

MDR-modulating activities were detected in a number of plant samples. The two most intersting series of compounds were obtained from *Cynanchum otophyllum* (Asclepiadaceae) and *Piper sarmentosum* (Piperaceae).

Two pregnane steroid glycosides were isolated from the methanolic extract of the root of the Yunanese liana, *Cynanchum otophyllum*, a Chinese medicinal plant. The first of these was identified as the known *C. otophyllum* metabolite otophylloside B, which has been previously reported to exhibit anti-epilepsy properties.[17] Otophylloside B has three sugar units linked to the steroid nucleus. The second active metabolite was identified as a novel dicymaroside analogue of otophylloside B, caudatin dicymaroside, in which the terminal oleandrose sugar is missing (manuscript in preparation). The structures and IC_{50}'s in the MDR drug accumulation assay of these two compounds are shown in Figure 10. On comparison with cyclosporin A in secondary assays, otophylloside B proved to have more potent drug potentiation properties and to be signficantly less toxic. These compounds were not progressed further as they were less amenable to synthetic approaches than the diketopiperazine series obtained from microbial screening.

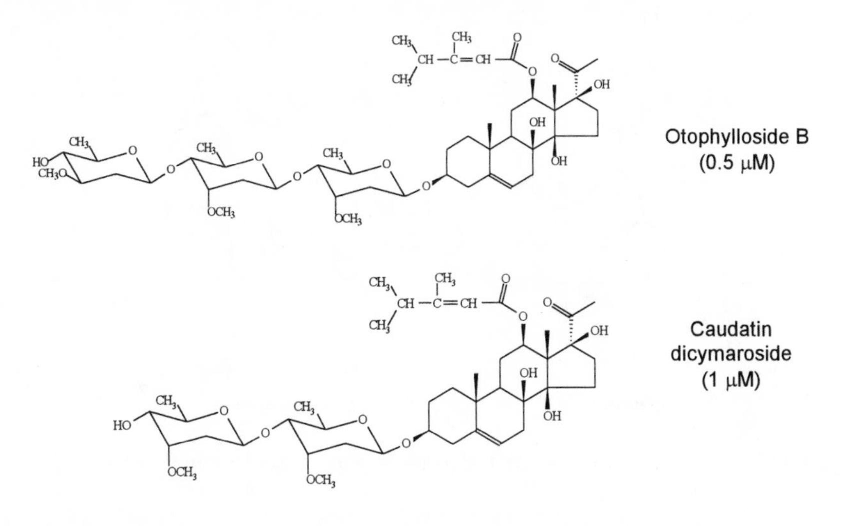

Figure 10 *Structures and IC_{50}'s in MDR drug accumulation assay of otophylloside B and caudatin dicymaroside*

Bioassay-guided purification of extracts of *Piper sarmentosum* (Piperaceae) led to the isolation of three novel related metabolites with MDR-modulating activities in the low micromolar range. Spectroscopic studies showed that these compounds, the piperdimerines, have interesting structures, each containing a *tert*-substituted cyclohexene ring, two methylyenedioxyphenyl units and two piperidine amide units (manuscript in preparation). The structures and IC_{50}'s of the piperdimerines are shown in Figure 11.

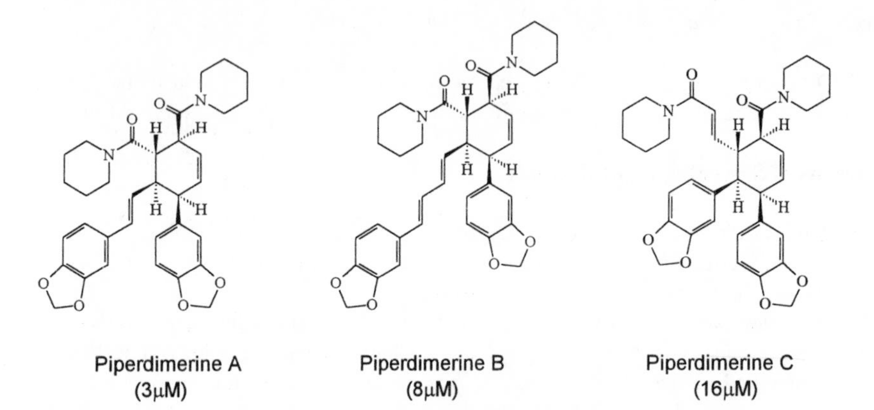

Figure 11 *Structures and IC_{50}'s in the MDR drug accumulation assay of the piperdimerines*

The piperdimerines can be considered as Diels-Alder adducts formed by the addition of two units of piperine as shown in Figure 12. This is likely given the position of the double bond in the cyclohexene ring and that the product is stereospecifically syn with respect to the molecule of piperine which behaves as the dienophile. Although piperdimerine A was active in secondary assays, it was also toxic, resulting in a very narrow therapeutic window.

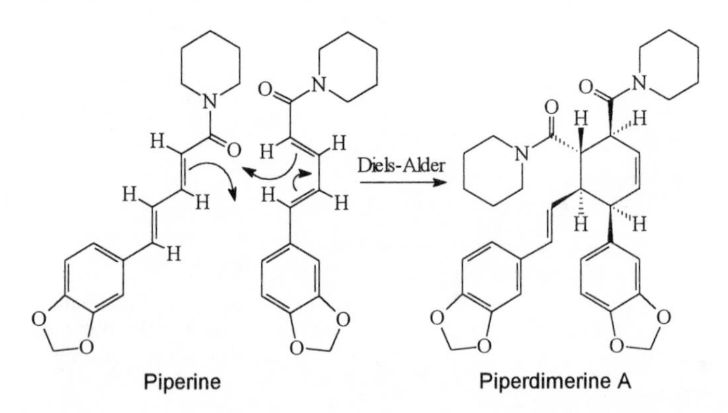

Figure 12 *Possible route of formation of piperdimerine A from two molecules of piperine*

5 CONCLUSIONS

Biological diversity continues to provide an important source of chemical diversity for drug discovery. Xenova's experiences show that compounds isolated from a range of microorganisms and plant species have novel biological activities and that the combination of plants and microorganisms provides access to an extensive range of chemical classes and novel structures. Lead compounds can be discovered by natural product screening which are not accessible by other technologies. These compounds often occur as families of related compounds which can provide valuable preliminary structure-activity relationships.

One of the major challenges to natural products drug discovery from other sources of chemical diversity is to reduce the time necessary for lead compound identification. For natural products the lead identification process usually involves assay development and screening, refermentation/resupply of plant material, assay-guided purification and structure elucidation. Xenova has recently re-engineered its drug discovery processes through implementation of new dereplication methods, stringent selectivity strategies and the powerful use of databases. It has reduced the time for discovery of lead compounds by over 50% so that drug discovery projects can be completed within one year. The timeliness of natural products discovery now compares favourably with other approaches.[18]

ACKNOWLEDGEMENTS

The authors are grateful to many members of Xenova staff, past and present, who contributed to the work described here including Jacquie Holloway, Ian Chetland, Mike Luscombe, Liam Evans, Dave Kau, Trevor Gibson and Mo Latif.

REFERENCES

1. J. Bérdy, *Adv. Applied Microbiol.*, 1974, **18**, 309.

2. A. W. Alberts, J. Chen, G. Kuron, V. Hunt, J. Huff, C. Hoffman, J. Rothrock, M. Lopez, H. Joshua, E. Harris, A. Patchett, R. Monaghan, S. Currie, E. Stapley, G. Albers-Schonberg, O. Hensens, J. Hirshfield, K. Hoogsteen, J. Liesch and J. Springer, *Proc. Natl. Acad. Sci. USA*, 1980, **77**, 3957.

3. J. F. Borel, C. Feurer, H. U. Gubler and H. Stahelin, *Agents Actions*, 1976, **6**, 468.

4. C. Reading and M. Cole, *Antimicrob. Ag. Chemother.*, 1977, **11**, 852.

5. J. C. Chabala, H. Mrozik, R. L. Tolman, P. Eskola, A. Lusi, L. H. Peterson, M. F. Woods, M. H. Fisher, W. C. Campbell, J. R. Egerton and D. A. Ostlind, *J. Med. Chem.*, 1980, **23**, 1134.

6. N. Neuss and M. N. Neuss, *The Alkaloids*, 1990, **37**, 229.

7. H. F. Stahelin and A. von Wartburg, *Cancer Res.*, 1991, **51**, 5.

8. M. E. Wall, 'Chronicles of Drug Discovery', D. Lednicer (ed.), ACS, Washington D.C., U.S.A., 1993, Vol. 3, p. 327.

9. L. Lenaz and M. D. De Furia, *Fitoterapia*, 1993, **64** (suppl.), 27.

10. M. Potmesil, *Cancer Res.*, 1994, **54**, 1431.

11. A. M. Ainsworth, M. I Chicarelli-Robinson, B. R. Copp, U. Fauth, P. J. Hylands, J. A. Holloway, M. Latif, G. B. O'Beirne, N. Porter, D. V. Renno, M. Richards and N. Robinson, *J. Antibiotics*, 1995, **48**, 568.

12. M. Blackburn, U. Fauth, W. Katzer, D. V. Renno and S. Trew, *J. Ind. Microbiol.*, 1996, **16**, in press.

13. C. A. Buschi, A. B. Pomilio and E. G. Gross, *Phytochemistry*, 1981, **20**, 1178.

14. D. L. Dreyer, *J. Org. Chem.*, 1970, **35**, 2294.

15. J. M. Ford and W. N. Hait, *Pharmacol. Rev.*, 1990, **42**, 155.

16. C. K. Pearson and C. Cunningham, *Trends Biotechnol.*, 1993, **11**, 511.

17. M. Quanzhang, L. Jianrong and Z. Qianlan, *Scientia Sinica (Series B)*, 1986, **29**, 295.

18. M. Moore, *J. Biomolecular Screening*, 1996, **1**, 19.

Bioactivity-Guided Isolation of Phytochemicals from Medicinal Plants

M. Iqbal Choudhary* and Atta-ur-Rahman

H. E. J. RESEARCH INSTITUTE OF CHEMISTRY, UNIVERSITY OF KARACHI, KARACHI-75270, PAKISTAN

1 INTRODUCTION

Phytochemicals were the only source of medicines for human and live-stock until the end of the last century. During the current century they have been largely replaced by synthetic pharmaceuticals. However, even to date, one-quarter of all prescription drugs are of plant origin despite the fact that less than 5% of plant species have been investigated. Indeed many of the synthetic medicines currently in clinical use have been developed from lead compounds originally derived from natural sources. It is widely accepted that natural product chemistry surpasses the kind of chemistry that synthetic chemists can ever imagine accomplishing in the laboratory. Phytochemical diversity in terms of structural novelty is unprecedented in laboratory synthesis.

Plants provide enormous chemical diversity. Advances in bioassay screening, isolation techniques and structural elucidation have shortened and facilitated the process of drug discovery from medicinal plants. It is now general practice among natural product chemists to use some type of bioassay to direct the progress of phytochemical investigation towards the pure bioactive isolates.

Scheme-**1** summarizes logical choices of bioassay approaches based on available knowledge about vegetative materials while scheme-**2** is a typical flow chart of a bioactivity-guided process for the isolation of bioactive phytochemicals from medicinal plants.

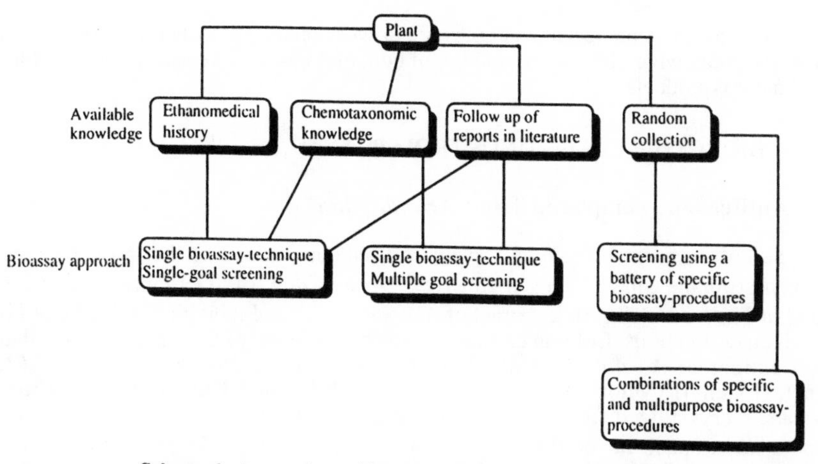

Scheme-1: Approaches to bioassay-directed natural product drug development

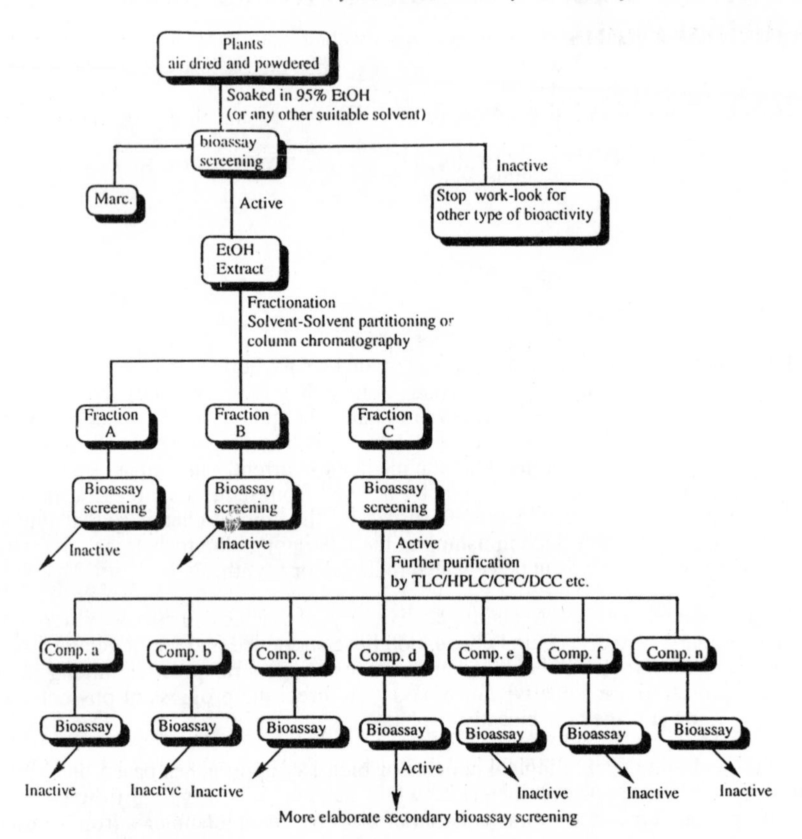

Scheme-2: A Typical bioactivity-guided isolation scheme.

Following are some recent examples of our work on the phytochemistry of medicinal plants and fungi which led to the isolation of both biologically interesting and structurally novel natural products.

2 BIOACTIVE COMPOUNDS FROM PLANTS

2.1 Antifeedant compound from *Arundo donax*

The boll weevil, *Anthonomus grandis* Boheman, is one of the most destructive cotton pests in the United States of America[1]. *Arundo donax* L (Germinae) is abundantly found in Thailand[2]. The crude extract of *A. donax* gave total inhibition of feeding of boll weevil on cotton bolls. Column chromatographic separation of the chloroform-methanol (1:1) extract of *A. donax* resulted in the isolation of tetramethyl-*N*,*N*-bis(2,6-dimethylphenyl) cyclobutane-1,3-diimine (**1**). Compound **1** was obtained as square crystals, recrystallized from hexane-ethyl acetate (9:1). The high-resolution electron-impact mass spectrum of this compound gave M^+ at *m/z* 346.2387 measured for molecular formula $C_{24}H_{30}N_2$, (calcd. 346.2409) which indicated eleven degrees of unsaturation. The IR spectrum showed characteristic absorption bands at 2945, 1584 and 14.51 cm^{-1} A ^1H NMR singlet at δ 1.25 integrating for 12 protons was consistent with four equivalent methyl groups. Another singlet at δ 2.10 integrating for 12 protons was consistent with four equivalent methyl groups attached to an aromatic ring and a multiplet at δ 7.00 (six protons for two trisubstituted benzene rings). The ^{13}C-NMR spectrum exhibited a signal for a methyl group at δ 18.0, signals at 22.1 and 22.2 for quaternary and methyl (attached to aromatic) carbons, respectively and three signals at δ 123.8, 126.2 and 128.3 corresponding to benzene ring carbons. Complete ^{1}H- and ^{13}C-NMR chemical shift assignments of **1** are presented in Table-**1**.

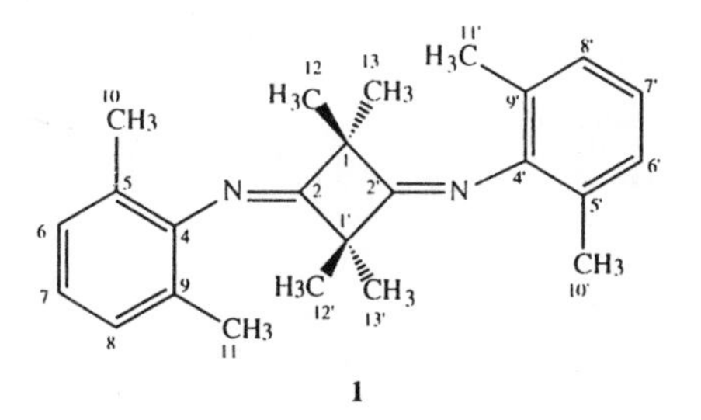

1

From the spectral data given above, **1** could be tentatively assigned as a structure which had two trisubstituted benzene rings with two methyl groups attached to each ring. Two of the degrees of unsaturation are due to the two C=N fragments which are indicated by the IR spectrum and the molecular formula $C_{24}H_{30}N_2$. The C=N fragments can be connected to benzene rings as shown in structure **1**. The above fragment affords a total of ten degrees of unsaturation. According the molecular formula and information above, the last degree of unsaturation can reasonably belong to a four-membered ring with two *gem*-dimethyl groups as shown in structure **1**.

To confirm this structure as tetramethyl-*N,N*-bis(2,6-dimethyl-phenyl) cyclobutane-1,3-diimine, X-ray crystallographic analysis was performed. Crystals of **1** were formed in the monoclinic space group $C_{2/c}$ with **a** = 11.918 (7), **b** = 11.204 (5) **c** = 15.745 (10)Å, and β = 95.08 (5°). All unique diffraction maxima were collected in range of $0^o \leq 2\theta \leq 55.0^o$ graphite monochromated M_o Kα radiation (0.71073 Å). Of the 2408 unique reflections 1515 (63%) had ([Fo])>5σ ([Fo]) and were judged observed. The structure was solved by multisolution direct methods and refined by full matrix least squares techniques[3]. The final discrepancy index for the observed data was 0.064. A computer-generated drawing of the final X-ray model is given in Figure-**1**. Hydrogen atoms are omitted for clarity. This is the first report of tetramethyl-*N,N*-bis(2,6-dimethylphenyl) cyclobutane-1,3-diimine[4] as a natural product and confirmation of its structure by X-ray.

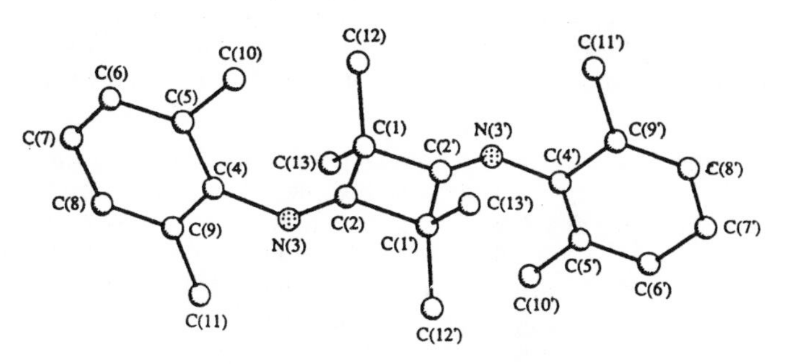

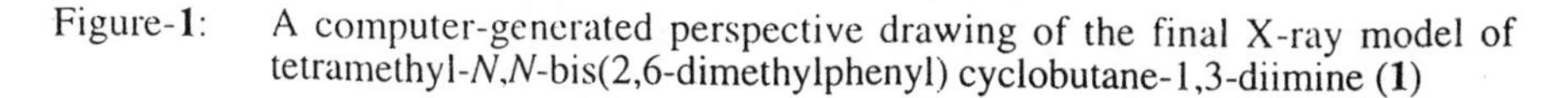
Figure-**1**: A computer-generated perspective drawing of the final X-ray model of tetramethyl-*N,N*-bis(2,6-dimethylphenyl) cyclobutane-1,3-diimine (**1**)

Compound **1** showed 54% inhibition of feeding against boll weevil at a dose of 0.5mg, as determined by agar plug bioassay feeding stimulant procedure developed by Hedin *et al.*[5]

2.2 An Anticholinergic Steroidal Alkaloid from *Fritillaria imperialis*

The genus *Fritillaria* comprises seven species belonging to the family Lilaceae[6]. It is an ornamental plant found in Kashmir at altitudes of 7000-9000 ft[7]. The bulbs of *Fritillaria* plants are bitter and find use in the treatment of sore throat, cough and haemoptysis[8]. The genus *Fritillaria* is a rich source of steroidal bases of different types. Phytochemical studies carried out by our research group on *Fritillaria imperialis*, collected from Turkey, which have resulted in the isolation of several new steroidal bases.

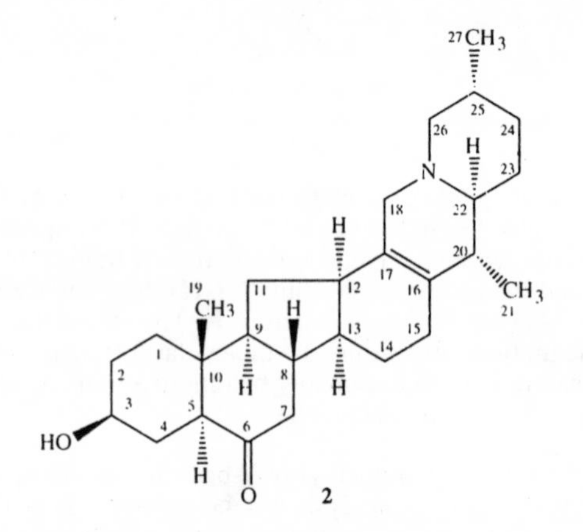

Ebeinone (**2**), $C_{27}H_{41}NO_2$ was isolated as a brown amorphous solid from the ethanolic extract of *F. imperialis* through various column and thin-layer chromatographies. The HREI MS of **2** showed the molecular ion at *m/z* 411.3089 corresponding to the molecular formula $C_{27}H_{41}NO_2$ and indicated the presence of seven degrees of unsaturation in the molecule. Base peak at *m/z* 112 arose by the cleavage of the C-17/C-18 and C-20/C-22 bonds, as commonly encountered in ceveratrum-type alkaloids[9]. Its UV spectrum showed terminal absorptions, while its IR spectrum afforded strong absorptions at 3614 (OH), 2909 (quinolizidine) and 1710 (C=O) cm^{-1}.

The ^{1}H-NMR spectrum ($CDCl_3$, 400 MHz) showed a singlet at δ 0.65 assigned to the C-19 tertiary methyl protons, whereas two doublets, integrating for three protons each, at δ 0.90 (J = 7.4 Hz) and 1.02 (J = 7.2 Hz) were due to the C-27 and C-21 methyl protons. A one-proton multiplet at δ 3.55 was ascribed to the C-3 proton, geminal to the hydroxyl group.

The ^{13}C-NMR spectrum ($CDCl_3$, 100 MHz) of **2** showed resonances for 27 carbon atoms. The signals at δ 132.2 and 131.0 were attributed to the tetrasubstituted olefinic carbons while the signal at δ 70.8 was assigned to the hydroxyl-bearing C-3 methine. Another signal at δ 211.1 was attributed to the ketonic carbonyl carbon. The HMQC spectrum served to establish the direct $^1H/^{13}C$ one-bond connectivity. Complete ^{13}C-NMR chemical shift assignments of **2** and $^1H/^{13}C$ one-bond shift correlation of protonated carbon atoms of **2** as determined by HMQC spectrum are shown in Table-**1**. The $^1H/^{13}C$ long-range coupling information was obtained from the inverse heteronuclear multiple bond connectivity (HMBC) experiment. This experiment was extremely useful for the confirmation of the position of double bond and ^{13}C-NMR chemical shift assignments of the quaternary carbon atoms (Table-**1**).

Table-**1**: $^{1}H/^{13}C$ Direct one-bond shift correlations of **1**, **2**† and ^{13}C-NMR chemical shift assignments of **8**#.

	1		**2**		**8**
Carbon No.	^{13}C δ	^{1}H δ	^{13}C δ	^{1}H δ	^{13}C δ
1	22.1 (-C-)	–	37.2 (CH_2)	1.35, 1.40	71.8 (CH)
2	* (-C-)	–	30.4 (CH_2)	1.80, 1.85	37.7 (CH_2)
3	–	–	70.8	3.55	66.0 (CH)
4	128.3 (-C-)[∂]	–	30.0 (CH_2)	1.40, 1.45	36.2 (CH_2)
5	126.2 (-C-)[∂]	–	56.6 (CH)	2.15	41.9 (CH)
6	123.8(CH)[∂]	7.00	211.1 (-C-)	–	32.9 (CH_2)
7	123.3(CH)[∂]	7.00	30.1 (CH_2)	1.85, 1.95	38.8 (CH_2)
8	123.3CH)[∂]	7.00	42.0 (CH)	1.65	36.1 (CH)
9	126.2 (-C-)[∂]	–	54.8	1.62	54.9 (CH)
10	22.2 (CH_3)	1.25	38.3 (-C-)	–	42.3 (-C-)
11	22.2 (CH_3)	1.25	31.6 (CH_2)	2.0, 2.10	207.3 (-C-)
12	18.2 (CH_3)	1.25	33.7 (CH_2)	2.4	136.4 (-C-)
13	18.2 (CH_3)	1.25	132.2 (-C-)	–	146.3 (-C-)
14	–	–	37.2 (CH)	2.60	44.65 (-C-)
15	–	–	22.3 (CH_2)	1.25, 1.30	29.8 (CH_2)
16	–	–	19.3 (CH_2)	1.90, 1.92	30.9 (CH_2)
17	–	–	131.0 (-C-)	–	85.64 (-C-)
18	–	–	46.1 (CH_2)	2.45, 2.55	12.2 (CH_3)
19	–	–	12.3 (CH_3)	0.65	17.1 (CH_3)
20	–	–	42.8 (CH)	1.72	40.3 (CH)
21	–	–	15.3 (CH_3)	1.05	10.8 (CH_3)
22	–	–	63.4 (CH)	2.52	66.6 (CH)
23	–	–	23.7 (CH_2)	1.45, 1.50	77.3 (CH)
24	–	–	29.6 (CH_2)	1.35, 1.40	27.4 (CH)
25	–	–	23.8 (CH)	1.72	31.6 (CH)
26	–	–	63.6 (CH_2)	2.95, 3.00	54.7 (CH_2)
27	–	–	17.6 (CH_3)	0.96	18.8 (CH_3)
28	–	–	–	–	–

†Determined by HMQC spectrum, #Multiplicity was determined by DEPT spectra, * signal was not visible, ∂signals are interchangeable

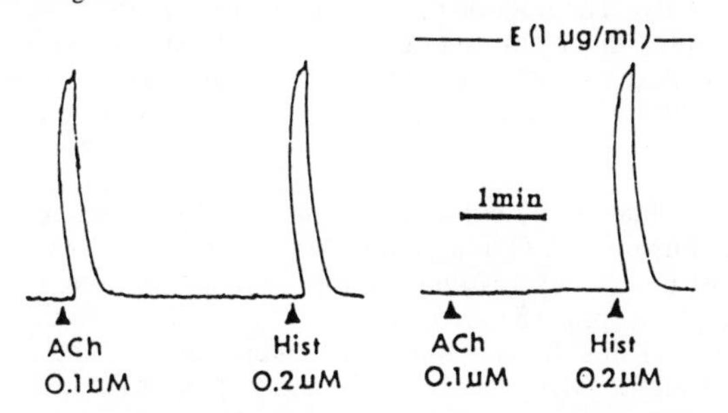

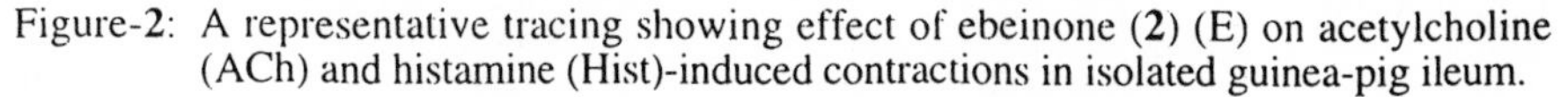

Figure-2: A representative tracing showing effect of ebeinone (**2**) (E) on acetylcholine (ACh) and histamine (Hist)-induced contractions in isolated guinea-pig ileum.

Compound **2** at concentrations of 1 μg/ml completely blocked contractile responses of acetylcholine while the histamine-induced contractions remained unaltered (Figure-**2**), suggesting the specific blockade of acetylcholine receptors by ebeinone.

2.3 Nematicidal Alkaloids from *Datura fastuosa*

Plant parasitic nematodes along with other pests cause significant crop losses in Pakistan[10]. According to a report, the root-like nematode was found to be responsible for a major loss of the banana crops in the Province of Sindh, Pakistan[11]. The commercially available synthetic nematicidal compounds are usually very expensive and often toxic and non-biodegradeable, causing environmental problems.

Datura fastuosa (Solanaceae) abundantly grows in Pakistan and the crude extract is widely used in the treatment of a variety of diseases such as ulcers, poisonous bites, pills, diarrhoea, asthma etc. in folk medicine. *D. fastuosa* is also used as an insecticide[12].

The dried plant material was extracted with ethanol. The extract was evaporated to a gummy residue (labelled as FO1), dissolved in water (labelled as FO2) and further extracted in organic solvents at different pH values. The resulting extracts were subjected to nematicidal bioassays on the larvae of root-knot nematode. The results of these experiments are summarized in Table-**2**.

Table-**2**: Effects of different fractions of *D. fastuosa* plant on larval mortality of root-knot nematode.

				% Mortality	
No	Extract	Conc.	24 Hrs	48 Hrs	72 Hrs
	Control		00.0	0.80	10.0
1	FO1	1000 ppm	89.0	100.0	100.0
2	FO2	"	90.0	100.0	100.0
3	FO3	"	90.0	97.0	99.0
4	FO4	"	00.0	0.10	07.0
5	FO5	"	04.0	68.0	73.0

The chloroform extract (FO3) obtained at pH 7.0 showed 99% nematicidal activity at concentration of 1000 ppm. The other most active extract was obtained at pH 9 labelled as FO5, while the extract in ethyl acetate at pH 9 and chloroform extract were found to have negligible activity.

The active extract FO3 was fractionated by column chromatography . The fractions were again tested for activity. The fraction containing the highest nematicidal activity was further purified using thick-layer chromatography to yield six compounds. All these compounds were examined for activity. Only one compound was found to have significant nematicidal activity. Spectral studies confirmed the compound to be tigloidine (**3**).

The active fractions obtained by column chromatography of the extract FO5 were combined and separated using HPLC in a system MeOH-buffer (0.2 M Na_2HPO_4) in a ratio of 25:75 at λ 230 nm. This resulted in the isolation of four known tropane alkaloids, apoatropine (**4**), tropine (**5**), hyoscyamine (**6**), and (-)-6β-tigloyloxytropane-3α-ol (**7**). Apoatropine (**4**) and hyoscyamine (**6**) were found to be inactive in the bioassay, while (-)-6β-tigloyloxytropane-3α-ol (**7**) and tropine (**5**) exhibited significant nematicidal activity.

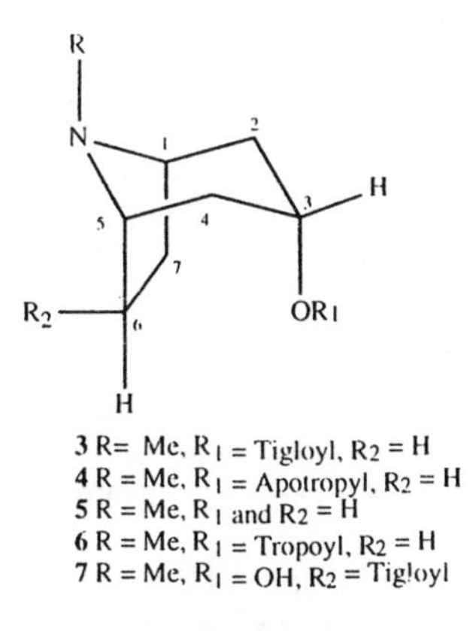

Table-**3**: Effects of different alkaloids isolated from *D. fastuosa* on larval mortality of root-knot nematodes.

				% Mortality	
No.	Compounds	Conc.	24 Hrs	48 Hrs	72 Hrs
. 1	Tigloidine (3β-tigloy-loxytropane) (**3**)	1000 ppm	50.0	60.0	84.0
2	(-)-6β-tigloyloxytro-pane-3α-ol (**7**)	"	44	77	83
3	Tropine (3α-hydroxytropane) (**5**)	"	04	10	47
4	Apoatropine (**4**)	"	00	01	08
5	Hyoscamine (**6**)	"	00	04	09

2.4 An Antihypertensive Alkaloid from the Rhizomes of *Veratrum album*

Veratrum species (Liliaceae), also called as white hellebore are found in central and southern parts of Europe[13]. These species occupy a prominent place among pharmacologically active[14] and insecticidal plants[15]. Our chemical studies on the *Veratrum album* Linn of Turkish origin have resulted in the isolation of a large number of new steroidal alkaloids. 1-Hydroxy-5,6-dihydrojervine (**8**), jervinone (**9**) and *O*-acetyljervine (**10**) are representative examples of bioactive steroidal alkaloids isolated from this plant by our research group.

1-Hydroxy-5, 6-dihydrojervine (**8**), $C_{27}H_{41}NO_4$, was isolated as a white amorphous solid from the ethanolic extract of *V. album*. The HREI MS of **8** showed the molecular ion peak at *m/z* 443.3020 which is in agreement with the molecular formula $C_{27}H_{41}NO_4$ (calcd. 443.3035). The base peak at *m/z* 126 and the peak at *m/z* 110 arose by the cleavage of the piperidine ring (ring E). Fragments appearing at *m/z* 114 ($C_6H_{12}NO$) and *m/z* 438 (M^+-CH_3) were also apparent in the EI MS of **8**. The UV spectrum of this compound showed an absorption at 249 nm attributed to the α, β-unsaturated cyclopentenone system. The IR spectrum displayed absorption bands at 3600, 1700 and 1620 cm^{-1} due to N-H, α, β-unsaturated carbonyl and C=C stretching vibrations respectively.

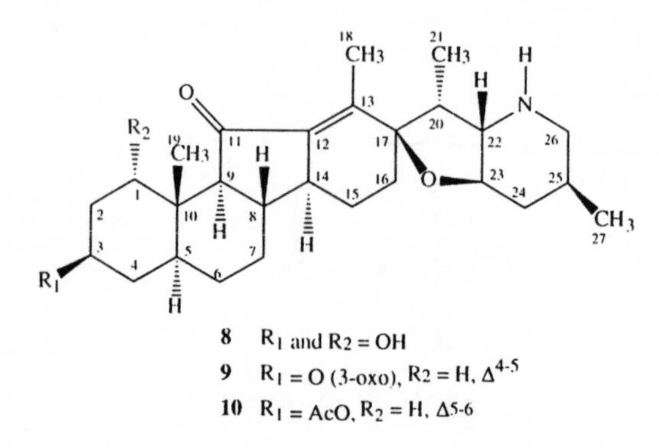

The ^{1}H-NMR spectrum of the compound **8** was similar to those of jervine and *O*-acetyljervine[16]. The spectrum showed two three-proton doublets at δ 0.92 and 0.96 ascribed to the C-27 and C-21 secondary methyl protons, respectively. Another doublet at δ 2.13 (J = 2.2 Hz) was attributed to the C-18 methyl protons showing homoallylic coupling with the C-14 methine proton. A broad multiplet at δ 41.0 ($W_{1/2}$ = 15 .0 Hz) was due to the C-3α proton, geminal to the hydroxyl group. A one-proton double doublet at δ 4.80 (J_1 = 4.6 Hz, J_2 = 2.4 Hz) was attributed to the C-1 proton, geminal to the hydroxyl group.

The ^{13}C-NMR spectrum (broad-band and DEPT) showed resonances of all twenty seven carbon atoms, comprising five quaternary, ten methine, eight methylene and four methyl carbons. The detailed chemical shift assignments of **8** are given in Table-**1**.

In anaesthetized rats, jervinone (**9**) (1-30 μg/kg) caused a fall in blood pressure in a dose-dependent manner. Table-**4** shows the combined effects of different doses of jervinone. The antihypertensive effect was very brief, returning to normal within one minute. The heart rate was not affected significantly except at higher doses (30 μg/kg) which produced a small degree of tachycardia. Noradrenaline (1 μg/kg) produced pronounced increase in the blood pressure and the pretreatment with jervinone (**9**) did not alter the vasconstrictor response to noradrenaline which rules out the possibility of the involvement of adrenoreceptor. Acetylcholine at doses of 1 μg/kg produced a decrease in mean arterial blood pressure comparable to that of jervinone (**9**) at 19 μg/kg. Pretreatment with atropine (1 mg/kg) completely abolished the antihypertensive response to jervinone (10 μg/kg) as well as to that of acetylcholine (1 μg/kg). *O*-Acetyljervine and 1-hydroxy-5,6-dihydrojervine also produced falls in blood pressure at doses of 10-300 μg/kg with a small degree of tachycardia. Antihypertensive effects of *O*-acetyljervine (**10**) and 1-hydroxy-5,6-dihydrojervine (**8**) were not modified by pretreatment of atropine, suggesting that the mechanisms of the antihypertensive effects of these compounds are different from those of jervinone or acetylcholine.

Table-**4**: Effect of compounds **8**, **9** and **10** on mean arterial blood pressure (MABP) in anaesthetized rats.

Dose (μg/kg)	1	3	10	30	100	300	No. of Obs (n)
8	-	-	-	17.66± 01.45	34.00± 03.46	55.95± 03.17	3
9	38.50± 03.50	49.35± 02.90	59.56± 02.15	75.00± 02.00	+	+	10
10	-	-	26.04± 01.02	34.02± 01.13	66.18± 01.81	+	3

3 STRUCTURALLY NOVEL PHYTOCHEMICALS

3.1 Novel Indazole Alkaloids from the Seeds of *Nigella sativa* L.

Nigella sativa Linn (Ranunculaceae) is an erect herbaceous annual plant found in South Asia and widely cultivated throughout Southern Europe, Central Asia, North Africa, Middle East, Pakistan and India[1]. *Nigella sativa* seeds are used for the treatment of various diseases, commonly believed to have carminative, stimulatory and diaphoretic properties. Our continuing studies on the chemical constituents of *N. sativa* seeds have yielded several new alkaloids. Nigellidine (**11**) is the second naturally occurring compound after nigellicine (**11**a) with an indazole ring.

Nigellidine (**11**) was isolated from the seed of *Nigella sativa* by column and thin-layer chromatography. The HREI MS showed the molecular ion at *m/z* 308.1527 corresponding to the molecular formula of $C_{19}H_{20}N_2O_2$, requiring eleven degrees of unsaturation. The strong UV absorptions (MeOH) at 230, 280 and 328 indicated extended conjugation in the molecule while the IR spectrum of the compound exhibited strong absorption bands at 3450 (OH) and 1620 (C=C) cm^{-1}. The ^{1}H-NMR spectrum of the methiodide salt displayed two three-proton singlets at δ 2.57 and 3.83 indicating the presence of $ArCH_3$ and OCH_3 groups, respectively. The ^{13}C-NMR spectrum showed resonances at δ 23.0 and 56.3 attributed to the methyl and methoxy carbons. Four methylene carbons resonated at δ 20.7, 21.3, 28.0 and 49.2 and were assigned to C-12, C-11, C-13 and C-10, respectively. The assignments were confirmed by DEPT, GASPE and HETERO COSY experiments (Table-**5**).

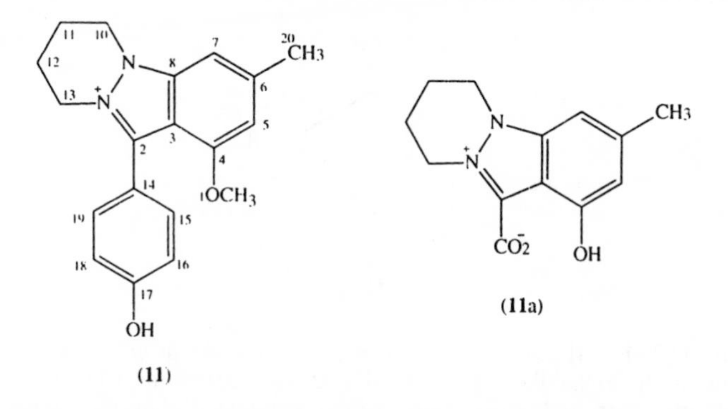

The structure of compound **11** was unambiguously determined by X-ray diffraction studies. In order to obtain suitable crystals, both the methyliodide and methylchloride salts of **11** were prepared and crystallized from aqueous methanol. The methylchloride salt gave better data and its crystals formed in the triclinic space group P_1 with a=9.334(3), b = 14.050(4), c = 14.862(4) Å, α= 81.83(2), β= 74.80 (2) and γ = 87.26(2). A reasonable density required four molecules in a unit cell and two units with the composition $C_{19}H_{21}N_2O_2Cl.H_2O$ formed an asymmetric unit. All unique diffraction maxima with $1\theta<105$ were collected using $2\theta:\theta$ scans and graphite monochromated CuKα radiation (λ = 1.5417Å). A total of 3931 unique reflections were collected and 2413 (61%) of these were judged observed ($F_o^2>3\sigma F_o^2$) and used in subsequent calculations. A computer- generated drawing of the final X-ray model of nigellidine is given in Figure-**3**. Both crystallographically independent molecules have the same conformation. All the atoms of rings B and C as well as C-10 and C-13 of ring A are in one plane . The methyl attached to O-1 *i.e.* C-22 (Figure-3) comes from the methylchloride used for the salt formation. The natural product exists as a zwitterion.

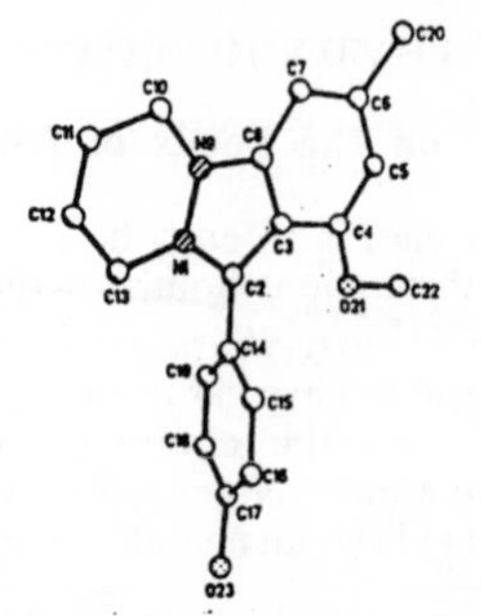

Figure-3: A computer-generated perspective drawing of the final X-ray model of nigellidine (**11**).

3.2 A Novel Isoprenylpolyketide from *Stachybotryis bisbyi*

Stachybotrys bisbyi (Srinivasan) is a soil fungus reported from various countries of East Europe, South Asia, and Egypt[18]. We isolated *S. bisbyi* from the seeds of *Oryza sativa* and initially grew it on potato dextrose agar plates. Extracts from ten day-old actinic fungus cultures grown in a liquid potato-based medium were assayed against a number of pathogenic bacteria. The crude extracts had significant activity against *Staphylococcus aureus* and this assay guided the subsequent isolation. Column chromatography of the combined organic extracts of *S. bisbyi* liquid culture yielded bisbynin (**12**) as a crystalline solid.

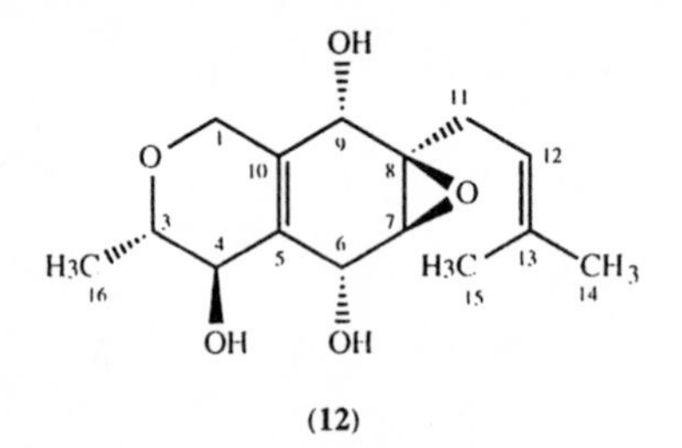

(12)

Bisbynin (**12**) had the molecular ion in the HRCI MS at *m/z* 283.1548 (M^{+}-H)- corresponding to the molecular formula $C_{15}H_{22}O_5$ ($\Delta = 0.03$ mmu). The UV spectrum of **12** showed only terminal absorption, which suggested that the five degrees of unsaturation required by the formula involved no extended conjugation. The ^{1}H-NMR spectrum of bisbynin (**12**) showed well resolved resonances for all twenty protons and ^{1}H-NMR chemical shift assignments of **12** is given in Table-**5**. The ^{13}C-NMR spectrum (broad-band decoupled, DEPT) showed resonances of all fifteen carbon atoms (Table-**5**).

The structure of bisbynin (**12**) was unambiguously defined by X-ray diffraction studies. The compound crystallized in the monoclinic space group $P2_1$ with **a** = 6.7867 (15), **b** = 8.781 (2), **c** = 12.248 (4) Å and $\beta = 97.32(2)^{o}$. One molecule of composition $C_{15}H_{22}O_5$ formed the asymmetric unit. The computer-generated drawing of the final X-ray model of bisbynin is given in Figure-**4**. The dihydropyran ring has a half-chair conformation while the central ring has a half-boat conformation with the hydroxyl groups at C-9 and C-6 in a pseudo-diaxial disposition. This arrangement leads to intramolecular hydrogen bonding between the hydroxyls (2.38 Å).

Table-5: $^1H/^{13}C$ Direct one-bond shift chemical shift correlations of **11**† and **12**†.

	11		12	
Carbon No	^{13}C δ	^1H δ	^{13}C δ	^1H δ
1	–	–	57.5(CH_2)	3.98,4.28
2	1615 (-C-)	--	--	--
3	145.1(-C-)	--	52.2(CH)	3.40
4	156.1(-C-)	--	55.9(CH)	3.84
5	106.9(-C-)	6.70	126.5(-C-)	--
6	144.1(-C-)	--	67.0(CH)	4.68
7	102.3(CH)	7.14	50.0(CH)	3.18
8	148.8(-C-)	--	51.9(-C-)	--
9	–	--	58.8(CH)	3.89
10	49.2(CH_2)	4.45	123.9(-C-)	--
11	21.3(CH_2)	2.30	21.9(CH_2)	2.31,2.81
12	20.7(CH_2)	2.20	109.2(CH)	5.20
13	48.0(CH_2)	4.57	120.0(-C-)	--
14	*	--	9.4(CH_3)	1.72
15	133.9(CH)	7.41	9.2(CH_3)	1.65
16	116.5(CH)	6.90	31.5(CH_3)	1.28
17	*	--	--	--
18	116.5(CH)	6.90	--	--
19	133.3(CH)	7.41	--	--
20	23.0(CH_3)	2.57	--	--
O-CH3	56.3	3.83	--	--

†Determinned by HETERO COSY spectrum *Signals were not visible

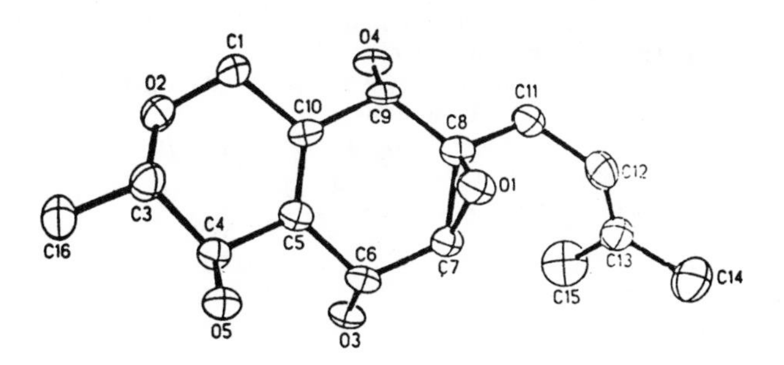

Figure-4: A computer-generated perspective drawing of the final X-ray model of bisbynin (**12**).

ACKNOWLEDGEMENTS

The research work was financially supported by Office of the Naval Research, U.S.A. through grant No. N00014-86-G-0229. We are also grateful to Prof. Bilge Sener (Gazi University, Ankara, Turkey), Prof. Jon Clardy (Cornell University, Ithaca, New York, U.S.A.) and Prof. W.H.M.W. Herath (Medical Research Institute, Colombo, Sri Lanka) for their collaboration. We wish to acknowledge our students, Dr. Athar Ata, Dr. Rahat Azhar Ali, Dr. Samina Naz, Mr. Afgan Farooq and Miss Durr-e-Shahwar, who were involved in various aspects of work mentioned in this paper.

REFERENCES

1. R. L. Ridgway, E. P. lloy and W. H. Cross, '*Cotton Insect Management With Special References to the Boll Weevil, Agriculture Handbook No. 589*', United State Department of Agriculture, 1993, 1.
2. D. H. Miles, K. Tunsuwan, V. Chittawong, U. Kokpol, M. I. Choudhary and J. Clardy, *Phytochemistry*, 1993. **34**, 1279
3. G.M. Sheldrik, '*SHELXTL Crystallographic Computing System*' (Nicolet Instrument Division, Madison, Wisconsin).
4. S. W. Pelletier, H. P. Chokski, H. K. Dasai, *J. Nat. Prod.* 1986, **49**, 829.
5. P.A. Hedin, A. C. Thompson and J. P. Minyard, *J. Econ. Entomol.*, 1966, **59**, 1811.
6. L. H. Bailey, '*Manual of Cultivated Plants*', MacMillan Company, New York, 1966, 218.
7. '*The Wealth of India*', CSIR, New Delhi, 1956, **4**, 63.
8. L. M. Perry, '*Medicinal Plants of East and South East Asia*', MIT Press, Cambridge, Massachusetts, 1980, 236.
9. R. Shakirov, R. N. Nuriddinov, S. Yu Yunusov, *Khim. Prir. Soedin.*, 1965, **1**, 384.
10. A. Ghaffar, '*Soilborne Diseases Research Center (Final Research Report, 1986-88)*', Dept. of Botany, University of Karachi, Pakistan, 1988, 110.
11. M. A. Maqbool, S. Hashmi and A. Ghaffar, *Problem of Root-knot Nematode in Pakistan and Strategy for their Control* ' in "*Advances in Plant Nematology*", (Ed. M. A. Maqbool, A. M. Golden, A. Ghaffar, L. R. Krusberg), NNRC, University of Karachi, Karachi-75270, Pakistan, 1988, 294.
12. R. C. Roark, '*Experts from Consular Correspondence Relating to Insecticides and Fish Posion Plants*', U.S. Dep. Agric. Bur. Chem. and Soils.
13. V. E. Tayler, L.R. Brady, J. E. Robbers, '*Pharmacognosy*', Lea and Febiger, Philadelphia, 1988, 238.
14. J. Fried, H. L. White, O. Wintersteiner, *J. Am. Chem. Soc.*, 1950, **72**, 4621.
15. D. G. Crosby, '*Naturaly Occurring Insecticides*', (M. Jacobson, D. G. Crosby Eds), Marcel Dekker, New York, 1971, 177.
16. Atta-ur-Rahman, R. A. Ali, T. Parveen , M. I. Choudhary, B. Sener and S. Turkoz, *Phytochemistry*, 1990, **30**, 368.
17. R. N. Chopra, S. L. Nayar and I. Chopra, '*Glossary of Indian Medicinal Plants*', CSIR Pulication, New Delhi, 1956, 176.
18. A. Kubatova, *Czech Mycology*, 1994, **47**, 151.

The *Dictionary of Natural Products* and Other Information Sources for Natural Products Scientists

John Buckingham and Stephen Thompson

CHAPMAN & HALL, 2–6 BOUNDARY ROW, LONDON SE1 8HN, UK

1 INTRODUCTION

The title of this conference makes it clear that the slant of our contribution should be towards an industrial audience, and we have therefore put together a paper with the emphasis on the practicalities of information sources available to natural products scientists in industry. There will be an inevitable bias towards the virtues of the *Dictionary of Natural Products* (DNP), since this is a database with which we work on a day-to-day basis, but we hope also to give a not-too-biased account of both the current deficiencies of DNP and of the virtues of other information products.

Chemists and other scientists working at the 'coalface' of natural products research are busy people. The rapid advances in isolation and screening techniques that have taken place during what might be termed the new wave of natural products research during the last few years, have generated a strong demand for new information handling methods to take care of the large amount of data that is generated both within each active company and in the public domain.

1.1 Usage of Natural Product Information Sources

To the best of our knowledge and experience, the demands for natural product information made on the total information system within each company can be classified under the following headings:

1.1.1 Dereplication This is a vital 'bread-and-butter' everyday function of the company. An indication of the importance of rapidly-accessible databases to this activity can be glimpsed, for example, by the fact that the late Jim French named his database of selected UV spectra of natural products DEREP in cognisance of the principal use to which it would be put. It is estimated that the value of each dereplication achieved within a research group is approximately $50,000.[1]

Dereplication requires a compound-comprehensive database that can be searched by structural fragments (including especially multiple fragments, e.g. four methoxy groups in the same molecule, as detected spectroscopically), with the aim of maximising the speed with which unhelpful isolations can be detected and eliminated. Ideally it needs cross-searching between spectroscopic peak data (especially UV, since this is the spectroscopic technique most routinely used in HTS) and other types of data, e.g. molecular weight range.

1.1.2 Chemical structure and subgroup-related searching This addresses questions such as 'what other natural products related to the compound of current interest are known, and where do they occur?' It needs a compound-comprehensive database that is authoritative and well-organised, gives an accurate (though not necessarily comprehensive) account of natural source, and an authoritative schema of interrelationships between related natural products. The extent to which the last requirement is fully satisfied by substructure searching is discussed below.

1.1.3 Taxonomically-related searching This addresses questions such as 'which organisms related to the one already investigated are likely to throw up interesting leads', or in simple terms 'where do we look next?', for which some kind of taxonomic dimension is required. The ideal database for this kind of search is occurrence-comprehensive as well as compound-comprehensive, and features a taxonomic relationship schema, although it is not necessary for it to have the full range of functionality that a true taxonomic database has.

This type of search may often be carried out by life scientists, taxonomic specialists working alongside or within and advising the isolation group.

1.1.4 Random browsing and idea generation The importance of this should not be overlooked. It requires a well-structured environment allowing navigation through various types of searching and a well-designed user interface that minimises the probability of the user losing, or losing sight of, valuable information.

It is implicit in the description of these types of search that the 'target audience' for new information products is now the end-user him(her)self as well as the professional information intermediary. The last few years have shown the shift from the latter to the former taking place at a greater or lesser speed in most research institutions, again with the pharmaceutical industry leading the way. The role of the intermediary has become more and more that of facilitator, familiar enough with the many new available products to be able to advise on their quality and suitability for the perceived purpose, making them available and ensuring that they work, providing appropriate training and induction for the end-user, but not always carrying out actual searching. This shift has mirrored the shift in information product design in response to the development of the end-user computing environment and the shift in the *kind* of complexity that products offer.

Effective use of the older products required good familiarity with command languages (for example, Messenger™ software for STN on-line databases) which tend to be subject-general, in other words applicable across the whole of chemistry or the whole of science. Mastery of these products involves, to a large extent, achieving sufficient familiarity with these commands to produce reliable results and avoid errors of incorrect search formulation that may make searches by non-experts less than reliable. (However the more recent introduction of new-generation front-end software such as STN Express has made searching by nonexperts much more user-friendly.)

The new wave of end-user based information products such as DNP have eliminated many of these technical complexities. However in their place they have added another range of complexities arising from the search strategies now possible. These tend to be subject-specific; the possible search strategies mirror the inherent complexities of the data itself, and imply greater subject specialisation on the part of the searcher, with corresponding ability to 'play' with the product so as to be able to rapidly reformulate the search parameters as results are obtained.

What then are likely to be the principal attributes of the 'ideal' information product for natural products specialists, in an era of end-user interactive searching? We suggest the following:

1. Comprehensive, certainly in recording all known natural product structures, and ideally in documenting all isolation records.
2. Searchable across all types of data that may be relevant for the natural products chemist, including spectroscopic data.
3. Accurate and well-maintained. Given the complexity of the subject, this probably means the involvement of a small team of subject specialists in building the product.
4. Topical.
5. User-friendly and with intuitive searching. Again, this is likely to mean that the search parameters have been designed with input from specialist natural product scientists.
6. Searchable both by text and substructure, and with rapid transfer of results between the two.

At this point the reader will no doubt expect us to demonstrate the ways in which DNP fulfils all of the above-mentioned requirements. It would be good to be able to do this at the present time, but we have to be honest and state that there are still certain shortcomings, which we are working hard to rectify. Before reaching this point, however, it would be appropriate to review briefly some of the other major information sources that are available to natural product scientists.

1.2 Electronic Natural Product Sources

1.2.1 Chemical Abstracts The virtues of CAS, including its newer manifestations such as SciFinder, are so well-known as to not bear repetition here and will be familiar to all members of the audience. We hope, therefore, that we will be forgiven by our friends at CAS if we mention here only a few deficiencies relevant to searching CAS for natural products information.

1. CAS does not flag natural products in the compound registry and many searches, especially substructure searches, throw up a large number of hits including a mixture of natural, synthetic and semisynthetic compounds. Of course, these extra hits may well be of interest but it is useful to be able to navigate first in a product that covers natural products only. Special strategies are necessary to search the CAS registry file for natural product information, using screens such as formula weight, carbon count, structure fragments, bioactivity and taxonomy,[2] but these are subject to a varying proportion of false drops and missed hit errors.
2. Compound registration is not 100% perfect; there was a period in the 1960s before the introduction of computerised compound registration when some natural products were incompletely or inaccurately registered, and natural products groups that were undergoing intensive research at that time (for example, the steroidal alkaloids) are not well documented in CAS.
3. In more recent years the brevity of CAS abstracts can cause difficulties in determining what natural products have been isolated in a particular paper and registered by CAS unless the reader has access to the paper itself.

4. CAS nomenclature since the 9th. Collective Index period (1973 -) is largely unintuitive for the natural products specialist and is driven, quite reasonably, by the need for a comprehensive scheme capable of unambiguous naming of all compounds. The nomenclature system in turn drives the systematic numbering scheme which may fluctuate between closely-related compounds and will mostly not correspond to the natural product specialist's preferred scheme.

5. Registration of compounds by CAS and the allocation of CAS registry numbers is associated with a large amount of redundancy. This is inevitable given the topicality of the CAS database and its cumulativity. Thus a compound, especially a natural product, may be registered and allocated a CAS number based on incomplete structural information. Subsequent reports in CAS may then register the structural identity to a greater degree of detail even though the identity of the compound has not changed. This leads to duplications and the more numerous 'quasi-duplications' describing the compound at different levels of structural, stereochemical or tautomeric detail. Once reported these remain in the system, although a true duplication may be resolved by CAS at a later date. One of the functions of a more heavily edited information source like DNP is to draw attention to such redundancy for the benefit of the user, without in any way making unwarranted and possibly inaccurate 'corrections' to CAS core information.

Having listed these points, it is important to re-emphasise that one of the functions of a specialist information resource such as DNP must be to 'mesh' in a meaningful way with the vast CAS database and to enable users to turn to CAS when appropriate and use it accurately and at minimum cost. Thus, for example, a function of DNP is to resolve for natural products the perceived ambiguities of CAS number allocation and to map from the DNP user-friendly nomenclature for natural products to the CAS name, either directly, or via the CAS number when the CAS name is excessively cumbersome.

1.2.2 Beilstein The Beilstein database too is probably well known to nearly all members of the audience. It shares with CAS the disadvantage of not having been constructed with natural products specialists principally in mind, with consequent user-unfriendliness of nomenclature and structure description.

One important feature of Beilstein, however, is the presence of a natural occurrence marker field (INP, Isolation from Natural Products) that can be incorporated into searches to restrict the hit set.

The high cost of searching on the Beilstein file is a drawback for the occasional user and accentuates the desirability, as with CAS on-line, of having some means of first formulating and refining queries using a product which does not impose time pressure, before carrying out a precise search that minimises expenditure. Crossfile searching between Beilstein, CAS and other files including NAPRALERT can be carried out on STN International.

1.2.3 NAPRALERT Napralert is an extensive database covering the isolation and pharmacology of natural products and bioactive extracts, assembled since 1975 by Norman Farnsworth, Chris Beecher and their group at the University of Illinois at Chicago. It is the world's most comprehensive resource on the medicinal aspects of biological species, currently incorporating over 120000 citations. It is essentially an abstracts database with added features such as an extensive index of pharmacological activity types linked to the abstracts.

The database is very powerful for searching all biological aspects, but it is currently not substructure searchable and chemical searching is possible only using the chemical names given in the original paper, which may be idiosyncratic.

Napralert is currently available for text searching on STN, or via the Napralert helpdesk at Chicago using fax or email.

The Napralert group are currently working with Chapman & Hall and our software developers to produce a CD-ROM version of the database, which will become available in 1997. Future developments may include closer collaboration with the Chapman & Hall database to provide substructure-based access to the Napralert data.

1.2.4 DEREP This is a UV-centred database of limited size (7000 compounds, approx. 4000 spectra) carefully assembled by the late James F. French on Microsoft FOXPRO™. It is based on literature data and covers the majority of commonly isolated natural products such as flavonoids and antibiotics, and including spectra for mixed antibiotic complexes. Its main function, as the name suggests, is to provide rapid dereplication of frequently encountered 'uninteresting' isolates. The future of DEREP is currently in doubt owing to Dr. French's death in early 1996.

1.2.5 Specialist Natural Products Databases. A number of smaller databases covering certain classes of natural product have developed over the years and are available commercially at differing cost levels.

Marine Natural Products are well served by two databases; the *Marine Natural Products Database* (4000+ compounds) from John Faulkner, California, and *Marinlit* (6000+ compounds) from Blunt and Munro in New Zealand. A comparison of these will not be made as we have not yet had the opportunity to evaluate Marinlit.

Antibiotics similarly have two databases available. The *Berdy Antibiotics Database* (23000 compounds including derivatives) provides a good range of data on most known antibiotics and related compounds. It includes UV spectroscopic data and isolation methods but is not structure searchable, currently being available only as a DOS-based product. *Actfund* is a Japanese database developed by the Umezawa group and documents all natural products isolated from actinomycetes since 1967. It is searchable by text only but includes UV data. It is now available as a Japanese Windows™ 3.1 product (20 MB hard disc required). It is expensive and appears to have few installations outside Japan.

A recent paper[2] reviews the use of most of these files in dereplication, and describes crossfile searching of STN files.

1.3 Printed Natural Product Sources

No natural products worker, particularly one hoping to get an accurate overall view of the complete literature of a certain group, should lose sight of certain invaluable printed resources.

Prominent among large encyclopaedic sources is the monograph *Konstitution and Vorkommen der Organischen Pflanzenstoffe (Exclusive Alkaloide),* compiled by W. Karrer, H. Hurlimann and E. Cherbuliez (Birkhauser Verlag, Basel). The Second Edition, published in 1981 as a main work in two volumes and two supplements (the second supplement in two volumes), contains a complete record of all plant isolations of natural products, except alkaloids, from the earliest days until 1966. This publication therefore provides an excellent resource complementing the CAS registry file which began in 1966, provided, of course, it is borne in mind that some of the structures given in Karrer have been revised since that date. It is published in German with English-language translation of compound names in supplement 2. Series discontinued.

Near-comprehensive treatment of the main groups of natural products was provided during the 1980 s by several specialist dictionaries published by Chapman & Hall such as the *Dictionary of Alkaloids* (1988), the datasets for which have been subsumed into the *Dictionary of Natural Products,* as described below, and are now updated as such.

Among review series, the most prestigious work covering natural products generally is *Progress in the Chemistry of Natural Products* (Springer), which began life in 1938 as *Fortschritte der Chemie Organische Naturstoffe.* The earlier volumes provided many encyclopaedic reviews of particular natural product groups but the more recent volumes, as well as being slimmer, lean more towards reviews of the biochemistry, pharmacology, etc. of smaller groups of natural products, or individual natural products, of current research interest. The function of providing brief critical literature reviews of current research of the isolation, structure determination and biological properties of natural products generally is now fulfilled with a high degree of success by *Natural Products Reports* (Royal Society of Chemistry, bimonthly).

Many groups of natural products are well served by review series in book form, and a good selection is provided from the Chapman & Hall list, notably the extensive treatment of the flavonoids in the multiauthored series edited by J. B. Harborne and others, *The Flavonoid* (1975), the coverage of which is extended by the volumes *The Flavonoids, Advances in Research, ...Advances in Research since 1980* and *...Advances in Research since 1986* (published 1982, 1988, 1994). A similar treatment for another specialised class is provided by R. H. Thomson's series *Naturally Occurring Quinones*, the most recent of which is *Naturally Occurring Quinones III (1987).* Other representative titles are *Lignans, Chemical, Biological and Clinical Properties* by D. C. Ayres and J. D. Loike (1990*)* and *The Biochemistry of the Stilbenoids* by J. Gorham (1994). All of these titles are 'chemically oriented' in that chemical structures and data form a large proportion of their content.

2 THE DICTIONARY OF NATURAL PRODUCTS (DNP)

The *Dictionary of Natural Products* is a sister-dictionary to the well-known *Dictionary of Organic Compounds* (DOC), which has a history of over 60 years. From the earliest days, the aim of DOC was to document in a selective and well-organised manner the most significant organic compounds, and from the First Edition in 1934 a substantial part of its coverage was devoted to reporting what was then known about the most important (and indeed, owing to the difficulty and slowness of isolating and purifying them in those days, this meant virtually all) natural products. In the 1930s many natural product structures were only partially, and often incorrectly, known, and DOC played a valuable role in collecting the available information, with references.

In the late 1970s Chapman & Hall developed a fielded database structure designed to assimilate and organise all of the data required for the new Fifth Edition of DOC (DOC5). At that time electronic publishing was still at a relatively early stage, and the primary purpose of the database remained for several years principally that of a typesetting resource for the printed Dictionary and its supplements. It was designed from the start with a highly fielded structure allowing maximum control of page layout in a typesetting environment.

An important part of the database design for DOC was the concept that it would be a data*bank* which would be capable of maintenance so as to represent an accurate snapshot of the state of development of organic chemistry at any one time, thus facilitating the publication of new Editions of the printed DOC as well as, potentially, electronic versions. This also implied the capability to produce specialised subset dictionaries for the use of researchers in particular areas of chemistry.

When we began to seriously consider the options for subset publications, several natural products areas immediately assumed prominence, firstly because of the recognition that an audience of subject specialists in the various subdisciplines existed, recognising themselves as 'alkaloid scientists' for example, and secondly, because DOC's leading role in natural products documentation over the years made such an expansion of the publishing programme appropriate in terms of our reputation and the expertise of our contributors and editorial staff. It was, however, obvious that considerable further compilation work was necessary to address the needs of the various subject specialists, which would not be met by a straightforward offprint of, for example, the alkaloid entries from DOC5.

Thus work began on a rolling series of printed dictionaries, including *Dictionary of Antibiotics and Related Substances* (published 1987), *Dictionary of Alkaloids* (1989), *Dictionary of Steroids* (including also a wide coverage of synthetic steroids) (1991) and *Dictionary of Terpenoids* (1991).

All of these compilations were as near possible comprehensive in terms of the natural products documented. Again using alkaloids as an example, DOC5 Main Work contained about 1800 entries covering the most important, well-known and structurally typical members of the class. At the time when work began on the *Dictionary of Alkaloids*, the Editorial Team's estimate of the number of known alkaloids put the figure at 6000-7000, and there did not seem to be any reliable way of arriving at an accurate figure short of compiling the Dictionary; the end total was 10000, admittedly some 3 years later than the estimates.

The amount of information collected in these specialised datasets about the occurrence, properties and primary literature was considerably expanded compared with DOC5, and is essentially comprehensive for the rarer natural products: however it still stopped well short of being a comprehensive record of occurrences for widely-distributed compounds (an example is Berberine for which approximately 4000 isolation papers are listed in CA).

The decision to implement electronic dissemination of the database coincided with the emergence of CD-ROM as an attractive delivery medium[3] and our realisation that we were within reach of assembling a comprehensive datasource covering all classes of natural product. During the period of approximately 1989-1991, updating of the datasets for the 'subset' dictionaries was combined with intensive literature work on assembling data for the remaining minor and not-so-minor (e.g. flavonoid) groups. The decision was taken to recognise in publication terms the fact that what had originated as a major expansion and development of a single dictionary, DOC, had by now matured into what were effectively two discrete enterprises. Thus were born sister-dictionaries, one of selected general organics, DOC, and a comprehensive dictionary of natural products, the *Dictionary of Natural Products* (DNP). The size of each dataset (approximately 300 MB including software) was wholly appropriate to publication on a single CD-ROM with space for a number of years' expansion.

DNP on CD-ROM, first published in June 1993, incorporates a flexible text-searching software package from Head Software International,[4] and the well-known PsiBase structure drawing and searching software from Hampden Data Services,[5] the latter modified to permit downloading of the initial screen search results from the CD-ROM to hard disc where the atom-by-atom search is carried out. An important feature of the software is the ease with which hitlist results can be passed back and forth between the two modules, as exemplified by the sample search shown below. DNP is rereleased six-monthly, each new release being a complete replacement for the previous disc. Additional search features have been added since 1993 in parallel with the expansion of the database, and in 1996 an upgraded interface to the Headfast software was implemented, making searching still more intuitive and adding a further tranche of search, display and output capabilities. There are a total of 19 available fields available for text searching of the data on the CD-ROM (Figure 1).

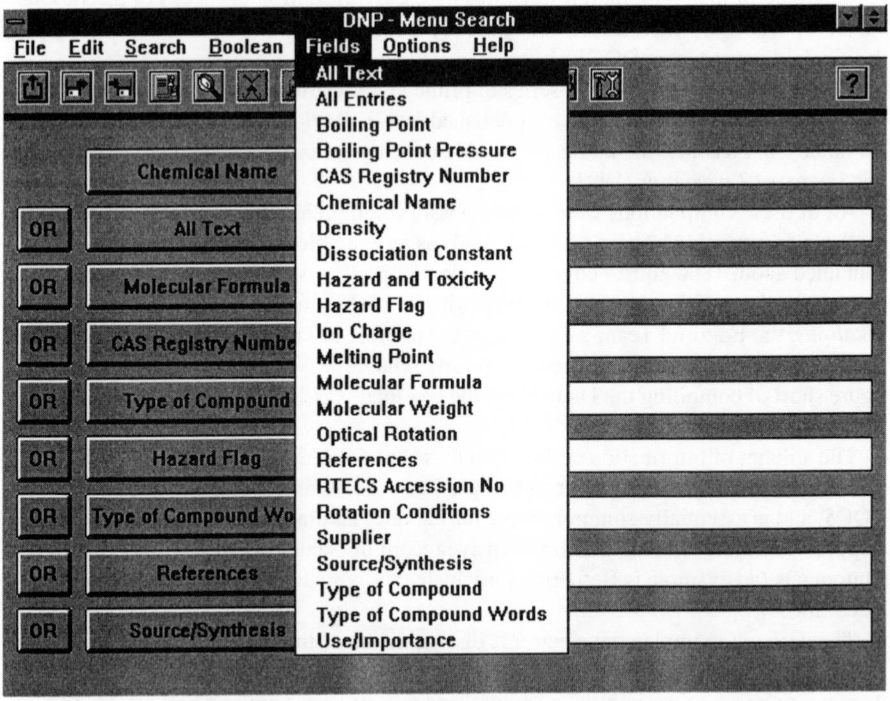

Figure 1. List of Searchable Fields in DNP

Statistics on the successive releases of DNP since the first release (version 1.1, June 1992) give an indication of the rate of growth of the database (Table 1). This growth is somewhat greater than the rate of characterisation of new natural products in the literature, because in the period since 1992, work has continued in identifying and adding to the database natural products not previously documented by us.[6]

Table 1. *Statistics on the Growth of the DNP Database, 1992-1996, as Represented by the Number of Searchable Connection Tables*

These figures do not represent the total of known natural products; the database has searchable structures for some derivatives that are not themselves natural products. The reduced total in the 1993 second release reflected reorganisation of the data and reconciliation of duplicates during the finalisation of the DNP printed edition main work. No new natural products were added to the database during this 6-month period.

1992 (release 1)	80000
1992 (2)	87984
1993 (1)	99791
1993 (2)	98432
1994 (1)	102674
1994 (2)	105047
1995 (1)	108189
1995 (2)	113453
1996 (1)	119550

An important function of the DNP database is to rationalise and organise the scattered literature on natural products by combining data from different sources on related natural products within the same entry, and in the process standardising the structural information so that ready comparisons within and between entries can readily be made. The high-quality structure diagrams (HQSDs) which are a vital feature of the hard-copy version are carried over into the CD-ROM version where they are present in a moveable window (Figure 2).

It is possible to argue that this approach is elaborate in an era of substructure searching in which a database can be rapidly and accurately searched provided the connection tables are accurate. We do not subscribe to this view, if only because the careful editorial study of the structure of the compound necessary for the drawing of a HQSD highlights many errors and inconsistencies in the literature. The conciseness of the database is a feature which all contributors and editors strive to maintain and which minimises the amount of inspection necessary during the various stages from initial query to meaningful results.

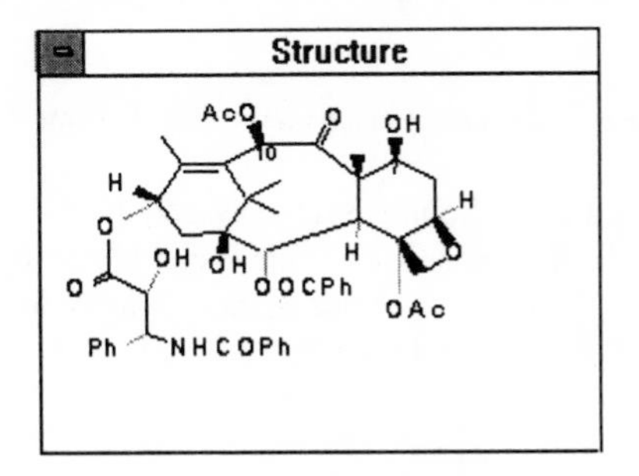

Figure 2. Example of a High Quality Structure Diagram

2.1 Carrying Out Searches with the CD-ROM Version of DNP

The object of this search is to identify all Natural Products with a Molecular Weight in the range 750 - 850 and which have been isolated from *Streptomyces* species. The results of this text search will be further refined using a substructure search to identify compounds containing a tetracycline fragment and a glycoside moiety which satisfy the search criteria.

The menu search facility is used to combine search terms. To enter a particular search term in a field, the term can be entered directly, or much more usefully, a browse index of all the terms in that field can be used. This ensures that all possible variations of the particular indexed term are found, and thus the search which will yield the maximum number of hits (Figure 3).

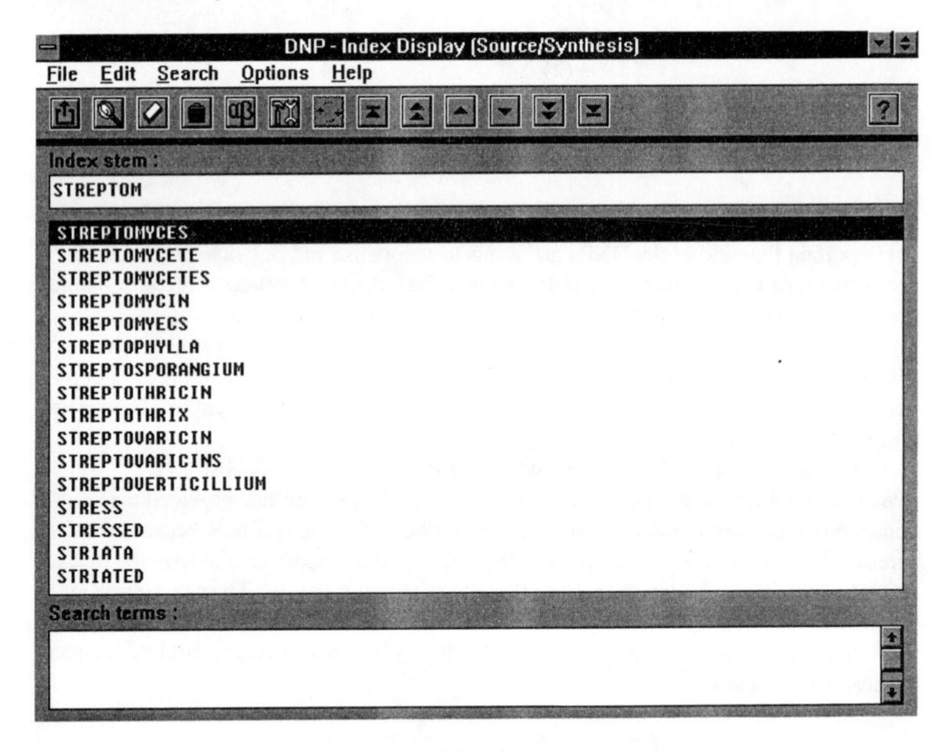

Figure 3. Entering Terms using the Browse Index

Isolation data is indexed under Source/Synthesis, and all possible variants for the term containing Streptomyces have been chosen. This is combined with the Molecular Weight term using Boolean logic (Figure 4). The use of the ~ symbol in the Molecular Weight field denotes searching in the range between 750 and 850.

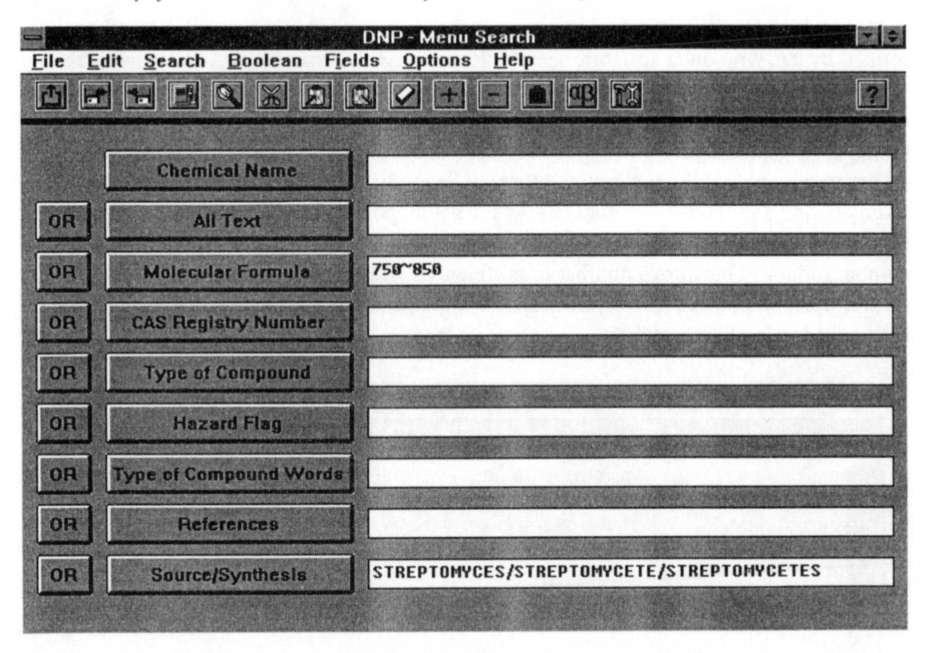

Figure 4. Combining Search Terms in Menu Search

The text search yields a hit list of 300 hits, listed alphabetically (Figure 5). Other options allow this list to be manipulated to display and be sequenced by Molecular Formula or CAS Registry number or printed as a list or as selected individual entries.

DNP - Summary Display (Text search)
File Format Sequence Help
Item 20 of 3292
Refine with Structure search
Name
Acivicin; (αS,5S)-form
Acivicin; (αS,5S)-form, 4S-Hydroxy
Aclacinomycin A
Aclacinomycin A; 9-Hydroxy
Aclacinomycin A; 13-Methyl
Aclacinomycin B
Aclacinomycin B; 9-Hydroxy
Aclacinomycin B; 9-Hydroxy, 4C-epimer
Aclacinomycin M
Aclacinomycin M; 10-Hydroxy
Aclacinomycin M; 13-Methyl
Aclacinomycin M; 4'''-Epimer
Aclacinomycin M; 4''-Epimer, 8-hydroxy
Aclacinomycin M; 4'''-Epimer, 13-Me
Aclacinomycin S
Aclacinomycin Y
Aclacinomycin Y; 10-Hydroxy
Aclacinomycin Y; 10-Hydroxy, 3B-deoxy

Figure 5. Alphabetically Sequenced Hit List

The entries for this hit list can be viewed at this stage or the search can be further refined by carrying out a structure search. Refining with Structure Search transfers the results of the text search to structure search allowing substructure searching on these hits or on the whole database. In this case the substructure search involves looking for tetracycline glycoside fragments. The required substructures are drawn from scratch (alternatively, one of the text search hits may be modified), in this case a tetracycline fragment and a pyran ring (Figure 6). As the actual point of attachment of the two fragments is not known, they are not drawn as being connected, again to maximise the number of hits. There are a number of preferences which may be changed by the user, for example, had an exact structure search been required, the search software can be set to look only for an exact match.

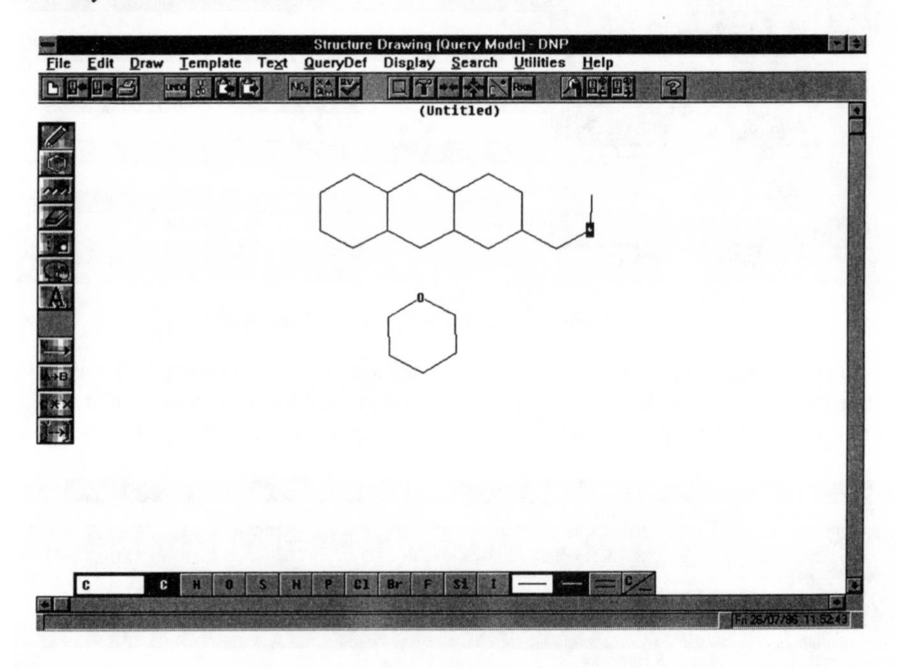

Figure 6. Constructing the Query Structure

The structure search engine refines the text results twice (using an initial screening search followed by an atom-by-atom search) to produce a second hit list which can be browsed in structure search mode, but returning to text search allows access to the text entries of the 54 hits that satisfied the combined text, numerical and substructure search criteria.

From any hit list, double clicking on one of the items gives access to the entry screen display. This was deliberately designed to resemble the printed dictionary page as much as possible, but with the added functionality of a moveable window for the HQSD, and editorially placed hotlinks providing cross-reference to other entries on the database. The information displayed in the entry includes CAS number, isolation data, physical properties (if known) and a bibliographic section consisting of 'tagged' references. The entry for Cinerubin A is shown (Figure 7).

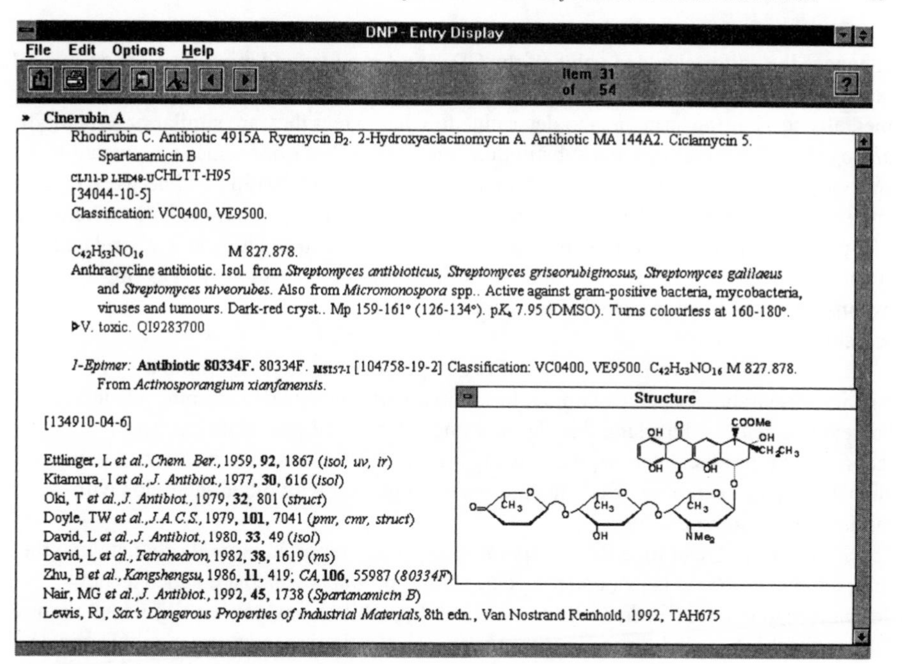

Figure 7. Text Entry of Cinerubin A

3 CONCLUSION

To what extent does DNP fulfil the attributes of an ideal information product as delineated earlier in this paper? We hope that the following will be adjudged by our existing users as an objective summary.

1. It is essentially comprehensive in its coverage of known natural products,[7] and future compilation will ensure that the asymptotic approach to 100% coverage is maintained. It is also essentially comprehensive in its incorporation of synonyms , notes on alternative numbering schemes and other vital items of 'orienting' data. A great deal of attention is paid to ensuring accuracy of recorded information, and DNP contains much editorial comment aimed at reconciling inaccuracies and inconsistencies in the literature. Compilation is carried out by subject specialists; for example, all alkaloid entries in DNP are compiled, edited and proofed by one specialist, with cross-checking and proofing support supplied by the in-house Chapman & Hall staff.

2. Flexible and intuitive searching is possible across all of the data types in both the text and structure modules. Dereplication-type searching of multiple structural fragments is straightforward. The rapid exchange of results between text (including numerical range) and substructure searching makes this the most intuitive database for natural products scientists wishing to browse and generate ideas across the whole field of natural product chemistry.

3. The organisation of data on closely-related natural products within each entry enhances the editorial added value of the CD-ROM product and minimises the amount of attention which the user has to devote to non-essential peripheral activities such as mentally reorienting structures to determine to what extent they are similar or structurally analogous. The structures and substitution patterns are described under nomenclature schemes which are biogenetically meaningful, self-consistent within and between entries and readily followed by the user. Access to the contents is also possible by searching on a Type of Compound classification code which brings together biogenetically related substances even where they have undergone structural modification such as ring expansion or contraction, which may make them difficult to find by substructure searching.

4. The usefulness of DNP for dereplication purposes is currently limited by the paucity of searchable spectroscopic information, particularly UV maxima. We are however currently addressing this shortcoming.[8] Once UV peak data has been incorporated into the database, the existing product design and search software will enable Boolean searching of UV maxima in combination with other datafields with minimal development effort.

5. The current lead time for incorporation of newly reported natural products into the fully edited CD-ROM is just under a year; natural products which are added as new derivatives to existing entries appear more rapidly, while a few new entries which require further checking and editing may take longer. We would like to improve on this and an increase in topicality of the database is anticipated in the second half of 1997.[8]

6. Since the database is not isolation-comprehensive, DNP is currently of restricted usefulness for taxonomically-based queries, although as with other types of query where it cannot at present provide the definitive hit set, it will always provide an accurate set of indicative results that can, if necessary, be used to formulate a more accurate search strategy in the very large databases. At present, CAS, Beilstein or Napralert give a more complete record of the occurrences of the more common natural products (i.e. those likely to be found in more than 3-4 species). Transfer to CAS searching is facilitated by the extensive coverage of CAS numbers in DNP, and the accurate provision of these even for complex natural products lacking trivial names is a useful function of DNP.

Future development of DNP may include links with other databases to enhance the coverage of the taxonomic dimension.

One extensive project has already been completed to provide a comprehensive validated taxonomic dataset for an important group of organisms. In conjunction with the ILDIS (International Legume Database and Information System), during the early 1990s a complete record was assembled of all isolations of natural products from the Leguminosae family, described using fully validated taxonomic nomenclature according to the international ILDIS consortium.[9] The resulting dataset documents approximately 20000 isolations of 6000 different natural products from 2700 different leguminous plants, and was published in book form as the *Phytochemical Dictionary of the Leguminosae*. At present for technical reasons[8] the results of this survey are not available in DNP but it is planned to read them into the database during 1997. ILDIS is a pioneer group in the evolution of consensus taxonomy, and the emergence of other international groups has been a slow process, but others are now forming.

7. Finally, a word about costs. Few companies are in a position to carry out unlimited searching of large databases virtually regardless of cost. For all information departments

the availability of an intuitive, flexible and infinitely available (subject to network licence agreement) resource such as DNP on CD-ROM is producing economies by helping to formulate search strategies accurately before going on-line, and by solving many queries without the need to do so at all. Beyond this, however, many companies making DNP and its sister-dictionaries available to end-users as one of their 'key information products'[10] are finding that there is a perceptible enhancement of research productivity as these tools reach the hands, and the PC screens, of the subject specialists.

4 ACKNOWLEDGEMENTS

DNP on CD-ROM represents the combined efforts of many contributors, contractors and Chapman & Hall members over a number of years. We would particularly like to mention Hilary Bradley, Tim Hamer, Steve Hawe, Bob Hill, Jack Lee, Vivian Lingard, Fiona Macdonald, Jane Macintyre, Nick Moon, Christine Pattenden, David Proctor, Pete Rhodes, Andy Roberts, Rachel Roberts, Ian Southon and Brian West, with apologies for any inadvertent omissions.

5 REFERENCES

1. Cordell, G. A., *Phytochemistry*, 1995, **40**, 1585
2. Corley, D. A. and Durley, R. C., *J. Nat. Prod.,* 1992, **57**, 1484
3. Since 1984 the Chapman & Hall chemical file was available for text-searching only on DIALOG (file 303).
4. Head Software International, Croudace House, 97 Godstone Rd., Caterham, Surrey CR3 6RE, UK
5. Hampden Data Services, Kingmaker House, Station Rd., New Barnet, Herts. EN5 1NZ, UK
6. The most prolific category of these is the plant glycosides, especially those not named trivially, which were in general not compiled for the database before about 1990. We estimate that the database now documents over 80% of these. For all other classes except polypeptides the figure is much higher and is almost certainly >98%, although even the odd alkaloid still continues to turn up.
7. We use here a relatively wide definition of the term, to encompass primary as well as secondary metabolites, but excluding polypeptides except those considered important enough by the scientific community to merit the allocation of a trivial name.
8. A major upgrade of the DBMS of the parent Chapman & Hall Chemical Database (as distinct from the CD-ROM production software) is scheduled for the first half of 1997.
9. ILDIS, c/o Dr. F. Bisby, Medical/Biological Building, University of Southampton, SO9 3TU, UK
10. We define a 'Key information product' not in terms of the sense of 'key' as a colloquial term for 'important', but rather in its more literal usage as 'unlocking'.

The Future Prospects of the Archaic Species *Ginkgo biloba*

Z. Pang, F. Pan, X. Q. Zhu and S. He*

NANJING BOTANICAL GARDEN, INSTITUTE OF BOTANY, JIANGSU PROVINCE AND ACADEMIA SINICA, PO BOX 1435, NANJING 210014, PR CHINA

ABSTRACT: The importance of the conservation of the diversity of the *Ginkgo biloba* L. is emphasized and the status of the exploration as well as the utilization of the *Ginkgo biloba* leaves in China are described. Several subjects worthy of research both now and in the future are also presented in this paper.

Key words: conservation, *Ginkgo biloba*, flavonoid, terpene lactone.

1 INTRODUCTION

Chinese medical and pharmaceutical sciences are essential constituents of the glorious Chinese civilization and its marvellous scientific heritage. China is extremely rich in terms of flora. Hence, the extent of utilization of medicinal plants in China is certainly among the best in the world.

According to recent investigations all over the country the total number of the varieties of medicinal plants is more than ten thousand. More than one thousand of these species are commonly used in Chinese medical practice. Natural medicine is regarded as the best medical option for mankind. For discovering new medicines we must recognize that human knowledge provides a rich heritage of ethnobotanical knowledge to draw upon. Exploration of the knowledge contained in ancient classical works and learned from folk experience can be facilitated by modern scientific and technological developments. To fulfil this task we must rely on the application of new technology and the co-operation of scientists of different disciplines. One of the most notable species generating tremendous interest is *Ginkgo biloba* L. which is listed as a rare species in volume one of the Chinese Red Data Book. It has a long cultivation history, of at least one thousand years, and has been recognized as a traditional Chinese medicine for the same length of time. At present it has been verified by modern medical and pharmaceutical technology as a very important raw material for a series of medicines. So it is a good example for us to discuss in terms of the conservation and prospects for utilization of medicinal plants.

2 DIVERSITY AND CONSERVATION OF *GINKGO BILOBA* L.

The origin of *Ginkgo biloba* L. can be traced back to the Carboniferous Period 230,000 thousand years ago. In the Mesozoic Era there were at least 14 genera in

Ginkgoaceae. In the Jurassic Period there were still more than 20 species in the genus. This is the so-called golden period of *Ginkgo*. There was only one species during the Tertiary Period. Ginkgo fossils have been discovered on many sites around the world except along the equator. So *Ginkgo biloba* L. is regarded as a living fossil. The *Ginkgo biloba* tree has a life span of between 2,000 and 4,000 years. The longevity of the ginkgo tree can be attributed to its remarkable ability to adjust itself to the wide variety of environments. The ginkgo tree has proven to be extremely resistant to insects and fungal diseases. A ginkgo tree even survived the nuclear explosion at Hiroshima.

In China the edible fruit of the gingko tree is considered as a precious nut and tonic food. This usage and concept is unique to China. For this reason Chinese people have studied the variations of gingko fruits and described and named a lot of cultivars based up on the shape, size and different features of these fruits. In the 18th century, gingko was introduced to Europe and then to North America. In those areas people did not use it for its fruits but as a landscape tree. Consequently, a number of varieties have been produced by horticulturists and forestrymen and also a great many cultivars for ornamental purposes have been recorded. In 1984, Santamour and He reported all the botanical and horticultural varieties and cultivars of *Ginkgo biloba*. The total number is 80. There are still more cultivars to be reported in China.

Although *Ginkgo biloba* has long been recognized as a medicinal plant there was no detailed research on the diversity of the species concerning active constituents in the leaves until the 1980's. If we protect the diversity of *Ginkgo biloba*, especially in *ex-situ* conservation, depending only on the information from the research fields of fruit and ornamental purposes, it is difficult to say that this will provide effective conservation of the diversity of ginkgo as a medicinal plant. This situation is also true of many other medicinal plants, because there is often little information about the variation of active constituents. This point could be considered as a unique element in the conservation of medicinal plants. For this reason, it is essential to research the genetic diversity of the species and the old trees are most precious to us. In China, there are several *Ginkgo biloba* trees having an estimated age of more than 3,000 years distributed in Shandong, Anhui, Hunan and Sichuan. As for *Ginkgo biloba* trees of more than one thousand years old, there are little more than 100 plants. It is reported that there are about 180 trees above an age of 500 years and for trees over one hundred years old there are about 200 thousand plants. It is extremely important to understand the genetic differences among the one thousand year old trees and between the thousand year old and the younger trees. Studies on some other medicinal plants have shown that not only plants grown in different geographical areas with different morphological characteristics could have different chemical constituents, but also plants with similar morphological features and grown on the same site may have different contents of chemical constituents. For conservation, we should collect all different variations including morphological and chemical ones. But for exploration and cultivation it seems that more studies on the medical consequences of these variations, especially those variations resulting in different chemical constituents, should be conducted.

Are there any remaining wild ginkgo trees? This is an interesting question. So far there is no evidence to identify any wild population. Hence the one thousand year old trees are particularly important.

3 EXPLORATION AND UTILIZATION OF *GINKGO BILOBA* LEAVES

2.1 Resources of Leaves

In spite of ginkgo nuts which are considered as a tonic food, the main medicinal resource is the leaves. China owns 70% of the total amount of *Ginkgo biloba* plants. Jiangsu is one of the main planting areas in the country and has more than 50% of the country's *Ginkgo biloba* leaf resources. But leaves are usually collected from different areas, in different seasons and by different methods. Sometimes, leaves are collected after the harvest of nuts. Therefore the content of the ginkgoflavone glycosides in leaves, which is one of the main group of active constituents for medicines, may vary from 0.7% - 1.7% depending on various ecological conditions, cultural practices and other genetic factors. In order to meet the demand for large amounts of ginkgo leaves of high quality for medicines, huge dense plantations for leaves have already been established around the country. Nevertheless, there is almost no research on the relationship between the content of active constituents and different variations as well as different provenances of the species.

As for variation in the content of terpene lactones, this is also unclear. In this case, a certain amount of leaves may contain a very varied quantity of active constituents. If the content of active constituents is uncertain then the effect and result of a certain weight of the material could vary widely. This is a problem not only for ginkgo leaves but also for most traditional medicinal plants.

2.2 Struggling for High Quality Extracts of *Ginkgo biloba* (GBE) Product

High quality resources are important. But high quality resources do not always result in high quality products (extracts). In Europe, the first pharmaceutical speciality ginkgo extracts were commercialized during the 1960s. Among them, leading the market, is a standardized extract GBE 761 manufactured by the German pharmaceutical company Schwabe and later also produced under license by Beaufour Laboratories in France. During the past decade a hot tide of GBE products has appeared in China. At present there are many industrial plants making GBE products.

Although the results in laboratories are good, in some cases even perfect, the quality of the products from industrial workshops is often variable. Most industrial products do not meet international criteria. A project studying the scientific process and modern technology for making GBE has been conducted in the Institute of Botany, Jiangsu province and Academia Sinica for more than eight years since 1988. Now the project has provided its third generation of products. Using the advice of the research group the extracts from the industrial plant affiliated with the Institute have achieved contents of ginkgoflavone glycosides $\geq$ 25%, and of terpene lactones $\geq$ 6%. The quality of the product meets the standard accepted by the international market. The yield of the process has reached 1.8-2.5% depending on the quality of the leaves. Based on this technology, a workshop with an annual production of tons of GBE has recently been built in Jiangsu province.

High quality extracts require high quality leaves. It has been found that the contents of flavone glycosides and bilobalide are affected seriously by harvesting time (Table 1).

Table 1 *Harvesting time and contents of active components*

Harvest time	FG %	BB %	TT %	BB/TT
16/VI	1.660	0.215	0.415	51.8
2/VII	1.610	0.205	0.409	50.1
16/VII	1.562	0.202	0.397	50.9
1/VIII	1.505	0.200	0.401	49.9
15/VIII	1.410	0.202	0.448	45.1
2/IX	1.370	0.190	0.417	44.8

Furthermore, the leaf quality also varies with the source and there is obviously variation among individuals. For example, the content of active components of high quality plants could be two times the content of the low quality plants.

It is worth mentioning that all leaf materials are produced by special dense plantation of *Ginkgo biloba* affiliated with the industrial plant under the particular advice of cultural practice. It shows that in the development of traditional medicinal plants a good connection between the raw material (resources) and the industrial products (extracts) is very important. This is also true in most other cases of the exploration and development of traditional medicinal plants.

2.3 Exploration and Utilization of Active Constituents in *Ginkgo biloba* L.

For over one thousand years, the tree has been part of the traditional Chinese medicine. Chinese herbal pharmacopoeias, both ancient and modern, contain reference to the extract of *Ginkgo biloba* leaves as being good for the heart and lungs. Up to now a great number of apolar and polar compounds have been isolated from ginkgo leaves. The extracts from the leaves contain active substances: flavonoids and terpene lactones, the structures of some of which are shown in Fig. 1. The actions of the flavonoids are reported to include: protection of capillaries against fragility; action as potent antioxidants; anti-inflammatory action; reduction of edema caused by tissue injury; free radical scavenging action.

The actions of terpene lactones include: inhibition of platelet aggregation and prevention of clots in blood vessels by blocking PAF receptors; reduction of oxygen radical discharge and other proinflammatory functions of macrophages; inhibition of lung bronchial constriction and hypersensitivity which can trigger asthma; inhibition of eosinophilimmune cell activation which can trigger blood vessel spasm.

The application range of GBE is mainly related to peripheral circulatory insufficiency due to degenerative angiopathy and cerebrovascular insufficiency with symptoms including vertigo, tinnitus, headache, short-term memory loss, hearing loss, vigilance and mood disturbance.

So far the total extract of ginkgo leaves has been used in the modern medicines or medical tablets for health. It still contains at least several kinds of flavonoids and terpene lactones. Actually, it contains an "active constituent complex". In the future the active constituents should be further separated and each could be investigated for some special effect on certain diseases, potentially leading to the development of new, more specific medicines.

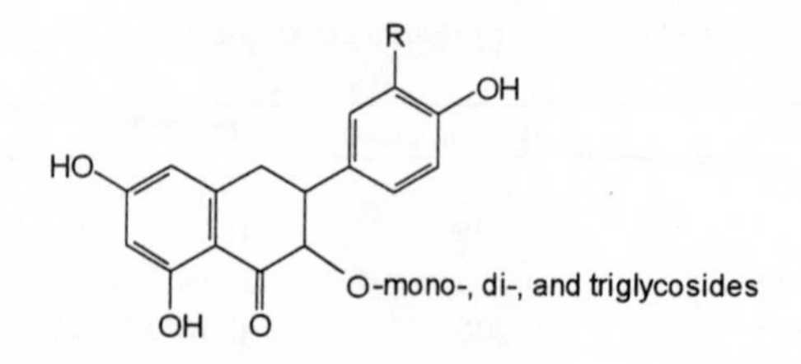

R	Name of compound
H	Kaempferol derivatives
OH	Quercetin derivatives
OCH_3	Isorhamnetin derivatives

Flavonoid glycosides from leaves of *Ginkgo biloba*

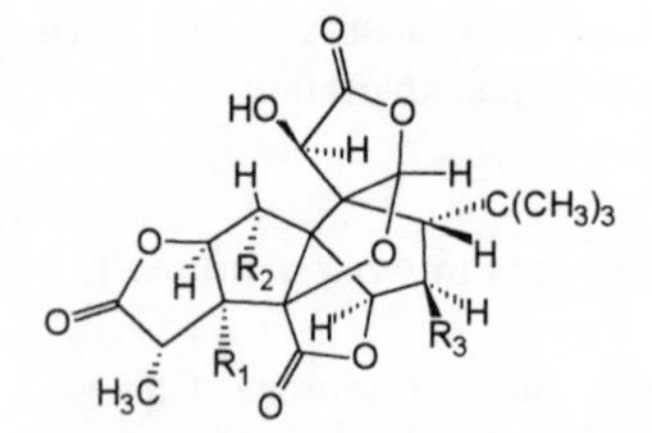

Ginkgolides (diterpenoid)

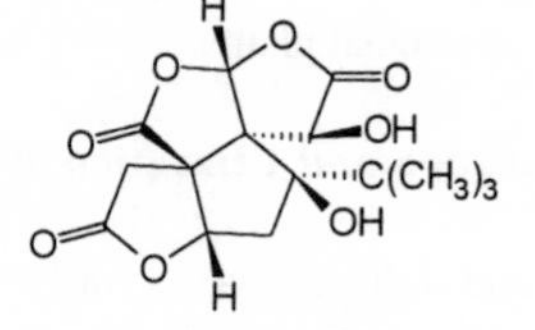

Bilobalide (sesquiterpenoid)

R_1	R_2	R_3	Ginkgolide
OH	H	H	A
OH	OH	H	B
OH	OH	OH	C

Figure 1 *Active compounds from Ginkgo biloba of leaves*

4 PROSPECTS FOR THE CONSERVATION AND UTILIZATION OF *Ginkgo biloba*

The main subjects for today and tomorrow are as follows:

(1) More effective measures should be taken to protect all old ginkgo plants, to prevent loss of any useful genetic factors. Genetic variation among the old trees should be studied with molecular biological technology.

(2) In order to collect larger quantity of leaves for extraction of the active constituents, it is necessary to select good varieties, which have higher leaf yields and higher contents of flavonoids and terpene lactones in their leaves.

(3) The effects of ecological conditions, cultural practice and post-harvest process on the content of active constituents should be further studied.

(4) More effective processes for making high quality GBE with modern technology and scientific principles.

(5) Attention should be further paid to the development and utilization of the known medicinal constituents in ginkgo leaves, such as the separation of the bilobalides for curing senile dementia, ginkgolide B for curing of asthma, etc.

(6) Studies should be also taken to investigate other chemical constituents in order to find valuable new medicinal substances in ginkgo leaves.

These research areas are also common to most traditional medicinal plants.

REFERENCES

1. S. A. He, Z. M. Cheng, The role of Chinese botanical gardens in the conservation of medicinal plants, in "Conservation of Medicinal Plants", edited by O. Akerele, V. Heywood and H. Synge, 1990, 229-237.

2. R. T. Major, *Science*, 1967, **157**, 1270.

3. P. Mickel and D. Hosford, Ginkgolides-Chemistry, Biology, Pharmacology and Clinical Perspectives, J. R. Prou Science Publisher, Barcelona, 1988, Vol. 1, 1.

4. K. Nakanishi, Pure Appl. Chem., 1967, **14**, 89.

5. L. X. Liang, Chinese Ginkgo, Shangdong Science and Technology Press, 1988, 311.

6. P. D. Tredici, The Ginkgos of Tian Mu Shan Conservation Biology, Vol. 6, No. 2, p202-209, J. Y. Wu, P. L. Cheng and S. J. Tang 1992.

7. J. Y. Wu, P. L. Cheng and S. J. Tang, Isozyme analysis of the genetic variation of *Ginkgo biloba* L. population in Tian Mu Mountain, *J. Plant Resource and Environment*, 1 (2): 20-23, 1992.

8. F. S. Santamour Jr., S. A. He, etc, Checklist of cultivated Ginkgo, *J. of Arborculture* 9 (3), 88-92, March 1983.

9. Collected Essays of the First National Ginkgo Symposium, Hubei Science and Technology Press, 1992.

The Ups and Downs of the Ethnobotanical Approach to Drug Discovery

A. I. Gray[1,*], A. M. Mitchell[2], N. Rodriguez Jujuborre[3], B. de Corredor[4], J. E. Robles C.[1,5], R. D. Torrenegra G.[5], M. D. Cole[6] and C. Cifuentes[5]

[1] DEPARTMENT OF PHARMACEUTICAL SCIENCES, UNIVERSITY OF STRATHCLYDE, GLASGOW, UK
[2] EXPEDICIÓN HUMANA, INSTITUTO DE GENETICA, PONTIFICIA UNIVERSIDAD JAVERIANA (PUJ), BOGOTÁ, COLOMBIA
[3] RESGUARDO MUINANE BORA, AMAZONAS, COLOMBIA
[4] DEPARTMENTO DE ANTROPOLOGÍA, UNIVERSIDAD NACIONAL, BOGOTÁ, COLOMBIA
[5] DEPARTMENTO DE QUÍMICA, PUJ, BOGOTÁ, COLOMBIA
[6] DEPARTMENT OF PURE AND APPLIED CHEMISTRY, UNIVERSITY OF STRATHCLYDE, GLASGOW, UK

1 INTRODUCTION

Drug discovery from natural sources generally takes two forms: *ethnobotanical*, the examination of plants used in ethnic or traditional medicine, or *random screening*, employing bioassay-guided high-throughput screening techniques. The latter method is the preferred approach by the majority of pharmaceutical companies. However, the traditional medicines and other substances sourced from nature by the indigenous people of South America have yielded many useful drugs including Cocaine, Quinine and Tubocurarine.[1] Does this mean that there are no more discoveries to be made?

We have set up a project entitled 'Pharmacognosy of the Americas' to study the plant products used by some indigenous communities of the Amazonian region of Colombia. The project forms a part of a larger programme 'Expedición Humana' started in 1988 by PUJ as an interdisciplinary study of the biodiversity of Colombia.

1.1 Expedición Humana

1.1.1 Expedición Humana 1988. This programme of research was set up as a four year plan by the Universidad Javeriana in Bogotá, Colombia, to coincide with the 500th anniversary of the 'discovery' of the Americas. The ultimate goal was to investigate the origin, evolution, and cultures of the ethnic groups living in Colombia to generate an understanding by the Colombian population of their heritage. At the same time the project was intended to be a study of the dynamic process of interaction between man and his environment. One of the principal objectives of the expedition was; "To study the different native groups in the country; to record their cultures, traditions and history." This part of the project was seen as a niche for our team of investigators whose expertise spanned anthropology, biochemistry, microbiology, pharmacy and phytochemistry and the project "Pharmacognosy of the Americas" was created.

1.1.2 Pharmacognosy of the Americas. This project was originally set up to study and record the traditional medicines used by the ethnic communities in the Caquetá region (Figure 1) of Colombia with the intention of expansion to include other communities in different regions of the country. The general objective of the study was to investigate the traditional system of preventative and curative medicine used in the Muinane and Uitoto cultures.

Figure 1 *Map of Colombia indicating Araracuara*, Caquetá Region.*

More specifically, i) to determine which plants are used by the "Abuelos-Sabedores" (the elder men of the community who are botanical doctors following the tradition of their tribal heritage); ii) to record the "mythical history", told by oral transmission, which forms the basis of the knowledge held by the ethnic groups with respect to their medicinal plants; iii) to examine the biological activity(s) of crude plant extracts; iv) to perform phytochemical studies on selected plants in an attempt to determine the active components; v) to conduct case studies of patients treated by the "Abuelos-Sabedores" using medicinal plants with information provided by patients, "Abuelos-Sabedores" and clinicians. In this paper we present some of our experiences of working with the Uitoto (dialect Nipode) and Muinane (dialect Muinane-Bora) people living in the Amazonas/Caquetá region.

2. RESULTS AND DISCUSSION

2.1 Field trips to Amazonas/Caquetá region

Expeditions were mounted in Bogotá with the primary destination of Araracuara, Caquetá Department (Figure 1). These trips were usually of at least one month duration and were scheduled in a variety of seasons such as the dry 'summer' period (January - March) and wet 'winter' (June - August) months. The work was carried out in cooperation with a number of "Abuelos-Sabedores" from the Muinane and Uitoto communities in the region. Many plant species have been collected that are used to treat wounds and some other skin diseases. The plants collected include several representatives of the family Burseraceae. This family is well known for the production of commercially important resin exudates used as incense, for example Myrrh (*Commiphora spp.*) and Frankincense (*Boswellia spp.*) found in Africa and Asia.[2] The medicinal actions of Myrrh (Commiphora Resin), among other things, as an antimicrobial, expectorant and antiseptic for use in the treatment of respiratory catarrh, common cold and topically on wounds and abrasions is well documented.[3]

2.1.1 Selection of plants for investigation. The burseraceous genera endemic to Colombia include *Bursera, Crepidospermum, Dacryodes, Paraprotium, Protium, Tetragastris* and *Trattinickia*.[4] They are known collectively in Colombia as 'incienso'

(Incense) but we have been introduced to them from the local dialect point of view where they are known as Uguko, Uguna, Xemena, etc. The different species are recognised by the "Abuelos-Sabedores" according to the 'energies' of each plant which we may translate as the organoleptic, e.g. smell, taste and form of resin produced, and macroscopic characteristics of the various parts of each plant. These trees produce resin, in many instances in copious amounts, upon mechanical damage to the bark. In some cases large specimens of resin were already available for collection, having been produced as a result of insect attack; the live larvae (grubs) were often encountered, apparently living out a part of their life cycle in the trunk / resin exudate of the tree. Insect attack was not always confined to the trees as some of us were bitten with astonishing regularity by every manner of ant, fly or mosquito! Some of the 'down' side of this form of investigation.

The plant species known to the Uitotos and Muinanes as 'Uguna' (possible *Tetragristis panamensis?*) and several other resin producing species, used by them as antiseptics, wound healing agents and incenses, have been investigated by us and found to have antibiotic activity against some plant and human pathogens.

2.2 Selection of Bioassays

Owing to the use of these resins by the indigenous people as wound healing agents and antiseptics, antimicrobial bioassays were employed to screen the extracts and fractions.

2.2.1 Microbial growth assay and bioautography.[5] Several bacteria such as *Escherichia coli, Staphylococcus aureus, Clavibacter michiganensis, Erwinia carotovora, Pseudomonas syringae, Streptomyces scabies* and fungi such as *Cladosporium cucumericum, Fusarium solani and Trichoderma viride.* The latter two plant pathogenic fungi are used (Figure 2, A and B) to illustrate the activity of three resins, Uguna, Xemena and Uguko, when compared to DMSO, the solvent used to disperse the compounds in the nutrient agar, and to the positive control Griseofulvin. The radial growth progression of a 'plug' of actively growing fungus, applied to the centre of the nutrient agar plate, was measured at two-day intervals. It appears that these three incense extracts have some measure of antifungal activity. Bioautographic examination[5] of Thin-Layer chromatograms of these incense extracts (Silica gel G, developed with CH_2Cl_2:MeOH mixtures as mobile phase) indicated that the more polar components have a higher degree of inhibition of fungal growth.

2.3 Isolation and Identification Techniques

In order to gain some insight into the type(s) of compound one is dealing with, *i.e.* how do we choose an appropriate chromatographic method for separation, we obtained Nuclear Magnetic Resonance spectra of the crude resins. For example, the ^{13}C NMR spectrum (J-mod / APT-type experiment) of Uguna resin in $CDCl_3$ (Figure 3) revealed the terpenoid nature of components with ketonic carbonyls *ca.* 220 ppm, aldehydic carbonyls *ca.* 205 ppm, acidic carbonyls *ca.* 180 ppm, olefinics/aromatics *ca.* 100 - 150 ppm, carbinolics *ca.* 75 - 85 ppm and the tell-tale aliphatics, a veritable forest of methyls, methylenes and methines!

2.3.1 Extraction. The resins were almost totally soluble in most organic solvents such as chloroform, ethylacetate, ethanol and methanol but only partially soluble in petroleum ether (b.p. 40-60°) and virtually insoluble in water. This solubility profile contrasts with Myrrh which contains up to 60% of water-soluble carbohydrate gum substances.[3]

2.3.2 Fractionation and Purification of components. Crude extracts were column chromatographed or subjected to vacuum liquid chromatography over Silicagel with CH_2Cl_2:MeOH mixtures or petroleum ether:EtOAc mixtures as mobile phase. Fractions were monitored by TLC, NMR and bioassays leading to the isolation of pure components (*cf.* Figure 4, the ^{13}C J-mod spectrum of the major component, (8), shows thirty carbons,

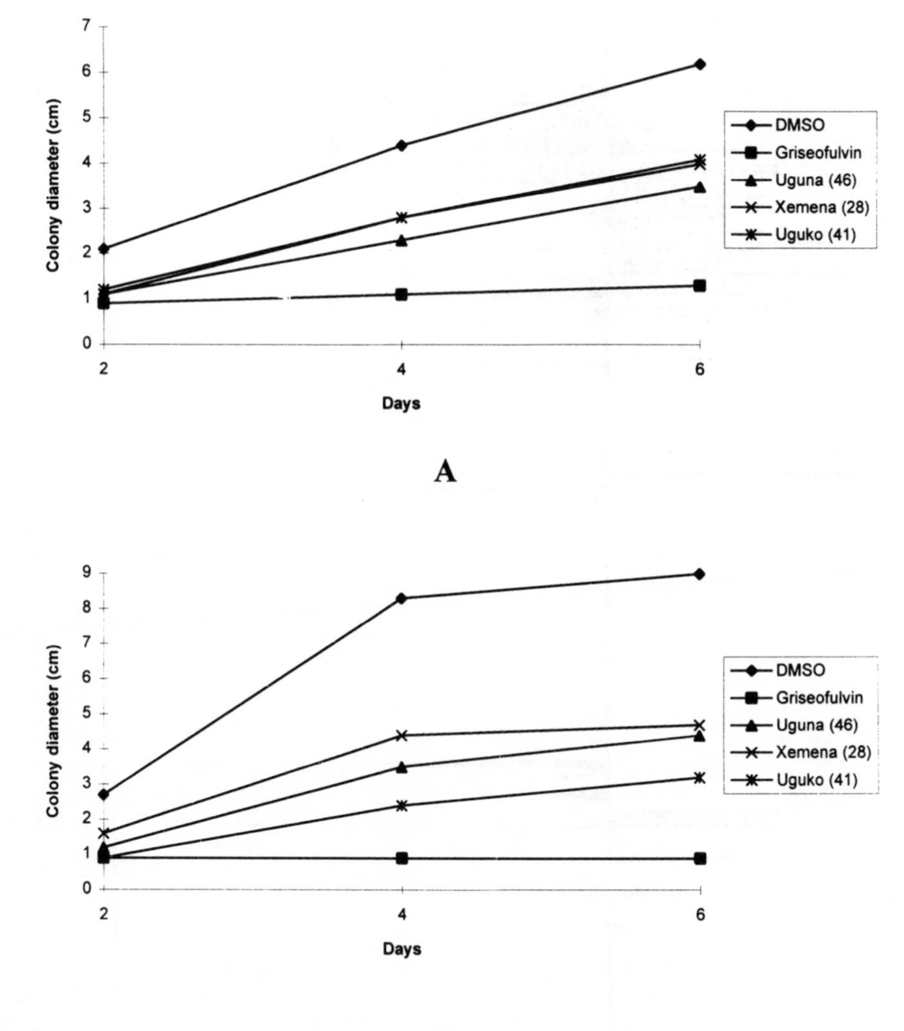

Figure 2 : *Colony growth vs. time for resins (5 mg/20ml agar) from three Burseraceous species with DMSO as solvent control and griseofulvin (200 μg/ml) as positive control on nutrient agar plates (9.1 cm diameter). Plugs (1 cm) of actively growing fungus **A** = **Fusarium solani** and **B** = **Trichoderma viride** were applied to the centre of the plates at the time of outset.*

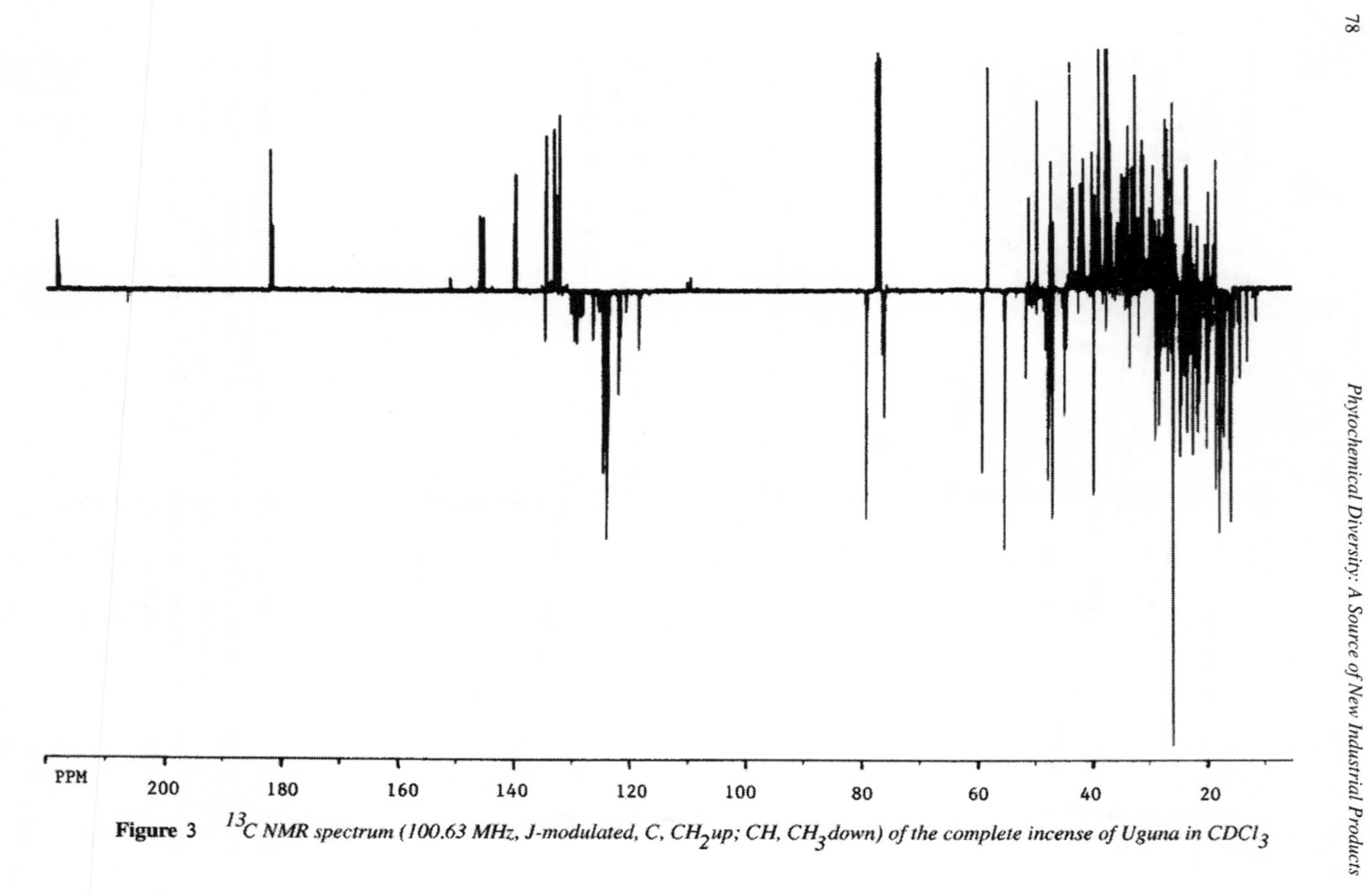

Figure 3 ^{13}C *NMR spectrum (100.63 MHz, J-modulated, C,* CH_2*up; CH,* CH_3*down) of the complete incense of Uguna in* $CDCl_3$

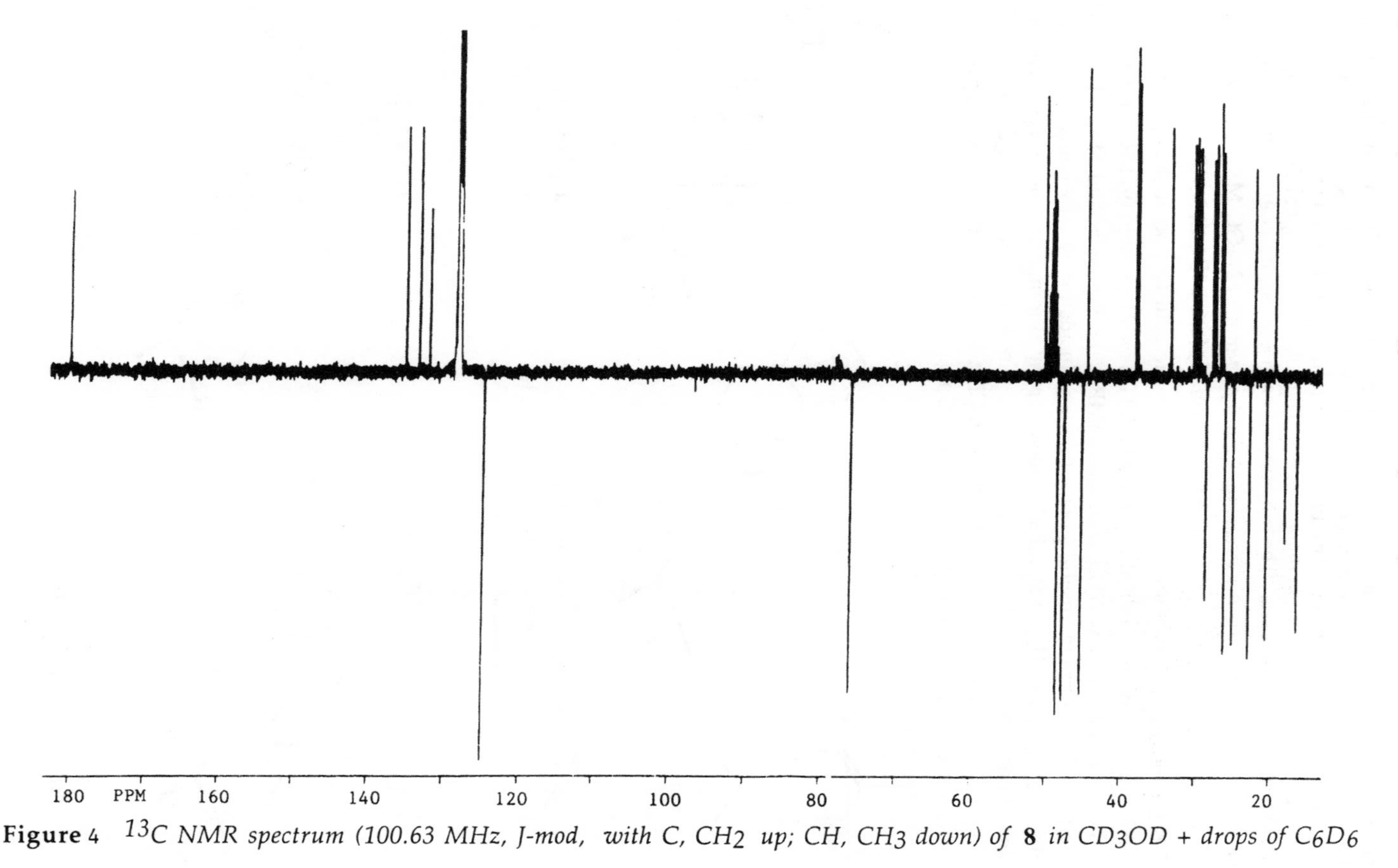

Figure 4 *^{13}C NMR spectrum (100.63 MHz, J-mod, with C, CH_2 up; CH, CH_3 down) of* **8** *in CD_3OD + drops of C_6D_6*

and this compound was taken for 2D NMR studies, e.g. Figure 5). In some cases mixtures of known substances were subjected to gas chromatography-mass spectroscopic analysis and compared with an MS database of compounds and with authentic samples.

2.3.3 Identification of components. The following compounds were identified as constituents of Uguna resin. The structure of each compound is given and method(s) of analysis(es) indicated. α-phellandrene (1) and *p*-cymene (2) [GC-MS], α-amyrenone (3) and β-amyrenone (4), α-amyrin (5) and β-amyrin (6) [GC-MS, IR, NMR and co-TLC with authentic samples], 3-oxotirucalla-8,24-dien-21-oic acid (7), 3α-hydroxytirucalla-8,24-dien-21-oic acid (8) and 3α-hydroxytirucalla-7,24-dien-21-oic acid (9) [IR, ^{1}H, ^{13}C, ^{1}H-^{1}H COSY, NOESY, ^{1}H-^{13}C-COBIDEC and HMBC (*cf.* Figure 5) and MS]. The HMBC experiment[6] is extremely useful in the structure elucidation of terpenoids and the strong methyl signals in the ^{1}H domain, if adequately separated from one another, can allow one to 'track' around the molecule (*cf.* Figure 5). In the case of the triterpene acids low solubility in $CDCl_3$ was overcome by dissolving them in CD_3OD where necessary and separation of signals was enhanced by adding a few drops of C_6D_6.

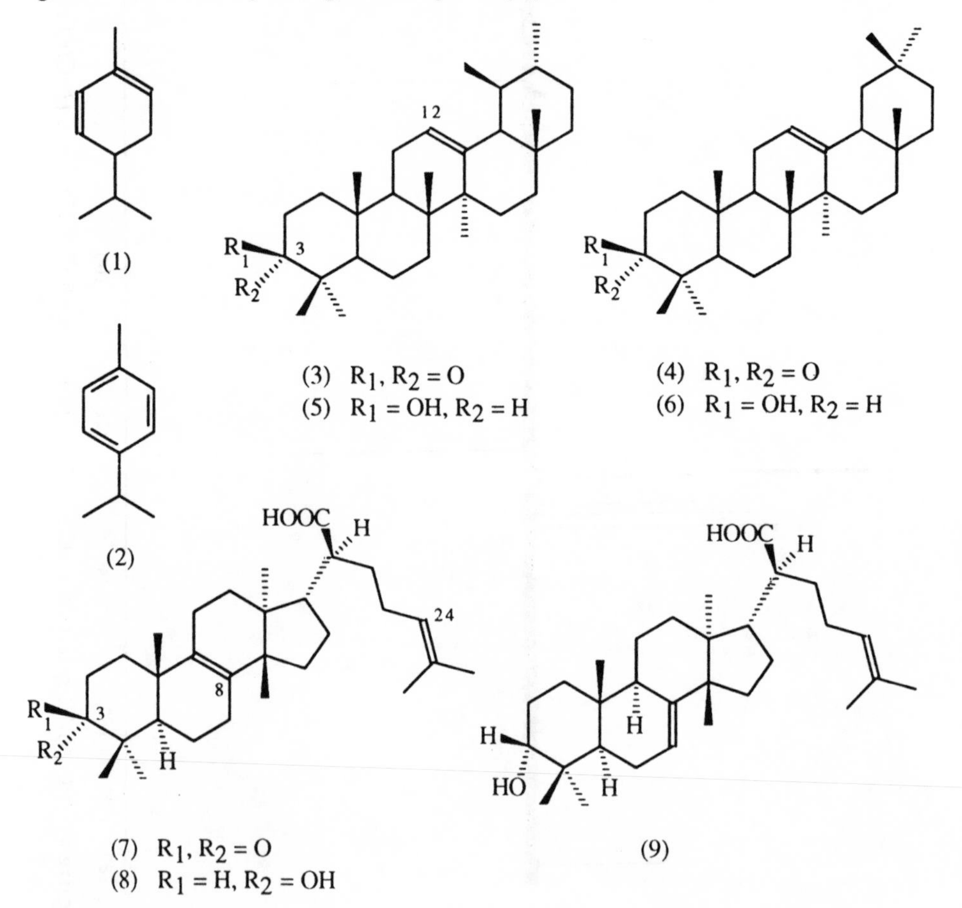

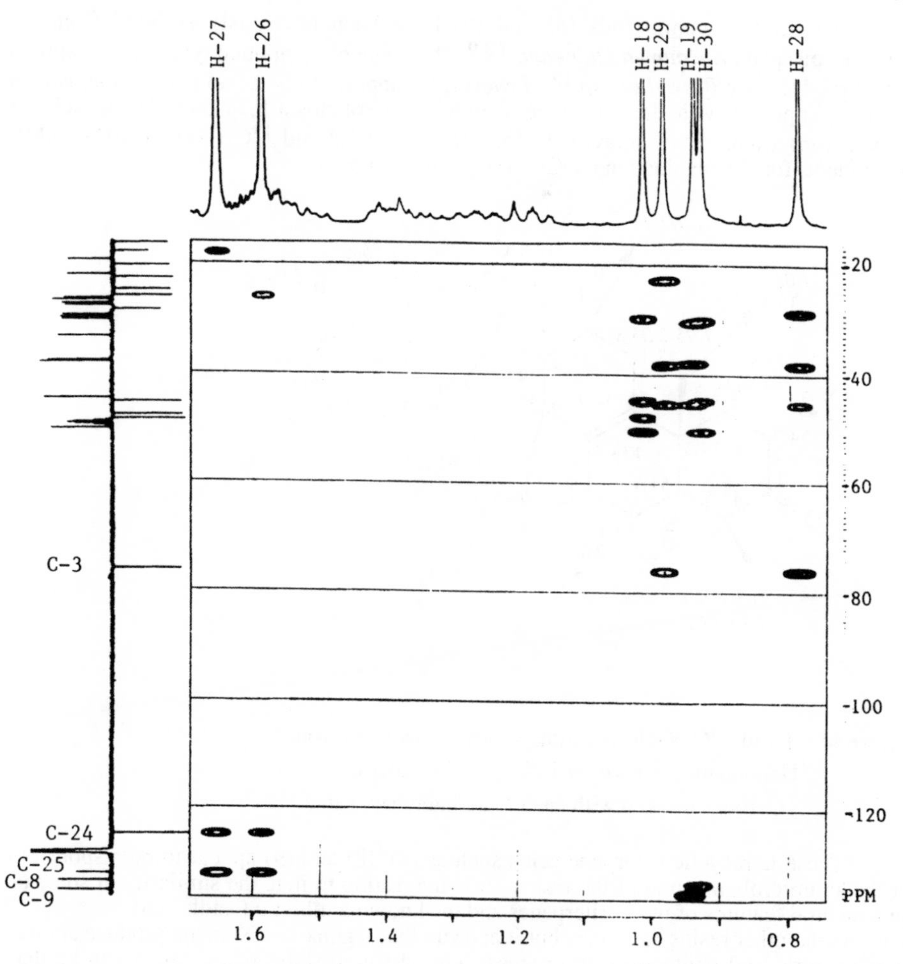

Figure 5 *Partial HMBC spectrum of* **8** *, isolated from Uguna incense, with optimisation of levels to show only methyl correlations to carbons (400MHz, D6 = 70msec for* J_{CH} *=7Hz, in* CD_3OD *+ drops of* C_6D_6*)*

The major compounds (8) and (9) have been previously isolated from the burseraceous plant *Aucoumea klaineana*.[7,8,9] The range of compound types is very similar to that produced by *Boswellia spp.*[10] However, (7) appears to be novel although an isomer of this compound with the lanostane skeleton, 3-oxolanosta-8,24-dien-21-oic acid, is known from a number of sources.[10] The ^{1}H (400 MHz) and ^{13}C (100.63 MHz) NMR assignments for (7), obtained in $CDCl_3$, are given in Figure 6.

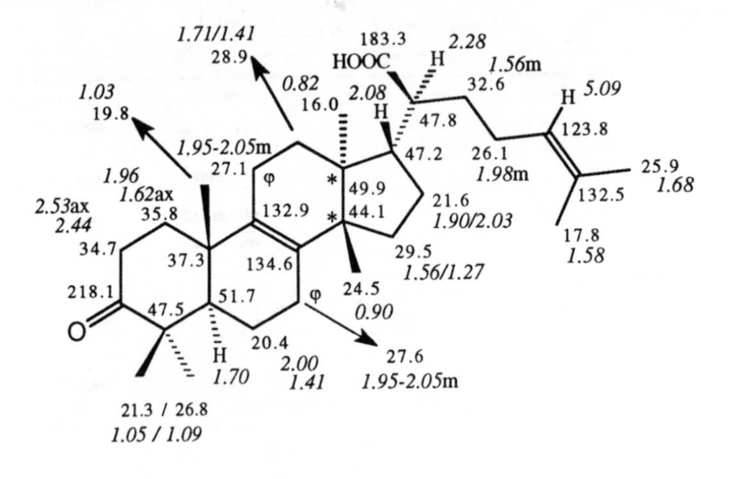

Figure 6 ^{1}H and ^{13}C NMR assignments for (7) isolated from Uguna resin. (^{1}H assignments in *larger italics*; ^{13}C smaller numbering. Carbons marked with φ or * are interchangeable.)

These tetracyclic triterpene acids such as (7), (8) and (9) appear to be responsible for the antimicrobial action of the resins. It is interesting to note the similarity of some of the structural features of these triterpenes and the known antibiotic fusidic acid.
Work on the other resins continues but it appears that Uguna and Xemena produce similar types of tetra- and pentacyclic triterpenes. One distinguishing feature seems to be that Xemena produces the aromatic ester benzylbenzoate (10) identified by GC-MS and NMR and comparison with an authentic sample. (10) is well known to orthodox medicine as an acaricide and pediculicide and is present in a number of natural oils[10] that have insect repellent properties. It is not surprising then that these resins are burned in the homes of the Muinanes and Uitotos "to clean the house". What of? Insects? Obviously the wood-boring insects that infest many of the species of Burseraceae do not find it repellent or have adapted to its presence in the resin.

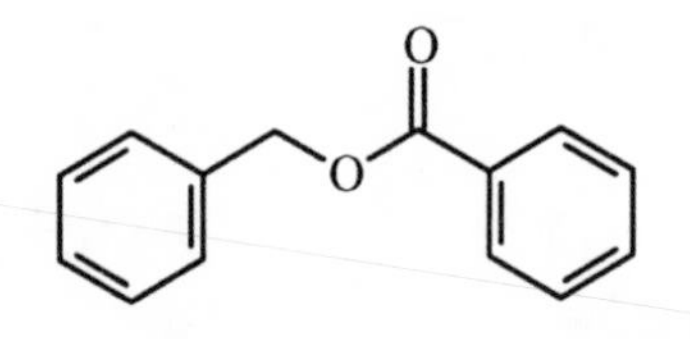

(10)

3 SUMMARY

The project 'Pharmacognosy of the Americas' undertook to investigate the traditional medicines used by the Muinane and Uitoto indigenous people of the Amazonian region of Colombia. We have investigated some burseraceous resins, especially Uguna (*Tetragastris panamensis?*), and found them to be very similar in chemistry and biological activity to (*Boswellia spp.* and others).[10] A number of terpenoid compounds (1-9) were isolated and characterised and one of them, (7), appears to be novel.

These Amazonian plant species may be viable commercial sources of the incense resins similar to Myrrh, Frankincense, Elemi, etc. and the indigenous people could harvest sustainably or cultivate the appropriate species. The problem of botanical identification remains, due to absence of flowers or fruit during the periods of the expeditions. The African species of Burseraceae growing in arid areas have also given taxonomists problems[11,12] since, for most of the time, the plant appears simply as a mass of thorny twigs!

Acknowledgements. The "Abuelos-Sabedores" and their relatives of the ethnic communities of Colombia without whose cooperation this work would not be possible. The authors also wish to thank the British Council, Bogotá (Academic Link Scheme), COLCIENCIAS, Colombia (Post-graduate sandwich studentship to J.E.R.C.) and the Royal Geographical Society for sponsorship. Ms T. Davidson, Dept of Pharmaceutical Sciences, University of Strathclyde for technical assistance with artwork.

References

1. W. C. Evans, 'Trease and Evans' Pharmacognosy', Bailliere Tindall, London, 13th Ed., 1989.
2. A. O. Tucker, *Econ. Bot.*, 1986, **40**, 425.
3. C. A. Newall, L. A. Anderson and J. D. Phillipson, 'Herbal Remedies, A Guide for Health-care Professionals', The Pharmaceutical Press, London, 1996, p.199.
4. J. Cuatrecasas, *Webbia*, 1957, **12**, 375.
5. K. Hostettmann, 'Assays for Bioactivity' in, Methods in Plant Biochemistry, P.M. Dey and J.B. Harborne, Eds., Academic Press, London, Vol. 6, 1991.
6. A. Bax and M. F. Summers, *J. Am. Chem. Soc.,* 1986, **108**, 2093.
7. A. M. Tessier, P. Delaveau and N. Piffault, *Planta Medica*, 1982, **44**, 215.
8. A. M. Tessier, P. Delaveau, N. Piffault and J. Hoffelt, *Planta Medica*, 1982, **46**, 41.
9. L. Guang-Yi, A. I. Gray and P. G. Waterman, *Phytochemistry*, 1988, **27**, 2283.
10. Dictionary of Natural Products, Chapman and Hall, London, CD-ROM, Release 4:1, 1995.
11. G. J. Provan, A. I. Gray and P.G. Waterman, *Flavour and Fragrance Journal*, 1987, **2**, 109.
12. G. J. Provan, A. I. Gray and P.G. Waterman, *Flavour and Fragrance Journal*, 1987, **2**, 115.

Modern Science and Traditional Healing

Thomas J. Carlson, Raymond Cooper, Steven R. King and Edward J. Rozhon

SHAMAN PHARMACEUTICALS, INC., 213 E. GRAND AVENUE, SOUTH SAN FRANCISCO, CA 94080, USA

1 HISTORICAL PERSPECTIVE

The pharmaceutical industry clearly has a goal and a challenge to discover, develop and deliver novel chemical and biological entities for the treatment of human diseases; and natural products have played, and will continue to play, an important role in this process. Tropical forest plant species have served as a source of medicines for people of the tropics for millennia. Many medical practitioners with training in pharmacology and/or pharmacognosy are well aware of the number of modern therapeutic agents that have been derived from tropical forest species. In fact, over one-hundred and twenty pharmaceutical products currently in use are plant-derived, and some 75 per cent of these were discovered by examining the use of these plants in traditional medicine.[1,2] Of these, as shown in Table 1, a large portion has come from tropical forest species. Three examples of plants are presented that (1) present a rich ethnomedical history, (2) offer a basis in ethnomedicine and traditional healing, and (3) have become the basis for multimillion-dollar drugs.

First is the willow, the botanical "parent" of aspirin. The willow (*Salix alba*) is a tree that grows in low-lying damp areas and along rivers in many European countries. The bark of this plant has been used for centuries by people in Europe to treat inflammation, pain and fever. In the 18th century, the information was formally documented with the Royal Chemical Society, and clinical administration of extract to people having fevers soon followed. At the end of the 19th century, developments in the German chemical industry led to the chemical and synthetic strategies for making today's aspirin based on the chemical found in the bark. Bayer Chemical Co. thus became the first to commercialize a synthetic drug based on an herbal remedy; this remains the largest selling drug of all time.

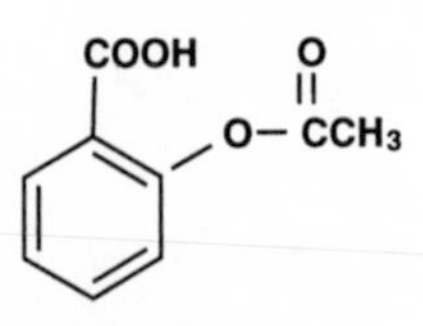

Aspirin From: Willow tree bark
Utilities: Anti-inflammatory, analgesic, antifebrile, prevention of heart attack

Table 1. Examples of clinically useful drugs from tropical rain forest plants and currently in use in the U.S.A.

Compound	Plant	Clinical Use
Bromelain	Ananas comosus (L.) Merrill (Bromeliaceae) (Pineapple)	Anti inflammatory; Proteolytic
Camphor	Cinnamomum camphora (L.) Nees & Eberm. (Lauraceae) (Camphor tree)	Rubefacient
Chymopapain	Carica papaya L. (Caricaceae) (Papaya)	Proteolytic; Mucolytic
Cocaine	Erythroxylum coca Lam. (Erythroxylaceae) (Coca)	Local anesthetic
Deserpidine	Rauvolfia tetraphylla L. (Apocynaceae) (Snakeroot)	Antihypertensive; Tranquilizer
L-Dopa	Mucuna deeringiana (Bort.) Merrill (Leguminosae) (Velvet Bean)	Antiparkinsonism
Emetine	Cephaelis ipecacuanha (Brot.) A. Richard (Rubiaceae) (Ipecac)	Amebicide; Emetic
Ouabain	Strophanthus gratus (Hook.) Baill. (Apocynaceae) (Twisted flower)	Cardiotonic
Papain	Carica papaya L.(Caricaceae) (Papaya)	Proteolytic; Mucolytic
Physostigmine	Physostigma venenosum Balf. (Leguminosae) (Ordeal Bean)	Anticholinesterase
Pilocarpine	Pilocarpus jaborandi Holmes (Rutaceae) (Jaborandi)	Para-sympathomimetic
Quinidine	Cinchona officinalis L. (Rubiaceae) (Yellow cinchona)	Antiarrhythmic
Quinine	Cinchona officinalis L. (Rubiaceae) (Yellow cinchona)	Antimalarial; Antipyretic
Reserpine	Rauvolfia serpentina (L.) Benth. ex Kurz (Apocynaceae) (Indian snakeroot)	Antihypertensive; Tranquilizer
Tubocurarine	Chondrodendron tomentosum R. & P. (Menispermaceae) (Curare)	Skeletal muscle
Vinblastine, Vincristine	Catharanthus roseus (L.) G. Don (Apocynaceae) (Madagascan periwinkle)	Antitumor and Antileukemic agents

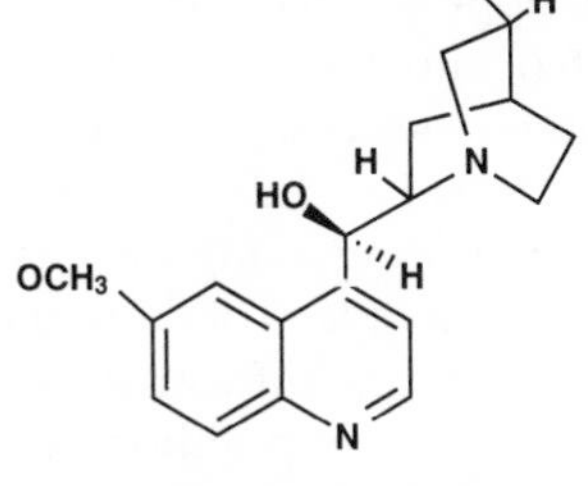

Quinine From: ***Cinchona spp.***
Utility: Antimalarial

As a second example, extracts from *Cinchona calisaya* and *C. officinalis* containing the antimalarial drug quinine have been known to healers in South America for hundreds of years. The cinchona bark was brought back to Spain in the 1500's, and an original sample has been preserved and is permanently on display at the Royal Pharmacy Museum in Madrid. The plant has been extensively studied by

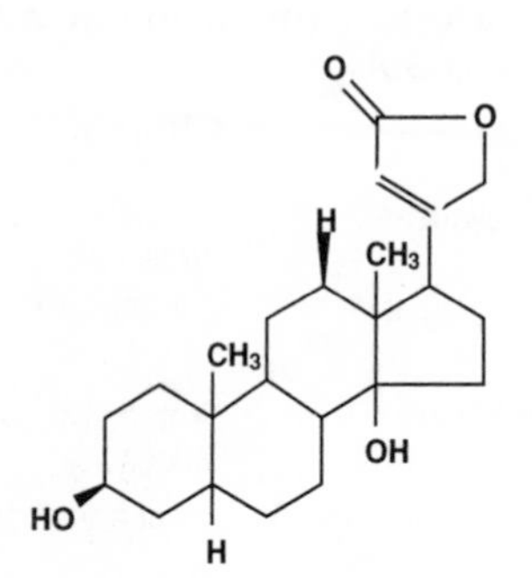

Digoxigenin (Aglycone of Digoxin)
From: Foxglove
Utility: Treatment of congestive heart failure

phytochemists, leading to the identification of the active constituent. Quinine and its derivatives have been and continue to be a major drug of choice in the fight against malaria.

Third, European herbalists were well acquainted with the properties of the digitalis plant (*Digitalis lantana,* and *D. purpurea,* commonly known as foxglove). This plant has been used for a long time to treat heart conditions. Digoxin is the main component prescribed for the treatment of congestive heart failure.

2 SCREENING APPROACHES

2.1 High throughput screening

The pharmaceutical industry is aggressively seeking new bioactive entities to treat the numerous disease targets. Tremendous investment is being pumped into this effort, particularly into research and discovery. Modern approaches to drug discovery include random high-volume screening of thousands of chemical entities. These are generated from chemical libraries, combinatorial approaches and natural products, all of which feed into various test systems. The technological advances of robotics for high-throughput, random screening in the 1980's gives the industry an ability to handle very large numbers of samples. This technology advance is now coupled with expanding the sources of screening entities. Thus one consequence has been a renewed interest in including novel and biodiverse tropical species; and this in turn has stimulated a renaissance of activities in the areas of plant natural product chemistry, pharmacognosy, and ethnomedical research. This approach includes the classic random collection of plants that are incorporated into the high throughput screening programs using a variety of mechanism-based assays with specific applications to numerous therapeutic areas. Any positive "hits" typically are subjected to an initial *in vitro* evaluation before proceeding along the chain of development. This methodology, that requires screening of tens of thousands of natural products and chemical entities, is well-suited to the infrastructure and philosophical approach to drug discovery of traditional, yet highly successful pharmaceutical companies.

2.2 An Ethnobotanical Approach

2.2.1 Philosophy

A different approach to new drug discovery involves a collection program for medicinal plants, with primary emphasis on the use of plants by indigenous people in the tropical regions of the world.[3,4] This approach, utilized by Shaman Pharmaceuticals since its inception, integrates a philosophy of looking for plant leads that already have been shown to be efficacious in humans. By isolating the biologically active compounds from these plants we thereby shortcut the long and expensive screening phase of drug discovery used by the conventional industry. It should be noted that the plant selection is very focused, based on ethnomedical use for a select disease target.

The two described approaches are shown in Figure 1. On the right-hand side, in a high-volume approach, crude extracts and samples are most often tested *in vitro*, using enzyme- and cell-based assays. *A priori*, no initial *in vivo* information is known. By contrast, in Shaman's approach (at left in Figure 1), the plants have already been used in humans, so a high percentage of the select number of actives is effective in animal models at the outset of the discovery process.

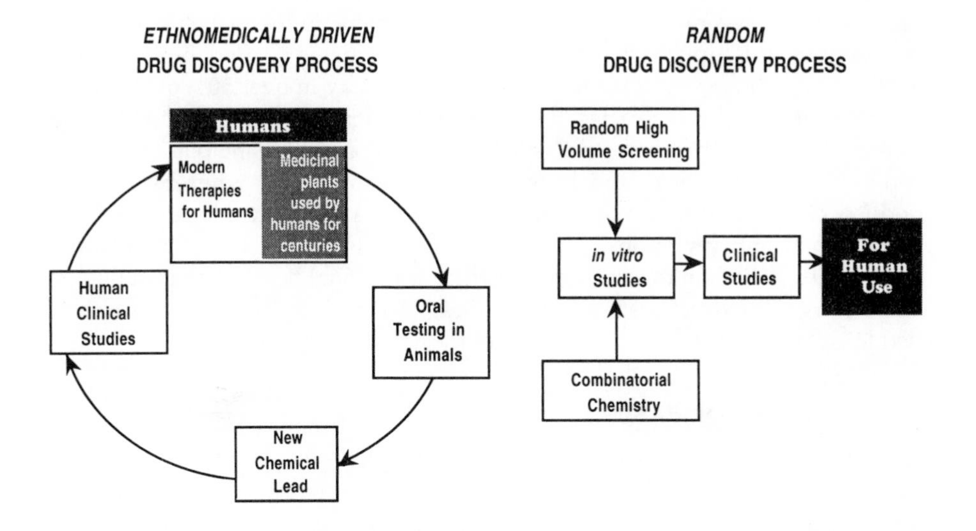

Figure 1. Different approaches to the drug discovery process

2.2.2 Plant Collection

The plant collection process is conceived over many months of planning, and by assessing those areas that may yield opportunities for collaboration with local traditional healers. We have a unique ethnomedical field research program that uses an ethnobotanist – physician team. Many weeks and months are spent with the local healers to understand their ways, their approaches and applications. A key phase early in our process takes place in the healer's village, where our ethnobotanist -- physician team carefully observes the craft of the healer, the choice and harvesting of the plant part and the way the healer prepares it. This phase is crucial to our own selection process and subsequent in-house extraction activities. We prepare our extract to closely resemble those preparations made by the healer. Many of the traditional preparations are water extractions.

2.2.3 Confirmation of Activity

In essence, the rain forest, its associated ethnomedical history, and the field research prioritization serve as the initial biological screen. The next step in our drug discovery process is to bring the plant in-house and confirm the pharmacological activity of the plant using an animal model that closely simulates the human disease for which the plant was selected. Using this *in vivo* testing approach, we have been able to rapidly identify active lead extracts and compounds for several important diseases in humans. In the diabetes program, for example, this approach has allowed us to efficiently find orally active preclinical candidates, and we have confirmed activity in over 50% of the plant extracts that have been selected for diabetes, by demonstrating blood glucose lowering activity in a diabetes mouse model. The most promising plant leads are then subjected to *in vivo*-guided fractionation campaigns, whereby natural products chemists use state-of-the-art chemical separation techniques to isolate the chemical entity responsible for the observed activity from its inactive components. We begin this campaign using a traditional healer's preparation or a classic organic extract preparation. Modern spectroscopic techniques are then used to elucidate the chemical structure of the active compound.

3 TARGET DISEASES

3.1 Diabetes

One of our targeted disease areas is diabetes, specifically type II non-insulin-dependent diabetes mellitus (NIDDM). We have selected this disease target because, first, it presents opportunities for improved drugs to better meet therapeutic demands; second, we can obtain valuable knowledge from traditional healers on plants they use to treat diabetes in their communities; and third, we have established an *in vivo* animal model to discover orally active molecules that lower blood glucose. Genetically modified mice (db/db and ob/ob mice) are used as animal models for NIDDM. The mice are dosed orally, and we monitor the effects of these extracts by measuring the blood glucose at various time intervals. Usually we administer extracts and pure compounds to mice by oral gavage and if necessary continue dosing at select intervals for 1 – 3 days. As a supplement to this *in vivo* approach, we also examine glucose transport through the use of a cellular *in vitro* assay based on 3T3-L1 adipocytes.[5]

3.2 Virology

Another program at Shaman which relies on the ethnomedical approach to new drug discovery is the antiviral program. We have focused on viral diseases that generally are inadequately treated with existing therapies; examples include respiratory syncytial virus disease, influenza, and several herpesviruses. As in the diabetes program, plants are selected and prioritized based on their use by the indigenous peoples for treating various symptoms of viral infection (e.g., for respiratory viral infections, runny nose, pharyngitis, cough, difficulty in swallowing and fever). Upon arrival at Shaman, plants are subjected to chemical extraction and the resulting extracts are tested for antiviral activity in *in vitro* (cell culture-based) and *in vivo* (animal-based) systems.

4 FIELD DATA AND LABORATORY CONFIRMATION OF SELECTED PLANTS

4.1 Diabetes

The genesis of the diabetes program relies on ethnobotanical data from the field. Our ethnomedical field research methodology includes discussing medical case presentations with healers. To gain a better understanding of the efficacy of these botanical medicines and how these plant extracts are effectively prescribed, our colleagues in a variety of host countries have conducted studies to measure blood glucose levels in patients with NIDDM who are treated with a botanical medicine. An example of one of these patients is given in Graph 1. All data points in this graph represent morning fasting blood glucose levels. Normal morning fasting blood glucose levels are 75 – 115 mg/dl. The patient in Graph 1 is a 47-year- old Latin American male who was diagnosed with NIDDM by a medical physician from his host country. Days 1–3 represent pre-treatment. On days 4-8 the patient was being treated. The introduction of the botanical medicine resulted in a significant drop in morning fasting blood glucose levels. By day 8, there was a drop of 90 mg/dl of blood glucose compared to the pretreatment levels on days 1, 2 and 3. Similar studies have also been conducted by tropical country colleagues that involve 6–10 patients. It should be noted that in these examples the patient's weight did not change significantly throughout the treatment period. These human data help prioritize the plants on which to focus pre-clinical and clinical studies.

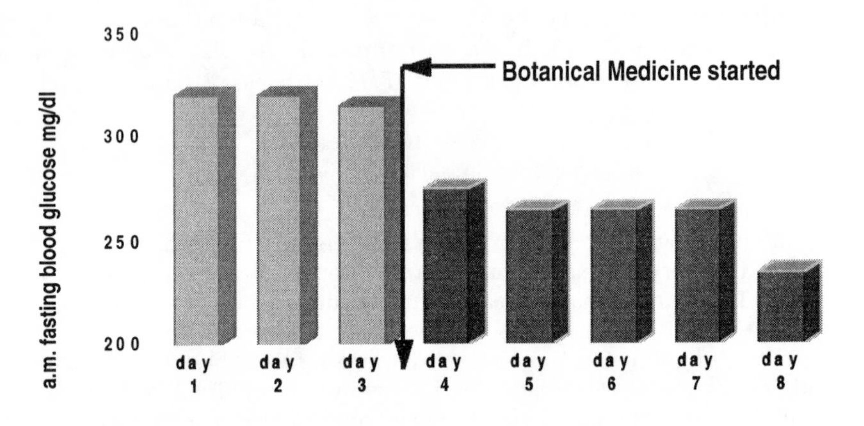

Graph 1. 47-year-old Latin American male with NIDDM

4.2 Virology

The direct assessment of the effects of antiviral plants on tropical patients is more difficult than the evaluation of the effects of antidiabetes plants. When evaluating viral infections, the field research physician is able to make a clinical diagnosis and follow the patient in the field throughout the treatment. In many instances, our field teams are able to assess whether or not a plant is efficacious in the patient. Selection of antiviral plants is based on discussions with the traditional healer as well as direct evaluation of their patients being treated with botanical medicines. Thus, from these discussions, interactions and observations, potential antiviral plants are assigned a priority value based on the research of the ethnobotanist-physician team (Table 2). Using an *in vitro* antiviral assay to test the inhibitory effect of these plants against respiratory syncytial virus (RSV) at Shaman, a direct correlation is observed between priority and activity. Plants in the highest priority categories (1.0–1.3) exhibit the greatest percentage of active plants, whereas plants in the lowest category, 3.0, exhibit the lowest frequency of active plants (Table 2). This correlation not only supports the ethnomedical approach to discovering new antiviral leads, but also confirms that the process for selection and prioritization of plants has significant benefit by directing our efforts to those plants that have the highest probability of containing antiviral compounds.

Table 2. Correlation between ethnomedically ranked plants for respiratory virus disease and activity against respiratory syncytial virus in the plant extract

RANK	ETHNOMEDICAL INDICATION	% MEETING ACTIVITY CRITERIA*
1.0-1.3	High-Medium Priority Viral Respiratory Infection	20
1.5	Low Priority Viral Respiratory Infection	15
2.0	Non Respiratory Viral, but Possibly Other Viral Indications	8
3.0	Non-viral Ethnomedical Use	8

* Antiviral activity, ED_{50} = 83 μg/ml & selectivity index $\geq$ 10.

We have also compared the frequency of isolating lead antiviral compounds from plants using Shaman's ethnobotanically driven approach and the high volume, random screening approach (Table 3, part 1). In this analysis, isolation of pure antiviral lead compounds, from plants indicated for three viruses, shows isolation frequencies ranging 1.6 to 8.2%. Compared to the frequency of isolating lead compounds from a fourth virus using the random screening approach (Table 3, part 2), the ethnomedically driven approach is 125 to 630 times more efficient (depending on which virus is used for comparison).

Table 3. Comparison of frequencies of isolating lead compounds using Shaman's approach and the random screening method

PART 1: SHAMAN'S EXPERIENCE

Area	Number of Plants Tested	Number of Active Compounds Isolated	% Isolation Frequency
RSV	97	8	8.2
FLU	123	2	1.6
CMV	231	5	2.2

PART 2: GENERAL INDUSTRIAL EXPERIENCE

Area	Number of Natural Products Tested	Number of Active Compounds Isolated	% Isolation Frequency
HSV	15,000	2	0.013

4.3 Frequency of confirming biological activity

Another recent analysis was to compare the frequencies for confirming antiviral activity using *in vitro* and *in vivo* primary assays for plants which were selected based on ethnomedical criteria. Using RSV again as the example, we found that we were more efficient in confirming the antiviral potential of a plant using the *in vivo*, animal-based assay than by using the *in vitro* assay (Table 4). Given that mice more closely resemble humans biologically than cell cultures, and that the animal model at least partially mimics human RSV disease, this finding seems sensible. Moreover, the criteria for selecting potential antiviral plants in the first place is based on a history of activity in humans, not cell culture systems. Similarly, the diabetes program was also more efficient when animal-based assays, rather than *in vitro* assays were used as the primary assay to confirm glucose lowering activity of plants (Table 4).

Table 4. Experimental frequencies for confirming biological activity in ethnomedically indicated plants

	Cellular	Animal
Antiviral	20%	29%
Diabetes	39%	57%

5 DEVELOPMENT OF ACTIVES

To date, several different classes of compounds have been isolated and their structures determined. Primary study of these compounds indicates significant lowering of blood glucose but the mechanism of action is not known at this time. Medicinal chemistry and synthetic efforts support these natural product leads in the event that the isolated natural product is available in a low yield and/or the plant source is not amenable to sustainable harvesting. In this situation, when feasible, a synthetic approach to the natural product is considered. An equally important medicinal chemistry mission is to use the isolated natural product as a template for further structural modification to reduce toxicity and/or improve potency. As a result of this process, new chemical leads can be generated from the initial orally active natural product lead. The medicinal chemistry program has been highly successful: two total syntheses originating from an antifungal natural product isolated from *Irlbachia alata* and *Anthocleista djalonensis* were completed and have been published[6,7], and a structural modification study originating from a natural product isolated from *Ambrosia chamissonis*[8] has been achieved.

6 SHAMAN'S RECIPROCITY STRATEGIES

Many modern medicines are plant-derived, but the origins of these pharmaceutical agents and their relationship to the knowledge of the indigenous people in the tropical forests is usually omitted.

The idea of compensating indigenous people for the use of knowledge about biological diversity is one based on fairness and equity. A logical means of compensating indigenous peoples for their role in a drug discovery process would be to accord them a share of the profits from the drug, once it is commercialized. However, because of the long period of time needed for commercial drug discovery and development (often ten years or more), such a mechanism for reciprocity requires a long waiting period before any benefit is realized by the indigenous peoples. Furthermore, in most instances, the indigenous knowledge gathered may not lead to a commercial product and thus, no benefit of any kind would come to the local people. From its inception, Shaman has been committed to the concept of reciprocal benefits: to developing new therapeutic agents by working with indigenous and local peoples of the tropical rain forests and, in the process, contributing to the conservation of biological and cultural diversity, or "biocultural diversity".[9-13]

Shaman, at its inception, founded the Healing Forest Conservancy as a nonprofit organization. The Healing Forest Conservancy is dedicated to conserve cultural and biological diversity and to sustain the development and management of the natural and biocultural resources that are a part of the heritage of native populations. The Conservancy was founded because no governmental organization existed to provide a formal and consistent process to compensate countries and communities for ethnobotanical leads which subsequently are developed into commercial product. The Conservancy ensures a mechanism for the species-rich tropical counties and the small-scale indigenous communities in tropical forests to be equitably compensated for their participation in the development of therapeutic agents. A number of pilot programs has been initiated and described in recent publications.[10,14-16]

7 CONCLUSION

Shaman Pharmaceuticals embraces the idea of creating an efficient drug discovery process through the integration of ethnobotany, modern medicine and natural products chemistry. Our approach integrates traditional healer information in the field followed by direct application of this knowledge base in our modern lab facilities. We believe we can continue to derive plant chemical entities using the knowledge of the traditional healer. To translate indigenous healing lore into modern medical science, we implement a valid methodology for adapting local use of medicinal plant preparations. Thus we combine the attributes of traditional and current knowledge, by choosing only those plants selected by traditional healers, then using state-of-the-art technology to isolate and purify the active principal and verify its activity in our in-house animal model.

Shaman has developed a pioneering technology platform, integrating the sciences of ethnobotany, ethnomedicine, medicine, modern separation science, medicinal chemistry, and primary *in vivo* screening. The process has led to the discovery of multiple orally active antihyperglycemic leads in our antidiabetes discovery program that are currently undergoing preclinical evaluation. We are currently entering Phase II clinical trials with Provir™, an oral product for the treatment of secretory diarrhea, and beginning pivotal Phase III clinical studies on Virend®, a topical antiviral agent for the treatment of genital herpes simplex virus.

Acknowledgments

The authors would like to acknowledge the contributions of Lisa Conte, Shaman's Scientific Strategy Team, outside investigators and the entire research, development and administration teams of Shaman Pharmaceuticals. We thank Dr. Silviano Camberos Sanchez for providing the patient data from Latin America. We also thank and acknowledge all the countries, indigenous groups, government and non-governmental organizations with whom we continue to have the pleasure of collaborating. These include indigenous communities, federations, governmental ministries, non-governmental organizations, and university scientists in the countries of Colombia, Ecuador, Peru, Guinea, Nigeria, Cameroon, Congo, Tanzania, Papua New Guinea, Philippines, Thailand, Indonesia, and many others. We would especially like to thank the indigenous federations and organizations of Aguaruna and Huambisa (CAH), the Ashanika Federation, Asociación Indigena de Desarrollo de la Selva Peruana (AIDESEP) and many other national and intentional organizations that work and collaborate with indigenous communities and their organizations. Activities of these organizations have provided regular monitoring, guidance and feedback about the ongoing activities of Shaman Pharmaceuticals.

References

1. N.R. Farnsworth, O. Akerele, A.S. Bingel. *Bull. World Health Org.* 1985, **63**, 965.

2. N.R. Farnsworth. Screening plants for new medicines. In 'Biodiversity', E.O. Wilson, ed., National Academy Press, Washington, D.C., 1988, p. 3.

3. R. Ubillas, S.D. Jolad, R.C. Bruening, M.R. Kernan, S.R. King, D.F. Sesin, M. Barrett, C.A. Stoddart, T. Flaster, J. Kuo, F. Ayala, E. Meza, M. Castanel, D. McMeekin, E. Rozhon, M.S. Tempesta, D. Bernard, J. Huffmann, D. Sonce, R.

Sidwell, K. Soike, A. Brazier, S. Safrin, R. Orlando, P.T.M. Kenny, N. Berova, and K. Nakanishi. *Phytomedicine.* 1994, **1**, 77.

4. S.R. King and M.S. Tempesta. From Shaman to human clinical trials: The role of industry in ethnobotany, conservation and community reciprocity. In 'Ciba Foundation Symposium', 1994, p. 197.

5. P. Cornelius, S. Enerback, G. Bjursell, T. Olivecrona and P.H. Pekala. *Biochem J.* 1988, **249**, 765.

6. D.E. Bierer, L.G. Dubenko, R.E. Gerber, J. Litvak, J. Chu, D.L. Thai, M.S. Tempesta, and T.V. Truong, *J. Org. Chem.* 1995, **60**, 7646.

7. D.E. Bierer, R.E. Gerber, S.D. Jolad, R.P. Ubillas, J. Randle, E. Nauka, J. LaTour, J.M. Dener, D.M. Fort, J.E. Kuo, W.D. Inman, L.G. Dubenko, F. Ayala, A. Ozioko, C. Obialor, E. Elisabetsky, T. Carlson, T.V. Truong, and R.C.Bruening. *J. Org. Chem.*, 1995, **60**, 7022.

8. D.E. Bierer, J.M. Dener, L.G. Dubenko, R.E. Gerber, J. Litvak, S. Peterli, P. Peterli-Roth, T.V. Truong, G. Mao, and B.E. Bauer, B. E. *J. Med. Chem.* 1995, **38**, 2628.

9. S.R. King. *Cultural Survival Quarterly* 1991,**15**, 19-22.

10. S.R. King. Conservation and Tropical Medicinal Plant Research. I n 'Medicinal Resources of the Tropical Forest', M.J. Balick, E. Elisabetsky, and S.A. Laird, S. A., eds., Columbia University Press, New York, 1996; p 63.

11. K. Moran. Ethnobiology and US Policy. In 'Sustainable Harvest and Marketing of Rainforest Products', M. Plotkin and L. Famolare, eds., Island Press, Washington, D. C., 1992, Chapter 5.

12. S.R. King and T.J. Carlson. *Interciencia* 1995, **20**, 134-139.

13. S.R. King, T.J. Carlson, and K. Moran. Biological Diversity, Indigenous Knowledge, Drug Discovery, and Intellectual Property Rights. In 'Valuing Local Knowledge: Indigenous People and Intellectual Property Rights', S. Brush, S. and D. Stabinsky, eds., Island Press, Washington, D. C., 1996, Chapter 8, p. 167.

14. K. Moran. 'Biocultural Diversity Conservation Through the Healing Forest Conservancy In Intellectual Property Rights for Indigenous Peoples, A Source Book', T. Greaves, ed., Society for Applied Anthropology, Oklahoma City, OK, 1994; p. 101.

15. S.R. King, T.J. Carlson, and K. Moran, K. *J. Ethnopharm.* 1996, **51**, 45.

16. K. Moran. Returning Benefits From Ethnobotanical Drug Discovery to Native Communities. In 'Biodiversity and Human Health', F. Grifo and J. Rosenthal, eds. Island Press. In press.

Bioactive Natural Products from Thai Medicinal Plants

C. Mahidol, H. Prawat and S. Ruchirawat

CHULABHORN RESEARCH INSTITUTE, BANGKOK 10210, THAILAND, AND CHULABHORN RESEARCH CENTRE, MAHIDOL UNIVERSITY, BANGKOK 10400, THAILAND

1 INTRODUCTION

In developing countries, medicinal plants continue to be the main source of medication. In China alone, 7,295 plant species are utilised as medicinal agents. The World Health Organization has estimated that for some 3.4 billion or 88% of the people in the developing world, plants represent the primary source of medicine.[1,2] A small percentage of the world's more than 250,000 flowering plants have so far been analysed for their possible medicinal uses. The most alarming cause for concern is that by the turn of this century, it is expected that some 25,000 species of plants will have ceased to exist due to excessive deforestation all over the world. This represents about five plant species a day between now and the year 2000.

It is thus a matter of utmost concern to public health and indeed to human life and welfare that urgent action is taken to prevent further diminution of actual and potential availability of medicinal and biological agents. Natural remedies that, although undocumented, may have been used for many thousands of years by the human race must be appropriately catalogued to ensure that vital ethnomedical information is not lost for ever.

Thailand is uniquely located to represent the fauna and flora which characterizes the biogeographic province of Indo-Burma. A number of the eastern Himalaya temperate taxa penetrate south into the northern mountains of Thailand while the southern part is evergreen forest thus making this area one of the richest floristic regions of the world.[3] It has been estimated that the vascular plants in Thailand include at least 10,000 species of about 1,763 genera from 245 families.[4] The numbers of alkaloid-containing plants are estimated to be only about 266 species of 176 genera in 67 families based on the Thai plants names and parts of the uncompleted flora of Thailand.[5] Thailand is endowed with a great diversity of indigenous medicinal plant species, but unfortunately only a small fraction of these have been scientifically investigated. The Thais have a long tradition of using medicinal herbs and plants in folklore medicine but many of the claimed curative properties have not been scientifically proven. In this review, work on the antimalarials isolated from the Thai plants will be highlighted.[6] The role of natural products for the treatment of malarial and other infectious diseases has been the subject of many excellent reviews.[7-13]

2 ANTIMALARIAL NATURAL PRODUCTS FROM THAI PLANTS

In the tropical world, protozoa are the cause of quite a number of diseases which inflict massive misery and death. Most prominent among the protozoal diseases is malaria which is most feared by the travellers to the tropics. It is not at all an exaggeration to say that malaria has been responsible for much of the human suffering and misery accompanying the process of social and economic development. About 270 million cases of malaria occur in the world every year and that 2.1 billion people (half the world's population) in some 100 countries now live in areas where there is a definite risk of falling sick from malaria.[14] Over the last ten years the malaria situation has been worsening in many areas of the world.

In most of the areas of Asia and Latin America where malaria now occurs, it had been considerably reduced or eliminated during the 1960s and 1970s. The situation has been worsening recently particularly in the frontier areas of economic development such as agriculture and mining in newly opened jungle, and in areas of warfare, illegal trading and migration of refugees. In many parts of the world where there are wars and other conflicts, or smuggling and movements of refugees across the borders, the disease has spread rapidly because no control and treatments could be undertaken.

The need to find new antimalarials is pressing due to the discovery of resistance of the human malarial parasite, *Plasmodium falciparum*, to the currently available common antimalarial drugs. Treatment has thus become both less effective and much more expensive. The problem is further aggravated by the resistance of vector anopheline mosquitoes to the most effective and least toxic insecticides which were used to kill them. It was once thought in the early 1950s that malaria had been conquered but by the late 1950s the parasite had been found to be resistant to chloroquine in South America and S.E.Asia. This was not at all unexpected because *Plasmodium falciparum* was reported to be resistant to quinine as early as 1910. Today, not only are there strains of *Plasmodium falciparum* that are resistant to chloroquine but also to other once useful drugs including amodiaquine, sulphadoxine-pyrimethamine, quinine and mefloquine. It has been recently stated that "If the present trend continues the affluent countries will soon arrive at the end of the drug reportoire".[15]

The potential of natural products as therapeutic agents in the treatment of malaria is enormous. Quinine (1) was introduced as an antimalarial more than a century ago. Quinine also served as a prototype antimalarial leading to the discovery of synthetic antimalarials based on quinoline compounds. 4- and 8-Aminoquinoline derivatives have proved their efficacy as antimalarials. Mefloquine (2), a quinoline methanol derivative , has recently been introduced as an antimalarial. Here again, Quinine is the prototype for the design of the drug.

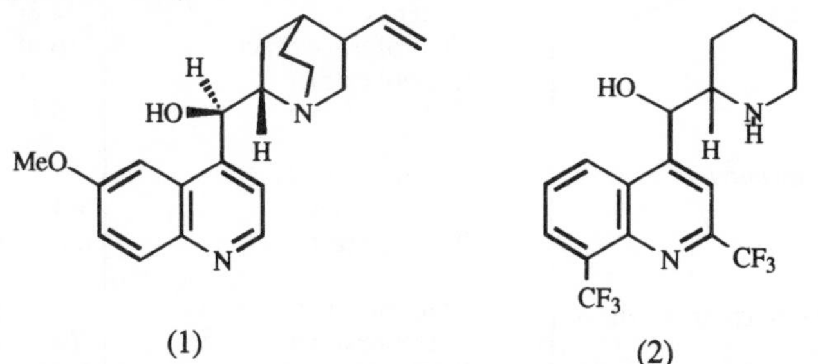

(1) (2)

The Chinese have a long history of use of a herb called *Qing hao (Artemisia Annua)* for the treatment of malaria. It was reported as early as in 1596 that chills and fever of malaria can be combated by *Qing hao* preparations. The pure compound was isolated in 1972 and named *Qinghaosu (QHS)* or *Arteannuin* meaning the "active principle of Qing hao", the name Artemisinin (3) is recommended by Chemical Abstracts.[16]

In Thailand, interest in the use of herbal medicine for the treatment of malaria has been long standing. Lately many groups of Thai scientists have been working on various plants to scrutinize the validity of these 'claims' as well as searching for new antimalarial agents from plants. During the second world war, thirty plant extracts used

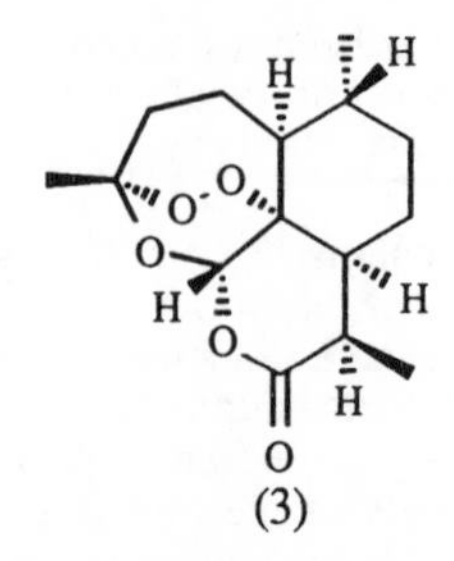

(3)

in herbal medicine were used to treat malaria. The treatments were applied to 543 patients with infections with *Plasmodium falciparum* (255 patients) and *Plasmodium vivax*. (288 patients).[17] Eight plant extracts were found to have a cure rate of higher than 50%. It was found that the plant of *Oroxylum indicum* showed a total cure rate as high as 65.7%. It is very interesting to note the definite preference for the plants of *Azadirachta* and *Nyctanthes arbor-tritis* for the treatment of those patients infected with *P. vivax* (80% cure rate and 33% for *P. falciparum*)

Scientists at the Division of Medical Research, Department of Medical Sciences, Ministry of Public Health and at AFRIMS (Armed Force Research Institute for Medical Science) have been interested in the search for antimalarial agents from Thai medicinal plants.[18] Screening of twenty Thai medicinal plants led to the identification of four plants i.e. *Brucea javanica, Picrasma javanica,Celastrus paniculatus* and *Tiliacora triandra* which gave good results for the *in vitro* testing (MIC of less than 25 μg/ml.)

The four plants which exhibited the minimal inhibitory concentration (MIC) of not more than 25 microgram per millilitre were then further studied by extraction with various solvents, each solvent extract was then again tested. The results are shown in the table.

PLANTS	SOLVENTS	MIC (μg/ml)
Brucea javanica	Hexane extract	94.80
	Chloroform extract	16.48
	Ethanol extract	32.80
	Water extract	19.48
Picrasma javanica	Chloroform extract	24
	Ethanol extract	64
	Water extract	no inhibition
Celastrus paniculatus willd	Chloroform extract	6
	Ethanol extract	79
	Water extract	153
Tiliacora triandra	Methanol extract	17
	Water-insoluble alkaloids	2
	Water-soluble alkaloids	22

The chloroform extracts gave the best result for the *in vitro* antiplasmodial activity for *Brucea javanica, Picrasma javanica* and *Celastrus paniculatus.* In the case of *Tiliacora triandra,* the water insoluble alkaloid fraction gave the best results.

The four active fractions were then subjected to various chromatographic techniques to isolate pure compounds for further testing.

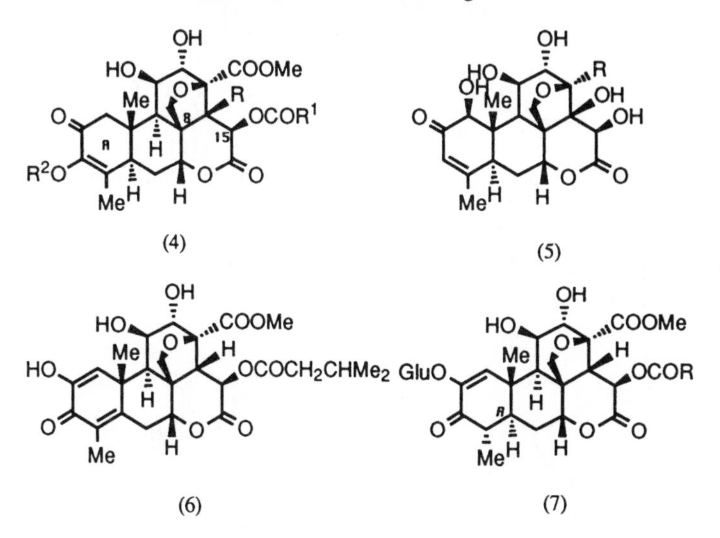

(4) (5)

(6) (7)

Twelve quassinoids have been isolated from *B. javanica* fruits and the structures have been elucidated. These quassinoids could be divided into four groups with mainly different functionalities in the A ring. In group I (4), there are seven compounds isolated with varying ester substituents, two compounds were isolated with structures as shown in group II (5) and only one compound was isolated having the structure as shown in group III (6) and finally two compounds with the glycoside groups were isolated as classified in group IV (7). The structures (8) of the five most active quassinoids are as shown.

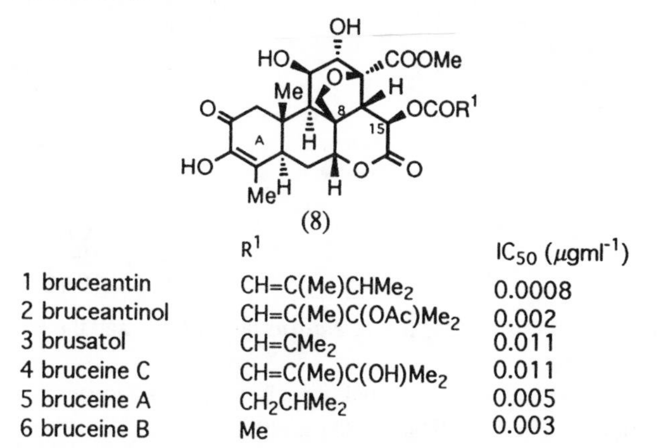

(8)

	R^1	IC_{50} (μgml^{-1})
1 bruceantin	CH=C(Me)CHMe$_2$	0.0008
2 bruceantinol	CH=C(Me)C(OAc)Me$_2$	0.002
3 brusatol	CH=CMe$_2$	0.011
4 bruceine C	CH=C(Me)C(OH)Me$_2$	0.011
5 bruceine A	CH$_2$CHMe$_2$	0.005
6 bruceine B	Me	0.003

Bruceatin exhibited a very impressive IC_{50} at 0.0008 μg/ml. It may be noted that the most active compounds are of type 1, differing only in the ester group.

Eurycoma longifolia, Ailanthus altissima and *Simarouba amara* have also been investigated and related compounds have been isolated and identified.[19-21]

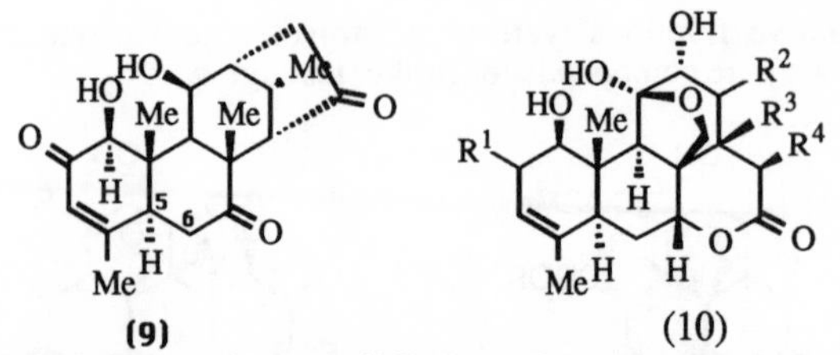

(9) (10)

Two structural types (9) and (10) have been isolated from these plants. *Eurycoma longifolia* produced both types of compounds while *Ailanthus altissima* and *Simarouba amara* produced only the compounds of structural type (10). The IC_{50} of the nine compounds isolated from these plants ranged from 0.004 μg/ml to 1.15 μg/ml.

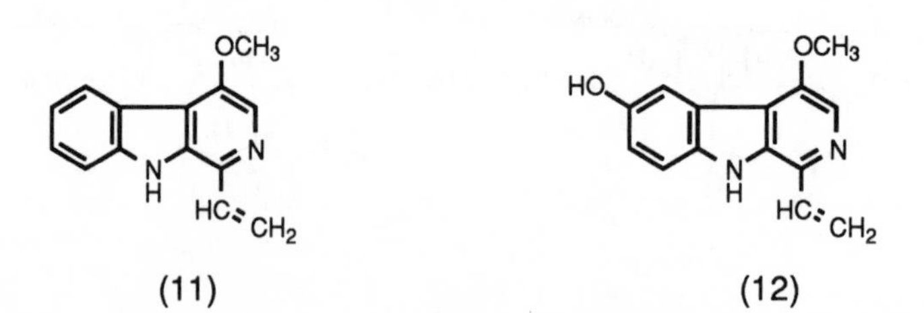

(11) (12)

Two carbazole alkaloids were isolated from *Picrasma javanica* and identified as 4-methoxy-1-vinyl-β-carboline derivatives (11 and 12, ID_{50} 2.43 and 3.28 μg/ml).[22]

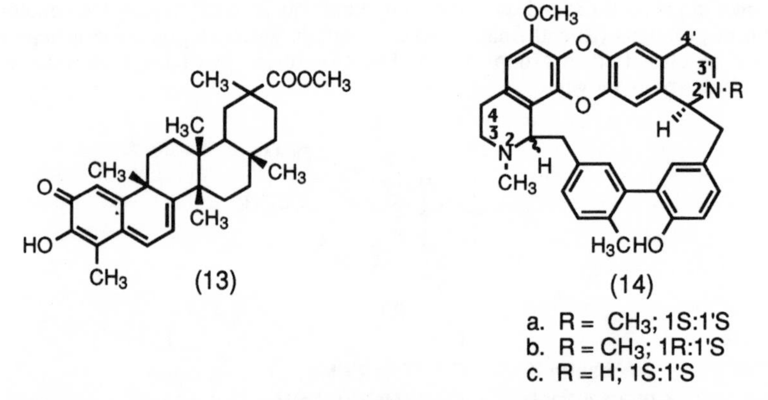

(13)

(14)

a. R = CH_3; 1S:1'S
b. R = CH_3; 1R:1'S
c. R = H; 1S:1'S

Pristimerin (13) was isolated from *Celastrus paniculatus*, and the ID_{50} was found to be 0.28 μg/ml.[23]

Three bisbenzylisoquinoline alkaloids were isolated from *Tiliacora triandra*, and identified as tiliacorinine (14a, ID_{50} 3.533 μg/ml) tiliacorine (14b, ID_{50} 0.675 μg/ml) and nortiliacorinine A (14c, ID_{50} 0.558 μg/ml). Two more alkaloids, coded alkaloid G and alkaloid H were also isolated but the structures are currently unknown. The ID_{50} of alkaloids G and H were found to be 0.344 and 0.916 μg/ml respectively.[24]

Azadiracta indica var. siamensis is another Thai plant that has been used for the treatment of malaria, interestingly the same practice was also adopted in Nigeria. Nimbolide, a bitter principle, was isolated as the main constituent of this plant and the structure was deduced to be pentacyclic lactone (15).

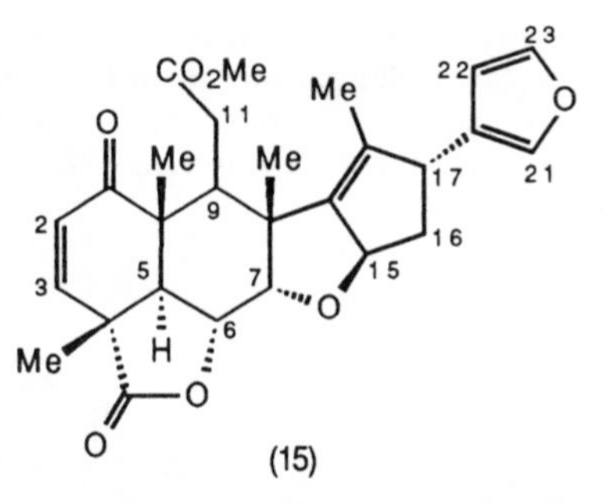

(15)

The terpenoid lactone nimbolide was found to inhibit *Plasmodium falciparum* in culture with a moderate potency.[25] The EC_{50} against the parasite line K_1 from Thailand was approximately 2.0 μM (0.95 μg/ml). The EC_{50} of the crude aqueous extract of *Azadirachta indica* var. siamensis (Sadao tree) was 115 μg/ml. and of the crude ethanol extract was 5.0 μg/ml. Since nimbolide is a major constituent in these extracts it could account substantially for their inhibitory activity. However, neither the crude extracts nor nimbolide showed any activity *in vivo* against *Plasmodium berghei* in mice either through ingestion or subcutaneous injection.

3 RECENT WORKS

3.1 Peroxide Compounds

As mentioned ealier, artimisinin has shown great promise for the treatment of malaria. It is most significant to note that artemisinin and its derivatives are at present the only group of compounds still effective for the treatment of some drug resistant strains of malaria in the Thai-Burmese and Thai-Cambodian border regions. It is foreseen that the 1,2,4-trioxane which is the integral unit of artimisinin, can act as a template for the development of further generations of antimalarial drugs especially through the design of novel synthetic 1,2,4-trioxane derivatives.[26]

Yingzhao (*Artabrotrys uncinatus*) is a rare perennial vine growing in the island of Heinan and coastal regions of Guangdong. The extracts of the roots of this plant have been used as a treatment for malaria for centuries. This lead resulted in the isolation and identification of the sesquiterpene peroxides, yingzhaosu A (16)[27] in 1979 and yingzhaosu C (17)[28]in 1988 which are responsible for the antimalarial activity. The synthesis of both yingzhaosu A[29] and yingzhaosu C[30] have recently been reported.

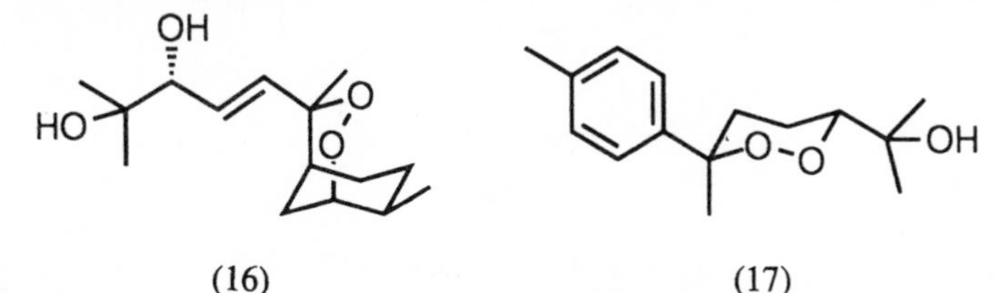

(16) (17)

Peroxide compounds have also been isolated from some plants of Thailand which have been used as folk remedies for malaria. Investigations[31] of *Amomum krevanh* Pierre, locally called "Kra-Waan" or "Round Siam Cardamon" in the international market has resulted in the isolation of diterpene peroxide (18). The compound was isolated from the hexane extract and exhibited *in vitro* inhibition of *P. falciparum* (EC_{50} = 0.17 mM. Also from another Thai plant, *Cyperus rotundus* Linn., a common weed distributed throughout the country, an antimalarial possessing the peroxide moiety was isolated. Activity-guided investigation of the extracts of the tubers of *Cyperus rotundus* led to the isolation and identification of the sesquiterpene endoperoxide, 10,12-peroxycalamenene (19)[32] which exhibited *in vitro* antimalarial activity with an EC_{50} value of 2.33 mM. The *in vivo* efficacy of these two peroxides remains to be tested.

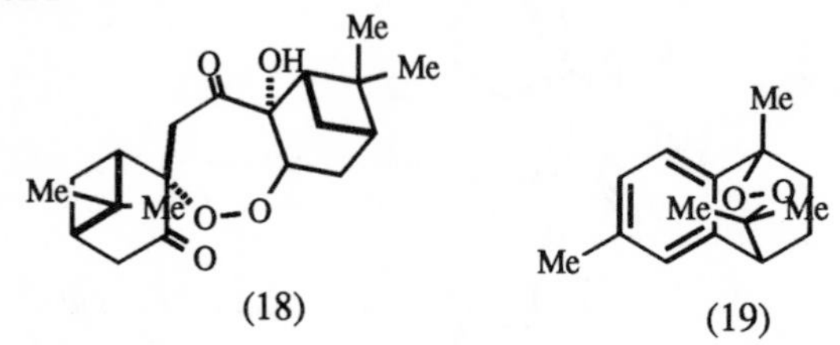

(18) (19)

3.2 Isoquinoline Alkaloids

3.2.1 Bisbenzylisoquinoline Alkaloids

As mentioned earlier, investigation of *Tiliacora triandra* yielded three antimalarials having the bisbenzylisoquinoline skeleton. Two other Thai plants which were alleged to have antimalarial properties have also been investigated and it has been found that these plants contain the bisbenzylisoquinoline derivatives which are responsible for these properties. *Cyclea barbata* Wall. was used in Thailand for the treatment of fever associated with malaria. Investigation[33] of this plant led to the isolation of tetrandine (20a) as a major alkaloid together with other related bisbenzylisoquinoline alkaloids. These alkaloids have EC_{50} values against *P.falciparum* of the order of 0.03-0.45 μg/ml. *Stephania erecta* Craib. has been used in Thai folk medicine as a skeletal muscle relaxant and an analgesic. A large number of bisbenzylisoquinoline alkaloids have also been isolated from this plant. The antimalarial activity of the bisbenzylisoquinoline alkaloids isolated from *Stephania erecta* [34] has been evaluated against *P.falciparum* chloroquine-resistant strain (W-2) and a chloroquine-sensitive strain (D-6). The EC_{50} was found to range from 0.045-0.31 μg/ml. Significantly, it was found that the 2-nor alkaloid was always more active in the test as compared with the di-N-Me derivative.

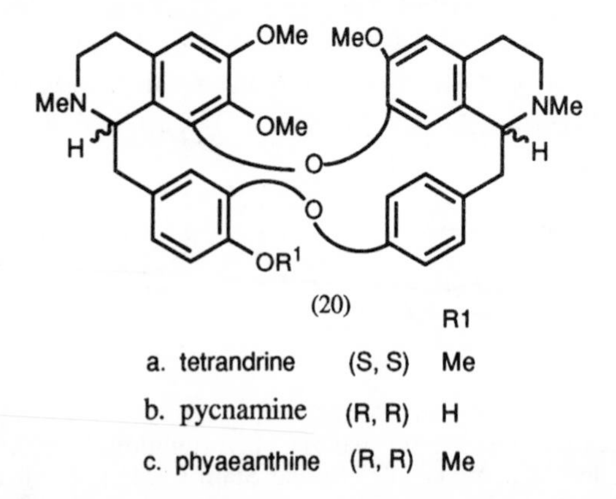

(20)

Eighteen species of higher plants used in folk medicine for the treatment of malaria and/or fever in Sierra Leone have been screened for antimalarial activity.[13] The active constituent was found to be pycnamine (20b), a bisbenzylisoquinoline alkaloid isolated from *Triclisia patens* as well as phyaeanthine (20c) and aromoline which exhibited EC_{50} ranging from 0.15-1.43 μg/ml. These findings lend scientific support to the use of these plants in traditional medicine for the treatment of malaria in two continents as well as stressing the importance of the role of traditional medicine as a guide for selection of plants to be studied.

3.2.2 Naphthylisoquinoline Alkaloids

The basic structure of the naphthylisoquinoline alkaloid consists of the linking of a naphthalene and a tetrahydroisoquinoline moieties. Being a biaryl system, these alkaloids can display atropisomerism due to the steric effect of the ortho group adjacent to the biaryl axis. Naphthylisoquinoline alkaloids have been isolated from the species *Ancistrocladaceae* and *Dioncophyllaceae*. The tropical plants species *Triphyophyllum peltaatum*, *Ancistrocladus abbreviatus* and *A. barteri* are well known in the traditional medicine of West Africa and are used for the treatment of fevers, malaria and other diseases. The extracts of these plants[35] and the pure alkaloids isolated from them have been tested and found to be active against a chloroquine-resistant strain (K1) and a chloroquine-sensitive clone (NF 54/64) of *P. falciparum in vitro*. The two most potent compounds were found to be dioncophylline B (21) (IC_{50} = 0.063 μg/ml for K1 and 0.224 for NF 54/64) and dioncopeltine A (22) (IC_{50} = 0.33 μg/ml for K1 and 0.021 μg/ml for NF 54/64).

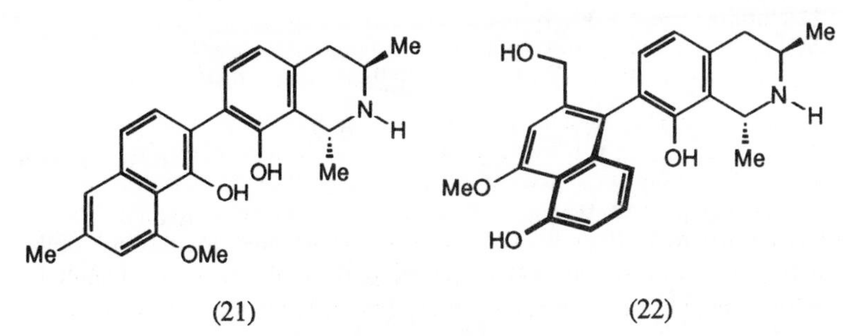

(21) (22)

Interestingly, in Thailand the roots of *A. tectorius* have also been used to treat dysentery and malaria. Investigations[36] of these plants have led to the isolation and identification of a new naphthylisoquinoline alkaloid (23). Recently a new naphthylisoquinoline, 4'-O-demethylancistrocladine (24) was isolated from Malaysian *A. tectorius*.[37]

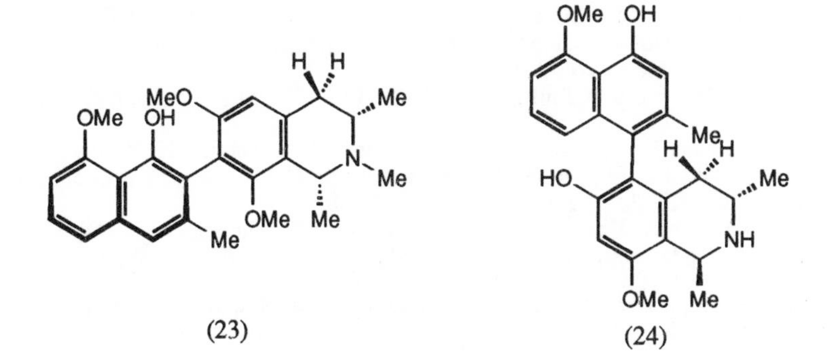

(23) (24)

Acknowledgment: We are grateful to the Thai Government and Petroleum Authority of Thailand for generous financial support of our research programmes.

REFERENCES

1. N.R. Farnsworth, O. Akerele, A.S. Bingel, D.D. Soejarto, and Z. Guo *Bull. WHO* 1985, **63**, 965.
2. G.A. Cordell, Proceedings of The Second Princess Chulabhorn Science Congress, In press.
3. P. S. Ashton, in "Biodiversity in Thailand"(edited by S. Wongsiri and S. Lorlohakarn) Prachachon Ltd. Co. Bangkok. 1989, p. 51.
4. T. Santisuk, in "Biodiversity in Thailand"(edited by S.Wongsiri and S.Lorlohakarn) Prachachon Ltd. Co. Bangkok. 1989, p. 81.
5. B. Tantisewie and S. Ruchirawat, in " The Alkaloids, Chemistry and Pharmacology" Volume 41 (edited by A. Brossi and G.A. Cordell), Academic Press 1992, p. 1. T. Smitinand, "Thai Plant Names" Funny Publishing Ltd. Bangkok., 1980.
6. C. Mahidol, P. Sahakitpichan and S. Ruchirawat, *Pure & Appl. Chem.*, 1994, **66**, 2353
7. J.D. Phillipson and M.J. O'Neill, Biologically active natural products, Proceedings of the Phytochemical Society of Europe, (edited by K. Hostettmann and P.J. Lea), Clarendon Press, Oxford.1987, **27**, 49
8. C.W. Wright and J.D. Phillipson, *Phytother. Res.*, 1990, **4**, 127.
9. J.D. Phillipson and C.W. Wright, *Planta Med.,* 1991, **57**, S53.
10. J.D. Phillipson and C.W. Wright, *J. of Ethnopharm.,* 1991, **32**, 155.
11. J.D. Phillipson and C.W. Wright, *Transactions of the Royal Society of Tropical Medicine and Hygiene* , 1991, **85**, 18.
12. J.D.Phillipson, C.W. Wright, G.C. Kirby, and D.C. Warhurst, "Phytochemical Potential of Tropical Plants (edited by K.R. Downum, J.T. Romeo, and H.A. Stafford), Plenum Press, New York, 1993, p. 1.
13. J.D. Phillipson, C.W. Wright, G.C. Kirby, and D.C. Warhurst, in "Phytochemistry of Plants Used in Traditional Medicine", Proceedings of the Phytochemical Society of Europe, (edited by K. Hostettman, A. Marston, M. Maillard, amd M. Hamburger), 1995, p. 95.
14. C. Laitman, *Tropical Diseases Research News* 1990, 31, 3.
15. W.H. Wernsdorfer, *Parasitol. Today* 1991, **7**, 297.
16. For reviews see: a. D.L. Klayman, *Science* 1985, **228**, 1049. b. S.S. Zaman and R.P. Sharma, *Heterocycles* 1991, **32**, 1593.
17. O. Ketusingh, in "Special Publication for the 60th Anniversary of Siriraj Hospital", 1950, 275.
18. T. Dechatiwongse, paper presented at a seminar on "Herbal Medicine and Malaria" November 1989; organizer: The Ministry of University Affairs.
19. J.D. Phillipson, M.J. O'Neill, and D.C. Warhurst, Proceedings of The First Princess Chulabhorn Science Congress 1987, International Congress on Natural Products, Vol. **2**, 55.
20. For a review see: J.D. Phillipson and M.J. O'Neill, *Parasitol. Today* ,1986, **2**, 355.
21. K. Pavanand, W. Nutakul, T. Dechatiwongse, K. Yoshihira, K. Yongvanitchit, J.P. Scovill, J.L. Flippen-Anderson, R. Gilardi, C. George, P. Kanchanapee, and H.K. Webster, *Planta Med.*, 1986, **2**, 108.
22. K. Pavanand, K. Vongvanitchit, H.K. Webster, T. Dechatiwongse, W. Nutakul, Y. Jewvachdamrongkul, and J. Bansiddhi, *Phytother. Res.*, 1988, **2**, 33.
23. K. Pavanand, H.K.Webster, K. Yongvanitchit, A. Kun-anake, T. Dechatiwongse, W. Nutakul, and J. Bansiddhi, *Phytother. Res.*, 1989, **3**, 136.

24. T. Dechatiwongse, P. Chavalittumrong, W. Nutakul, *Bull. Dept. Med. Sci.* 1987, **29**, 33.
25. S. Rochanakij, Y. Thebtaranonth, C. Yenchai, and Y. Yuthavong, *Southeast Asian J. Trop. Med. Pub. Hlth.*, 1985, **16**, 66.
26. G.H. Posner, C.H. Oh, L. Gerena, and W.K. Milhous, *J. Med. Chem.*, 1992, **35**, 2459.
27. X.T. Liang, D.Q.Yu, W.L.Wu, H.C. Deng, *Acta Chim. Sin.*, 1979, **37**, 215.
28. (a) L. Zhang, W.S. Zhou, X.X. Xu, *J. Chem. Soc., Chem. Commun.*, 1988, 523; (b) L. Zhang, W.S. Zhou, X.X. Xu, *Sci. China, Ser.* B, 1989, **32**, 800.
29. X.X. Xu, J. Zhu, D. Z. Huang, and W. Zhou, *Tetrahedron Lett.*, 1991, **32**, 5785
30. J. Boukouvalas, R. Pouliot, and Y. Frechette *Tetrahedron Lett.*, 1995, **36**, 4167.
31. S. Kamchonwongpaisan, C. Nilanonta, B. Tarnchompoo, C. Thebtaranonth, Y. Thebtaranonth, Y. Yuthavong, P. Kongsaeree and J. Clardy; *Tetrahedron Lett.*,1995, **36**, 1821.
32. C. Thebtaranonth, Y. Thebtaranonth, S. Wanauppathamkul and Y. Yuthavong, *Phytochemistry,* 1995, **40**, 125.
33. L.-Z. Lin, H.-L. Shieh, C.K. Angerhofer, J.M. Pezzuto, G.A. Cordell, L. Xue, M.E. Johnson, and N. Ruangrungsi, *J. of Nat. Prod.*, 1993, **56**, 22.
34. K. Likhitwitayawuid, C.K. Angerhofer, G.A. Cordell, J.M. Pezzuto, and N. Ruangrungsi, *J. Nat. Prod.*, 1993, **56**, 30.
35. G. Francois, G. Bringmann, J.D. Phillipson, L.A. Assi, C. Dochez, M. Rubenacker, C. Scheider, M. Wery, D.C. Warhurst, and G.C. Kirby, *Phytochemistry* 1994, **35**, 1461.
36. N. Ruangrungsri, V. Wongpanich, P. Tantivatana, H.J. Cowe, P.J. Cox, S. Funayama, and G.A. Cordell, *J. Nat. Prod.,* 1985, **48**, 529.
37. A. Montagnac, A.H.A. Hadi, F. Remy, and M. Pais, *Phytochemistry*, 1995, **39**, 701.

Novel Biologically-Active Alkaloids from British Plants

R. J. Nash[1], A. A. Watson[1], A. L. Winters[1], G. W. J. Fleet[2], M. R. Wormald[3], S. Dealler[4], E. Lees[4], N. Asano[5] and H. Kizu[5]

[1] INSTITUTE OF GRASSLAND AND ENVIRONMENTAL RESEARCH, ABERYSTWYTH SY23 3EB, UK
[2] DYSON PERRINS LABORATORY, SOUTH PARKS ROAD, OXFORD OX1 3QY, UK
[3] BIOCHEMISTRY DEPARTMENT, SOUTH PARKS ROAD, OXFORD OX1 3QY, UK
[4] BURNLEY GENERAL HOSPITAL, CASTERTON AVENUE, BURNLEY BB10 2PQ, UK
[5] FACULTY OF PHARMACEUTICAL SCIENCES, HOKURIKU UNIVERSITY, KANAZAWA, JAPAN

1 INTRODUCTION

In the last two decades attention has focused on tropical plants as sources of new natural products for potential pharmaceutical and agrochemical use but temperate species have been largely ignored. The fact that many plants in the U.K. deserve further investigation was highlighted by a recent study[1] of potato leaves and tubers which revealed two polyhydroxylated *nor*tropane alkaloids (calystegines) not previously described from this species despite the routine analyses conducted on potatoes by many laboratories worldwide. The calystegines are hydrophilic alkaloids which are not picked up by standard glycoalkaloid analyses. They are potent inhibitors of mammalian glucosidases and galactosidases[2,3] and have been found in over 70 potato varieties analysed at IGER and they are also present in the edible parts of aubergines (egg plant), sweet potatoes and chilli peppers. Alkaloids which inhibit glycosidases have aroused considerable interest in recent years as potential anti-viral, anti-cancer, anti-diabetic agents and as agrochemicals.[4-7] As a result of the potato study it was decided to screen native British plant species for hydrophilic alkaloids and glycosidase inhibitors as part of an on-going programme aimed at identifying novel biologically-active compounds. In the last three years the application of a combination of modern analytical tools, glycosidase screens, traditional isolation techniques and chemotaxonomy to a very limited number of species has enabled dozens of new alkaloids to be isolated as well as identifying alkaloids previously only thought to occur in tropical plants. While many of the compounds are still being evaluated for biological activity and many of the plants described here are still undergoing further investigation, in this paper we will outline the techniques used and some of the new structures discovered in the plant families Convolvulaceae, Solanaceae, Hyacinthaceae and Campanulaceae.

2 TECHNIQUES USED

2.1 Extraction and Initial Preparation for Glycosidase Inhibition Assays

Fresh or dried plant material (leaf) was extracted in 70% aqueous ethanol and left for 15 hours. The filtrate was applied to the cation exchange resin Dowex 50X8 (H^+ form) and after washing with water the alkaloids and amino acids were displaced with 1M ammonia solution and evaporated to a small volume before freeze drying.

2.2 Screen for Glycosidase Inhibition

Over 250 plant extracts prepared as above were tested for inhibition of glycosidase activities from crude homogenates of bovine brain material collected from abattoirs using *p*-nitrophenyl- (α-glucosidase)[8] and methyl-umbelliferyl substrates[9] (Sigma-Aldrich). The activities screened were α-glucosidase, β-glucosidase, β-galactosidase, β-*N*-acetylglucosaminidase and β-*N*-acetylgalactosaminidase. Over 20% of the plants analysed showed inhibition of glycosidases (usually of specific activities) and examples of the results obtained are shown in Table 1. A full paper will be published with details of these assays. Pure alkaloids were also tested on a range of glycosidases using both synthetic and natural substrates.

2.3 Gas Chromatography

Gas chromatography of trimethylsilyl-ether derivatives (prepared using Sigma-Sil-A) achieved good separations of all the plant glycosidase-inhibiting alkaloids using capillary columns such as BPX5 (SGE) but packed columns such as OV-1 also give good resolution.[10] The GC was coupled to a mass spectrometer (Perkin-Elmer QMass 910). An example of the type of separation which can be achieved is given by Asano *et al.*, 1996.[11]

2.4 HPLC with Pulsed Electrochemical Detector

The column was a Dionex CS3 Ionpac used in a Dionex DX500 system with an ED40 Pulsed Electrochemical Detector (with gold electrode). The solvent was 100mM hydrochloric acid with post column addition of 300mM sodium hydroxide. The method was modified from that reported by Donaldson *et al.*[12] and was found to be very sensitive (nannogram detection range) for glycosides of alkaloids and was able to separate most of the previously known plant glycosidase-inhibiting alkaloids and glycosides.

2.5 Isolation of Alkaloids

In general the alkaloids were fractionated using cation (Amberlite CG120 NH_4^+ form and CG50 NH_4^+ form) and anion exchange resins (Dowex 1 OH^- form) with fractions analysed by enzyme inhibition assays, GC-MS and HPLC. Chromatography on neutral aluminium oxide (Merck 90) using acetone with increasing proportions of water was also found to be useful. The isolation techniques used for specific compounds described here are detailed with the structural elucidation data in full papers.

3 EXAMPLES OF ALKALOIDS IDENTIFIED

3.1 Convolvulaceae

Calystegine A_3 (1), calystegine B_1 (2) and calystegine B_2 (3), along with some unidentified calystegines, were reported initially from roots of the species *Convolvulus*

Table 1. *Example of the Assessment of the Presence of Glycosidase Inhibitors in Common British Wild Plants*

Inhibition of Enzyme Activity (Bovine Brain)

Plant species	α-glucosidase	β-glucosidase	β-galactosidase	β-N-acetyl-galactos-aminidase	β-N-acetyl-glucos-aminidase
Convolvulaceae					
Convolvulus arvensis (Lesser Bindweed)	98%	78%	50%	5%	11%
Calystegia sepium ssp. silvatica (Larger Bindweed)	28%	50%	0%	30%	25%
Solanaceae					
Atropa belladonna (Deadly Nightshade)	90%	94%	40%	20%	15%
Solanum dulcamara (Bittersweet)	70%	65%	25%	20%	10%
Campanulaceae					
Campanula rotundifolia (Harebell)	80%	85%	44%	59%	65%
Hyacinthaceae					
Hyacinthoides non-scripta (Bluebell)	100%	70%	70%	33%	26%
Caryophyllaceae					
Silene dioica (Red Campion)	20%	40%	0%	70%	93%
Leguminosae					
Lotus corniculatus (Birdsfoot-trefoil)	25%	50%	0%	91%	88%

All of the extracts above were of aerial vegetative material. Strong inhibition of enzyme activities were given at a concentration of 100mg/ml but the screening results shown are of samples diluted at least 1:10 which indicate the specificity of the inhibition profiles.

arvensis and *Calystegia sepium*.[13,14] In our studies (Table 2) the alkaloid profiles of young leaves of both species appeared very similar and showed the presence of these alkaloids and also calystegine B_4 (4)[9] and calystegine C_1 (5)[3] as major alkaloids in *C. arvensis* (C_1 a minor alkaloid in *C.sepium*) and small amounts of calystegines A_5 (6) and B_3 (7)[3] (absent in *C. sepium*). The alkaloid concentrations decline as the leaves age.

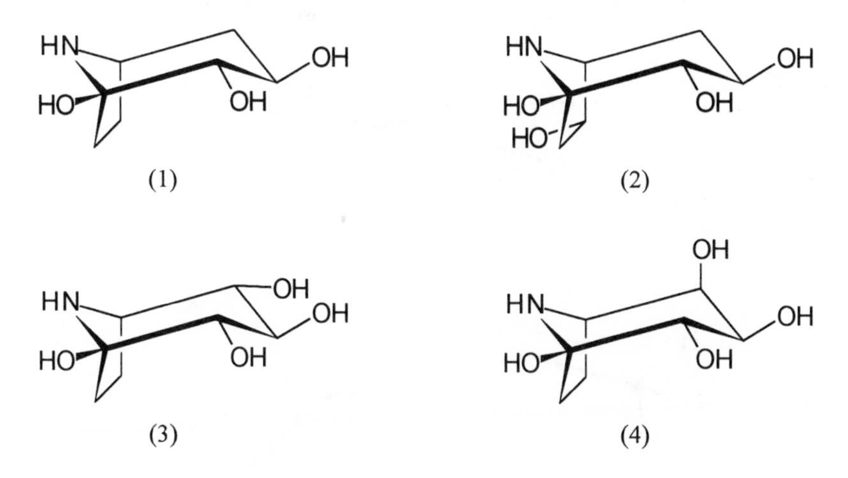

3.2 Solanaceae

The Solanaceae in the U.K. also contain a wide range of calystegines (Table 2). It is of interest that the Solanaceae species were found to contain glycosides of calystegines A_3 and B_1. The glycoside of calystegine A_3 is under structural elucidation but the B_1-glycoside has been identified as the 3-O-β-D-glucopyranoside (8).[15] Potatoes, reported to contain calystegines A_3 and B_2,[1] on cold storage also develop the 4-O-α-D-galactopyranoside of calystegine B_2 (9). Enzymatic synthesis and biological activities of the calystegine glycosides will be reported elsewhere. The Convolvulaceae species analysed contained no glycosides. The pyrrolidine alkaloid DMDP (10), patented as a potential nematocide,[7] was also detected as a minor component in one potato variety.

Table 2. *Calystegine Alkaloids in the Convolvulaceae and Solanaceae*

Species	A_3	A_5	B_1	B_2	B_3	B_4	C_1	A_3-glycoside	B_1-3-O-β-glucoside
C. arvensis	+	trace	+	+	trace	+	+	-	-
C. sepium	+	trace	+	+	-	+	trace	-	-
A. belladonna	+	trace	+	+	-	+	trace	+	+
S. dulcamara	+	+	+	+	+	-	+	+	-

Although calystegines have also been reported in *Morus alba* (Moraceae),[16] they were not detected here in any plant families other than the closely related Convolvulaceae and Solanaceae.

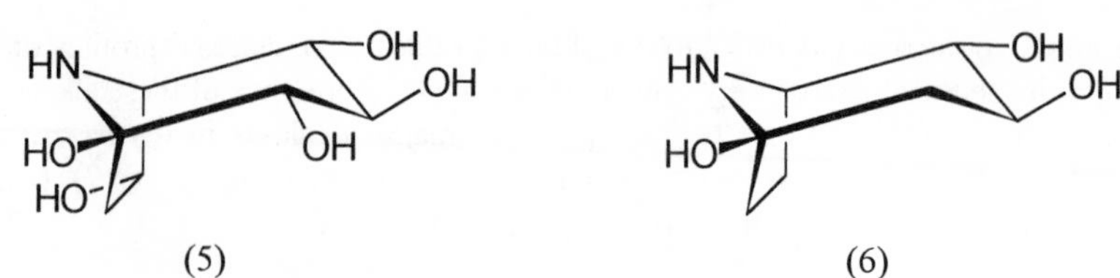

(5) (6)

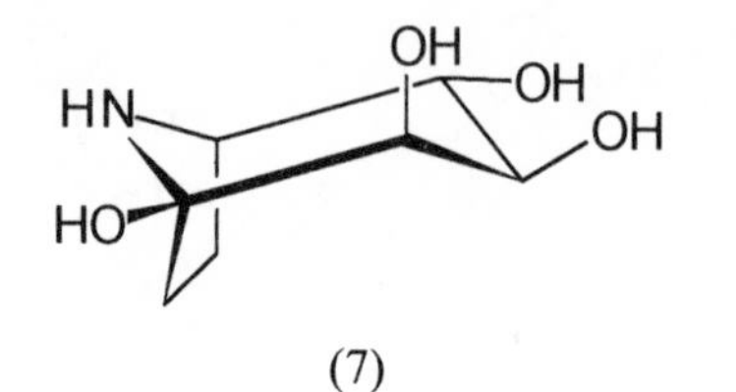

(7)

(8)

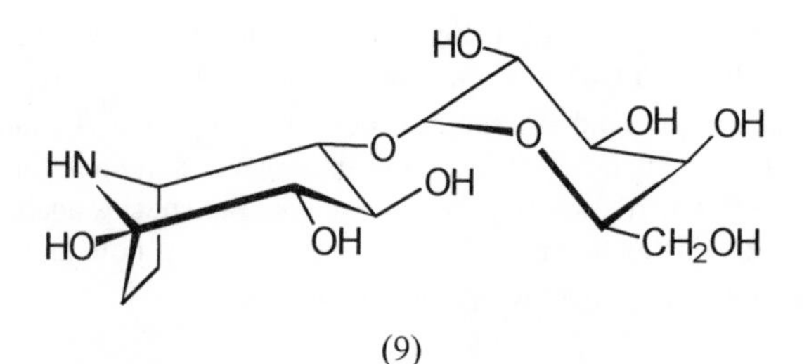

(9)

3.3 Campanulaceae

The inhibition of glucosidases by *Campanula rotundifolia* (Table 1) was found to be due largely to a high concentration (up to 2% dry weight in leaves/stems) of DMDP (10) which is the major alkaloid in all parts of this species. Several minor alkaloids which appeared to be unusual were also detected by GC-MS and HPLC and related species were analysed in the hope that one of these might prove a good source of the other alkaloids. *Campanula trachelium* (Nettle-leaved Campanula) was found to also contain DMDP as the major alkaloid but with increased concentrations of the other alkaloids. However, the best source of one of the new alkaloids proved to be *Campanula rapunculoides* (Creeping Campanula) from which the phenyl-pyrrolidine alkaloid (11) (or its enantiomer) was isolated. The structural elucidation and glycosidase inhibitory activity will be reported in a separate paper. *Wahlenbergia hederacea* (Ivy Campanula) also contains alkaloids related to 11 and this is currently under investigation.

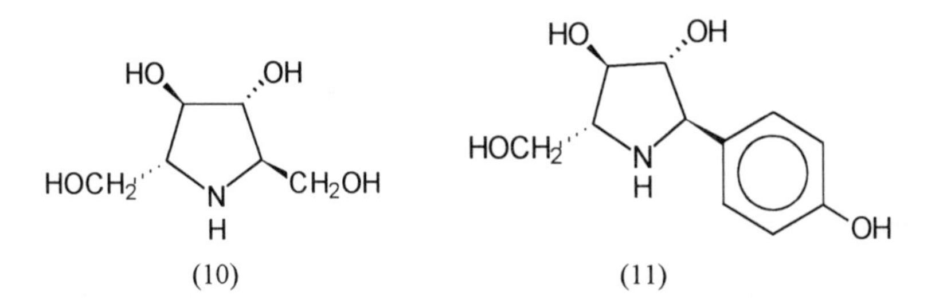

(10) (11)

3.4 Hyacinthaceae

Hyacinthoides non-scripta leaves, seed pods and bulbs also contain high concentrations of DMDP which is the major alkaloid at all growth stages.[17] In addition to DMDP, this plant also contains a number of novel alkaloids, many of which appear to be part of a biosynthetic pathway leading from DMDP to novel pyrrolizidine alkaloids, such as 12, which will be reported in full later. The new alkaloid α-homo-DMDP (13) (the stereochemistry cannot be defined at present at C-6) is the second most abundant alkaloid and this also occurs as an apioside (14).[17] α-homo-DMDP is a more potent inhibitor (K_i=1.5μM) of almond β-glucosidase than DMDP (K_i=10μM) but it is a weaker inhibitor of yeast α-glucosidase (K_i=54μM) compared to DMDP (K_i of 7μM).

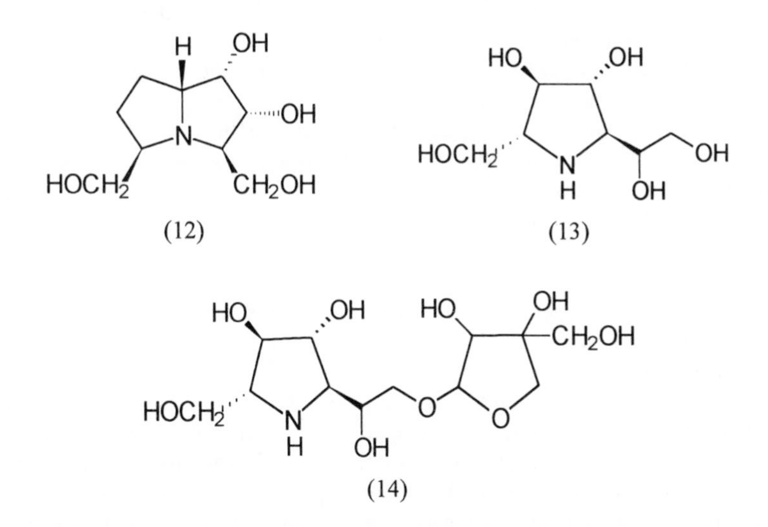

GC-MS analysis of an extract of *Scilla verna* (which is closely related to *H. non-scripta*) seed pods showed a profile containing no DMDP but a mixture of five novel polyhydroxylated-methylpiperidine alkaloids and a new pyrrolizidine alkaloid (15). 15 does not inhibit any glycosidases tested so far but full structural elucidation and activity tests of *S. verna* alkaloids will be reported shortly.

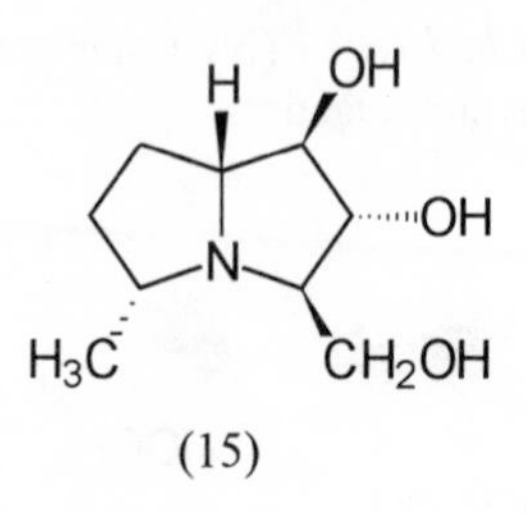

(15)

3.5 Caryophyllaceae and Leguminosae

The apparent specific inhibition of hexosaminidases shown by plants in these families (Table 1) are of interest because in humans the activity of these enzymes increases in blood serum in disease states such as diabetes mellitus, hepatic disorders and cancers including leukaemia.[18,19] Hexosaminidases may also have a role in the invasiveness of tumour cells. The natural inhibitors could therefore produce novel target compounds for synthesis of potential therapeutic agents.

4 DISCUSSION AND CONCLUSIONS

While we have reported only a small proportion of the alkaloids discovered so far in our survey it is clear that polyhydroxylated alkaloids of various structural types are common in British plants. It has been of interest to see that the pyrrolidine alkaloid DMDP, initially reported from tropical genera such as *Lonchocarpus* and *Derris*,[20] appears to be a common secondary metabolite in temperate species which also produce a wider range of derivatives of DMDP than those reported from tropical species. It may be of significance to animal (and perhaps human) nutrition and health that British plants appear to contain such a surprising range of compounds affecting specific cattle brain glycosidase activities. The trihydroxylated mannosidase-inhibiting indolizidine alkaloid swainsonine found in forages growing in Australia, China and the U.S.A. is known to cause a neurological disorder in cattle due to decreased α-mannosidase activity in lysosomes.[21,22] The effects are neuronal vacuolation, axonal dystrophy and loss of cellular function.

Glycosidase-inhibiting nitrogen analogues of mono- and disaccharides have aroused much interest in recent years as potential chemotherapeutic agents for use against cancers, HIV and other infections and as pesticides. For example, swainsonine, when given orally at low concentrations has a significant antimetastatic effect which appears to be due largely to augmentation of the immune system.[5,23] Many of the potential medical applications of these alkaloids seem to be due to inhibition of specific glycosidases involved in formation of glycoprotein oligosaccharide chains. Therefore, the new compounds discovered in this survey are being used as models for part of a synthetic chemistry programme aimed at producing selective glycosidase inhibitors for chemotherapeutic use. The discovery of an increasing number of glycosides and other conjugates of this group of alkaloids could create a much greater range of biological activities and applications. Recently it has been suggested that inhibitors of rhamnosidases may affect the growth of mycobacteria, such as those causing tuberculosis, by interfering with cell wall formation.[24,25] Interestingly, the *Scilla verna* piperidine alkaloids are isomers of synthetic compounds produced as rhamnose mimics[26] and one also inhibits a rhamnosidase (naringinase). (The activities will be published with the structural data). This study has concentrated on polyhydroxylated alkaloids, and in particular glycosidase inhibitors, but it could well be that hydrophilic members of other structural classes of secondary metabolite may also be awaiting discovery in common temperate species.

References

1. R. J. Nash, M. Rothschild, E. A. Porter, A. A. Watson, R. D. Waigh and P. G. Waterman, *Phytochemistry*., 1993, **34**, 1281.
2. R. J. Molyneux, Y. T. Pan, A. Goldmann, D. A. Tepfer and A. D. Elbein, *Arch. Biochem. Biophys.*, 1993, **304**, 81.
3. N. Asano, A. Kato, K. Oseki, H. Kizu and K. Matsui, *Eur. J. Biochem.,* 1995, **229**, 369.
4. V. A. Johnson, B. D. Walker, M. Barlow, T. J. Paradis, T. C. Chou and M. Hirsch, *Antimicrobial Agents and Chemotherapy*, 1989, **33**, 53.
5. P. E. Goss, J. Baptiste, B. Fernandes, M. Baker and J. W. Dennis, *Cancer Res.,* 1994, **54**, 1450.
6. M. A. Fischl, L. Resnick, R. Coombs, A. B. Kremer, J. C. Pottage, R. J. Fass, K. H. Fife, W. G. Powderly, A. C. Collier, R. L. Aspinall, S. L. Smith, K. G. Kowalski and C. B. Wallemark, *J. Acq. Immun. Def. Syn.,* 1994, **7**, 139.
7. A. N. E. Birch, W. M. Robertson, I. E. Geoghegan, W. J. McGavin, T. J. W. Alphey, M. S. Phillips, L. E. Fellows, A. A. Watson, M. S. J. Simmonds and E. A. Porter, *Nematologica*, 1993, **39**, 521.
8. A. M. Scofield, P. Witham, R. J. Nash, G. C. Kite and L. E. Fellows, *Comp. Biochem. Physiol.*, 1995, **112A**, 187.
9. P. Chow and B. Weissmann, *Carbohydr. Res.*, 1981, **96**, 87.
10. R. J. Nash, W. S. Goldstein, S. V. Evans and L. E. Fellows, *J. Chromatogr*. 1986, **366**, 431.
11. N. Asano, A. Kato, H. Kizu, K. Matsui, A. A. Watson and R. J. Nash, *Carbohydr. Res.*, 1996, **630**, in press.
12. M. J. Donaldson, H. Broby, M. W. Adlard and C. Bucke, *Phytochem. Anal.,* 1990, **1**, 18.
13. D. Tepfer, A. Goldmann, N. Pamboukdjian, M. Maille, A. Lépingle, D. Chevalier, J. Dénarié and C. Rosenberg, *J Bacteriol.*, 1988, **170**, 1153.
14. A. Goldmann, M. L. Milat, P. H. Ducrot, J. Y. Lallemand, M. Maille, A. Lépingle, I. Charpin and D. Tepfer, *Phytochemistry*, 1990, **29**, 2125.
15. R. C. Griffiths, A. A. Watson, H. Kizu, N. Asano, H. J. Sharp, M. G. Jones, M. R. Wormald, G. W. J. Fleet and R. J. Nash, *Tetrahedron Lett.*, 1996, **37**, 3207.
16. N. Asano, K. Oseki, E. Tomioka, H. Kizu and K. Matsui, *Carbohydrate Res.,* 1994, **259**, 243.
17. A. A. Watson, R. J. Nash, M. R. Wormald, D. J. Harvey, S. Dealler, E. Lees, A. Asano, H. Kizu, A. Kato, R. C. Griffiths, A. Cairns and G. W. J. Fleet, *Phytochemistry*, 1997, in press.
18. B. Woynarowska, H. Wikiel, M. Sharma, N. Carpenter, G. W. J. Fleet and R. J. Bernacki, *Anticancer Res.,* 1992, **12**, 161.

19. M. Horsch, L. Hoesch, G. W. J. Fleet and D. M. Rasi, *J. Enzyme Inhibition*, 1993, **7**, 47.
20. A. Welter, J. Jadot, G. Dardenne, M. Marlier and J. Casimir, *Phytochemistry*, 1976, **15**, 747.
21. S. M. Colegate, P. R. Dorling and C. R. Huxtable, *Aust. J. Chem.*, 1979, **32**, 2257.
22. B. L. Stegelmeier, R. J. Molyneux, A. D. Elbein and L. F. James, *Vet. Pathol.*, 1995, **32**, 289.
23. P. E. Goss, M. A. Baker, J. P. Carver and J. W. Dennis, *Clin. Cancer Res.*, 1995, **1**, 935.
24. J. C. Estevez, M. D. Smith, M. R. Wormald, G. Besra, P. J. Brennan, R. J. Nash and G. W. J. Fleet, *Tetrahedron Asymm.*, 1996, **7**, 391.
25. J. C. Estevez, J. Saunders, G. S. Besra, P. J. Brennan, R. J. Nash and G. W. J. Fleet, *Tetrahedron Asymm.*, 1996, **7**, 383.
26. J. P. Shilvock, J. R. Wheatley, B. Davis, R. J. Nash, R. C. Griffiths, M. C. Jones, M. Muller, S. Crook, D. J. Watkin, C. Smith, G. S. Besra, P. J. Brennan and G. W. J. Fleet, *Tetrahedron Lett.*, 1996, **37**, 8569.

Molecular Diversity and Selective Targeting of Enzymes in Tetrapyrrole Biosynthesis

Peter M. Shoolingin-Jordon, Josie F. A. Callaghan, Kwai-Ming Cheung, Paul Spencer and Martin Warren

BIOCHEMISTRY AND MOLECULAR BIOLOGY, SCHOOL OF BIOLOGICAL SCIENCES, SOUTHAMPTON UNIVERSITY, SOUTHAMPTON SO16 7PX, UK

1 INTRODUCTION

The selective toxicity of natural compounds stemming from the remarkable diversity of biological systems is exploited by the pharmaceutical and medicinal chemistry industries, two aspects being of central importance - molecular diversity and molecular specificity. The process of molecular redesign further broadens and exploits natural molecular frameworks. Such an approach is becoming ever more important as organisms adapt and develop resistance to both natural and synthetic compounds. The importance of identifying subtle species differences at the molecular level is rapidly becoming the key to developing selectivity with pesticides, herbicides and bacteriocides etc. This paper discusses tetrapyrrole biosynthesis, with specific reference to early stages of the pathway where significant interspecies differences exist that could be exploited in the design of novel and specific inhibitors.

Tetrapyrroles are crucial in a range of key biological reactions and include the haems, sirohaems, chlorophylls, corrins and related compounds[1, 2]. They have been adopted in living systems as versatile metal chelating agents for a range of redox reactions and play a vital role in the metabolism of oxygen and oxygen-containing derivatives. Many tetrapyrroles play a crucial part in energy metabolism, particularly photosynthesis and respiration, and disruption of these essential processes has a major impact on cellular viability. Even though evolutionarily ancient and present in almost all living systems, the tetrapyrrole biosynthesis pathway exhibits surprising metabolic diversity. The following account highlights some of these differences and deals with how knowledge at the molecular level can assist in the development of selective inhibitors with significant impact on one of the most ancient and important of biochemical pathways.

2 OVERVIEW OF THE PATHWAYS FOR THE BIOSYNTHESIS OF TETRAPYRROLES AND THEIR POSSIBLE ORIGINS

It is generally well accepted that the biosynthesis of tetrapyrroles first arose before oxygenic life existed and that the most ancient reactions would have led to the formation of the basic macrocyclic ring system, uroporphyrinogen III[1, 2]. This arises entirely from the carbon skeleton of glutamic acid *via* the intermediate 5-aminolaevulinic acid, through the pathway shown in Scheme 1. Methylation of uroporphyrinogen III on the a and b rings yields the branch-point intermediate, dihydrosirohydrochlorin (now called precorrin-2) from which vitamin B_{12} and the

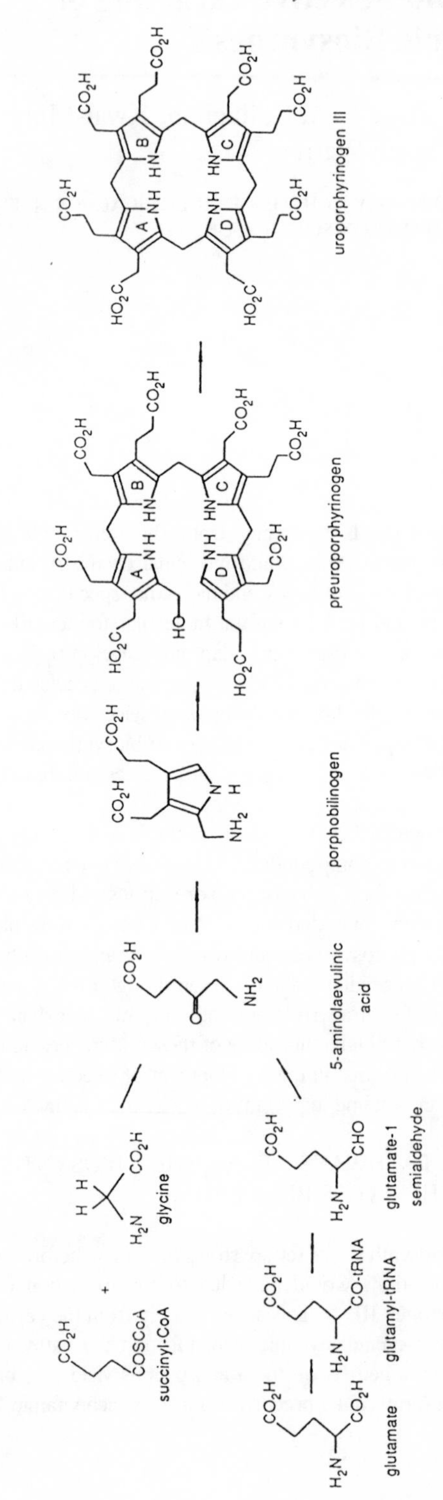

Scheme 1 *Biosynthesis of uroporphyrinogen III by the C_5 and glycine pathways*

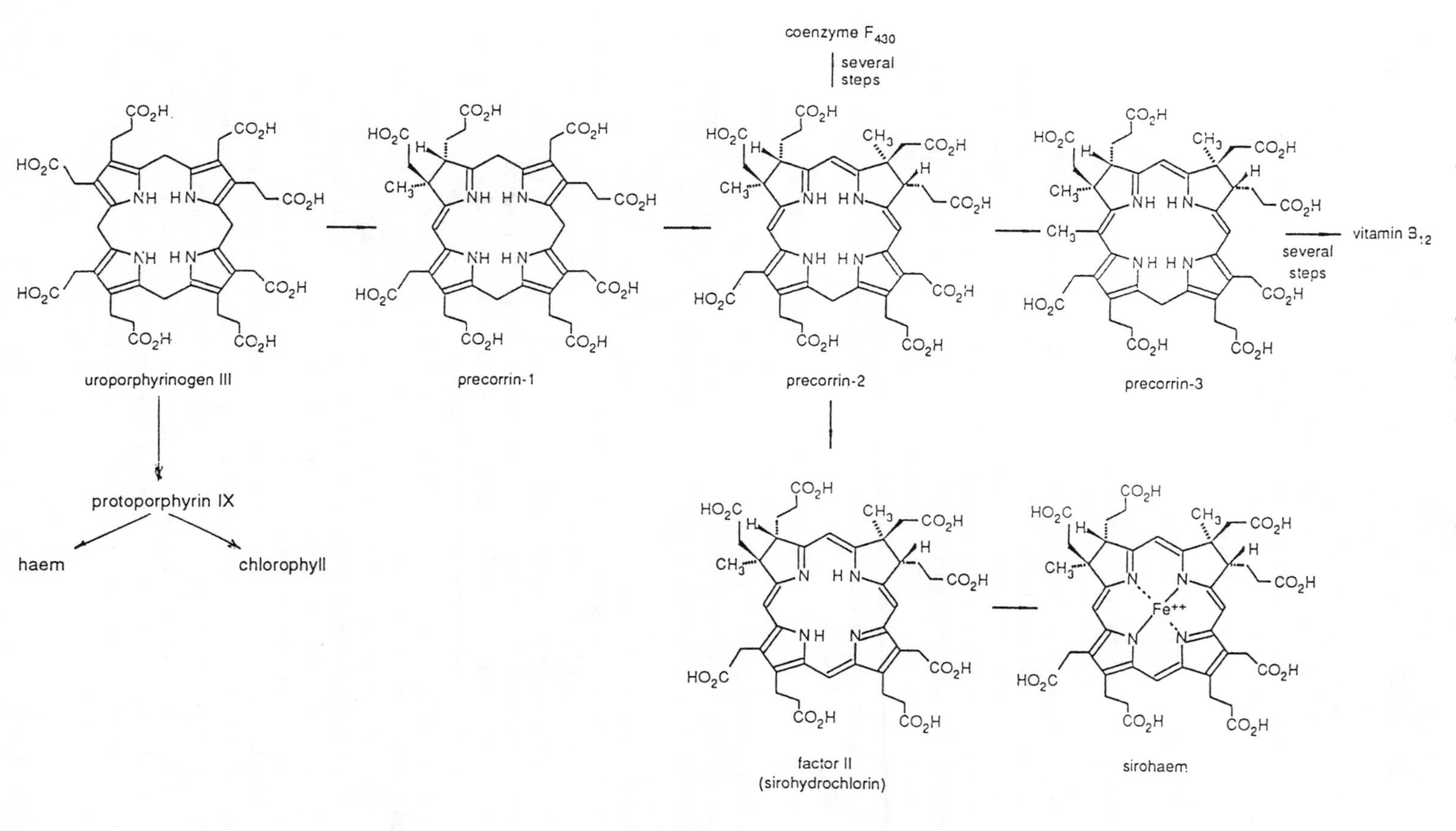

Scheme 2 *Biosynthesis of tetrapyrroles from uroporphyrinogen III*

Ni cofactor, F_{430}, from methanogenic bacteria, both arise (Scheme 2)[2]. Sirohaem, the prosthetic group of the sulphite and nitrite reductase enzymes, also utilizes precorrin-2 as an immediate precursor. Later in evolution, uroporphyrinogen III was further exploited by decarboxylation and oxidation resulting in protoporphyrin IX[2,3], from which the haem and chlorophyll branches of the pathway evolved (Scheme 2). The arrival of oxygenic life resulted in the evolution of an alternative means for the generation of 5-aminolaevulinic acid in animals and some photosynthetic bacteria in which glycine and the TCA cycle intermediate, succinyl-CoA, condense in a single enzymic step (Scheme 1).

3 THE BIOSYNTHESIS OF 5-AMINOLAEVULINIC ACID

3.1 Glutamate 1-semialdehyde aminotransferase

In plants and the majority of bacteria, 5-aminolaevulinic acid is biosynthesized from the intact carbon skeleton of glutamate in three stages, termed the C_5 pathway[4]. Firstly, glutamate is activated by coupling to $tRNA_{glu}$ to give glutamyl-$tRNA_{glu}$, an intermediate common to protein synthesis. Glutamyl-$tRNA_{glu}$ is then reduced by glutamyl-tRNA reductase to yield the reactive α-aminoaldehyde, glutamate 1-semialdehyde. Finally, glutamate 1-semialdehyde is transformed into 5-aminolaevulinic acid in a novel transamination catalyzed by the enzyme glutamate-1-semialdehyde aminotransferase[5]. The three-step C_5 pathway contrasts sharply with the single stage synthesis of 5-aminolaevulinic acid in animals and both routes offer the possibility for the design of selective inhibitors.

The first stage of the C_5 pathway, namely, the activation of glutamate as the ester of $tRNA_{glu}$, is common to protein synthesis. It is thus doubtful whether selective inhibition of the tetrapyrrole pathway can be accomplished without indiscriminate effects on protein synthesis. Similarly the targeting of the reductase enzyme, despite its uniqueness to the tetrapyrrole biosynthesis pathway, presents some problems because of the complex nature of the substrate $tRNA_{glu}$.

The most obvious target for selective inhibition of the C_5 pathway is the glutamate 1-semialdehyde aminotransferase enzyme[6]. Investigations with this enzyme have established that it exists in a pyridoxamine 5'-phosphate form that donates the amino group to glutamate 1-semialdehyde **(1)** to give an enzyme bound 4,5-diaminovalerate intermediate **(2)** (Scheme 3). The amino group at the 4-position of this intermediate is subsequently removed to restore the pyridoxamine 5'-phosphate form of the enzyme and to yield 5-aminolaevulinic acid. Several potential inhibitors of this enzyme have been designed and prepared in our laboratory. One of the most interesting is the lower homologue of glutamate 1-semialdehyde, aspartate 1-semialdehyde **(3)** (4-amino-5-hydroxy-dihydrofuran-2-one). This compound, like glutamate-1-semialdehyde, also exists in a stabilized cyclic form. The aminotransferase accepts this α-aminoaldehyde as a substrate and amino transfer ensues yielding 3,4-diaminobutyrate **(4)**. The enzyme is, however, unable to catalyze the second half of the reaction and the diamine is released leaving the enzyme in its pyridoxal 5'-phosphate form. The reduced form of glutamate 1-semialdehyde (5-hydroxy-4-aminovalerate) **(5)** is also a specific inhibitor of the C_5 pathway (Scheme 3). Again the aminotransferase enzyme appears to recognize this compound and the enzyme is inhibited. In contrast, the isomeric 5-aminolaevulinic acid analogue, 4-hydroxy-5-aminovaleric acid **(6)**, is not well recognized by the enzyme. Inhibitors acting against glutamate 1-semialdehyde (Scheme 3) clearly provide a blueprint for the design of herbicides or bacteriocides that would be likely to have little impact on haem biosynthesis in the animal kingdom.

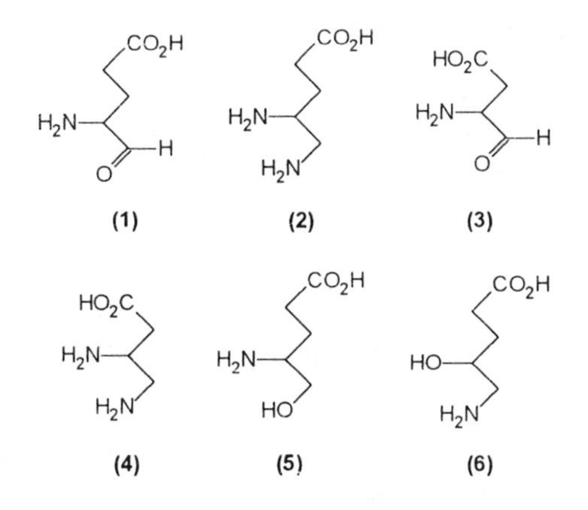

Scheme 3 *Compounds that interact with glutamate 1-semi aldehyde aminotransferase*

3.2 5-Aminolaevulinic acid synthase

The glycine "pathway" also has its specific inhibitors by virtue of the uniqueness of the enzyme 5-aminolaevulinic acid synthase. This pyridoxal 5'-phosphate dependent enzyme accepts only glycine **(7)** as the amino acid substrate and catalyses the condensation with succinyl-CoA to give a putative enzyme-bound intermediate 2-amino-3-ketoadipic acid **(8)** (Scheme 4). This is then decarboxylated to yield 5-aminolaevulinic acid[7, 8]. The enzyme catalyses the stereospecific exchange of the H^{Re} atoms of both glycine and 5-aminolaevulinic acid, presumably using the same catalytic base in both cases. The amino acid, aminomalonic acid **(9)** and its derivatives show strong competitive inhibition in the low micromolar range. The monoester **(10)** is an even better inhibitor than the dicarboxylic acid since it partially resembles the 2-amino-3-ketoadipic acid intermediate (Scheme 4). Interestingly aminomalonic acid behaves as a powerful inhibitor against both the bacterial and mammalian enzymes and exploitation of this difference may provide a useful basis for the design of pesticides that have little effect on plants.

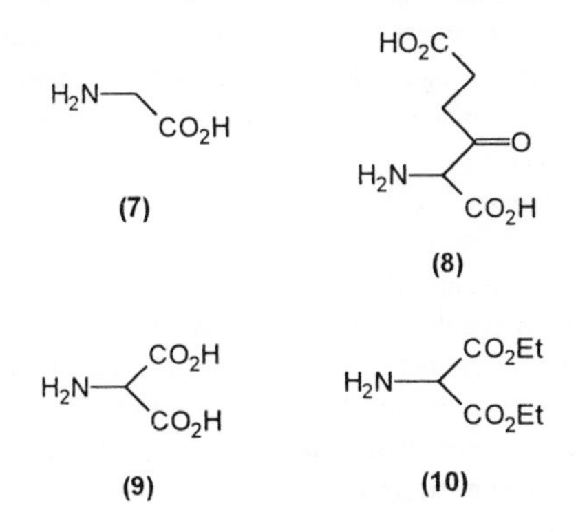

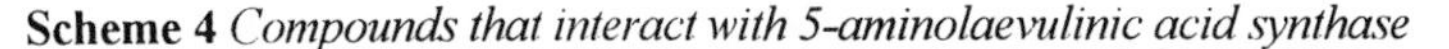

Scheme 4 *Compounds that interact with 5-aminolaevulinic acid synthase*

4 THE BIOSYNTHESIS OF PORPHOBILINOGEN

The biosynthesis of the tetrapyrrole building unit, porphobilinogen, from two molecules of 5-aminolaevulinic acid is catalyzed by the ubiquitous enzyme 5-aminolaevulinic acid dehydratase (porphobilinogen synthase). The reaction is the first of three enzyme stages found in all living systems that result in the biosynthesis of uroporphyrinogen III, the tetrapyrrole from which all other tetrapyrroles arise[1, 2] (Scheme 1). Dehydratases are all octameric in structure with similarities in amino acid sequences that suggest conserved three dimensional structures. Despite these similarities, the enzymes from animals and plants have completely different metal requirements[1, 9] and this is reflected in their responses to heavy metals. The effects of inhibitors, with structures related to intermediates *en route* to porphobilinogen, are also influenced by the nature of the bound metal. The animal dehydratases are zinc-dependent metalloaldolases that are exquisitely sensitive to lead and the effects of lead poisoning on lowering the levels of blood 5-aminolaevulinic acid dehydratase in humans are well documented[10]. The inhibition by lead occurs as a result of the substitution of lead for zinc in the metal binding domain that contains several reactive cysteine residues. These cysteine residues are readily oxidized to disulphide bonds with the concomitant loss of the zinc ion. In sharp contrast, the dehydratases from plants and some bacteria are magnesium-dependent metalloenzymes with aspartic acid metal ligands[11], a low sensitivity to lead and a remarkable stability to oxidation.

The differences in metal ion requirement not only provide a means to inhibit the animal and plant enzymes selectively but also result in major differences in the behavior of the two enzyme classes with intermediate analogue inhibitors (Scheme 5). 5-Aminolaevulinic acid dehydratases have long been known to be susceptible to inhibition by γ-ketoacids related to the substrate. Such inhibitors interact with a reactive active site lysine residue[12] through the formation of a Schiff-base. The Schiff base can be trapped by reduction with $NaBH_4$ leading to complete inactivation of the enzyme. We have studied two compounds related to putative enzymic intermediates, succinylacetone (SA) **(11)** and 3-acetyl-4-oxo-heptanedioic acid (AOHD) **(12)** (Scheme 5). SA has long been known to inhibit the human dehydratase with a K_i in the micromolar range. AOHD resembles a partially assembled intermediate made from two molecules of 5-aminolaevulinic acid linked together **(13)**. SA and AOHD are far more powerful inhibitors against the plant enzyme, possibly due to the dual binding of the inhibitors both as a Schiff-base and as a magnesium metal ligand. The selective inhibition of the plant enzyme for these two compounds may reflect the preference of magnesium for "hard" oxygen ligands. The interaction of these inhibitors with both the A-and P-sites could explain their high selective affinity for the plant enzyme.

The novel 5-aminolaevulinic acid analogue, 4-amino,3-oxobutyric acid (AOB) **(13)** (Scheme 5), shows a special affinity for the pea enzyme being 2000 times more effective as an inhibitor compared with the zinc-dependent enzyme from *E. coli*. This inhibitor does not form a Schiff base with the P-site of the dehydratase active site but interacts specifically with the A-site as a competitive inhibitor. This is the first specific A-site inhibitor to be characterized and represents a possible model compound for the design of future compounds as specific inhibitors of the plant enzymes. Further insight into the mode of interaction of the enzyme with its substrate and with inhibitors will arise from its X-ray structure, currently under investigation in our laboratory.

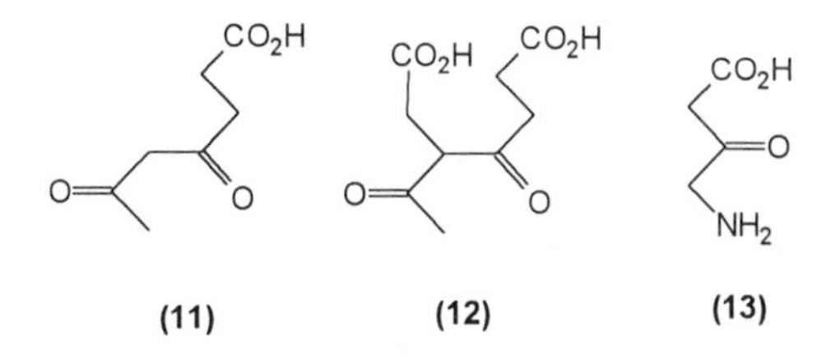

Scheme 5 *Specific inhibitors of 5-aminolaevulinic acid dehydratase*

5 THE ASSEMBLY OF THE TETRAPYRROLE MACROCYCLIC RING

Two enzymes are required to transform porphobilinogen into uroporphyrinogen III. Firstly, porphobilinogen deaminase assembles a tetrapyrrole intermediate, preuroporphyrinogen, by the sequential deamination and polymerization of four molecules of porphobilinogen. Preuroporphyrinogen is then cyclised, with rearrangement of the d ring, to yield uroporphyrinogen III by the enzyme uroporphyrinogen III synthase (Scheme 1). Preuroporphyrinogen is a highly unstable hydroxymethylbilane and rapidly cyclises without rearrangement in a non enzymic reaction to give the non physiological isomer, uroporphyrinogen I.

Porphobilinogen deaminase is typical of a polymerase since it builds the tetrapyrrole product by elongation of a novel enzyme primer, the dipyrromethane cofactor, which is attached covalently to the enzyme through a thioether linkage. The X-ray structure of the deaminase from *E. coli* reveals a protein with three domains and highlights the position of the cofactor, deep in a cleft between domains 1 and 2 (Figure 1)[13]. There are few differences between the deaminases from various organisms (there is 60% similarity in the primary structures of human and *E. coli* deaminases) and the possibility of species selective inhibition is therefore rather remote. The deaminase catalytic cycle involves the stepwise elongation of the dipyrromethane cofactor through enzyme intermediate complexes, by the sequence shown in Scheme 6. This generates a hexaxapyrrole species, ES_4, that liberates preuroporphyrinogen by hydrolysis and regenerates the cofactor, still attached to the enzyme[14]. We have investigated several porphobilinogen analogues as possible inhibitors and have found some interesting and novel results. Porphobilinogen is remarkably specific for its substrate and few pyrrole analogues show any promise as inhibitors[15]. However, opsopyrrole dicarboxylic acid (porphobilinogen without the aminomethyl side-chain) proved to be a weak, competitive inhibitor for the enzyme. More interestingly were porphobilinogen derivatives in which the free α-position had been substituted by a halogen. Thus α-bromoporphobilinogen[16] acted as a chain-terminating, suicide inhibitor, the enzyme recognizing the inhibitor as a substrate and catalyzing the normal deamination and coupling reactions. The presence of the bromine atom at the α-position blocks the further addition of either substrate (S) or α-bromoporphobilinogen (B) and results in complete inactivation of the enzyme because of the following termination complexes (E -> EB, ES -> ESB, ES_2 -> ES_2B and ES_3 -> ES_3B) as shown in Scheme 7. The ES_3B complex is unstable and generates the free enzyme and the product analogue, α-bromopreuroporphyrinogen.

Inhibitors of uroporphyrinogen III synthase are restricted to isomers of preuroporphyrinogen and a *spiro*-lactam related to the proposed *spiro*-intermediate[17] The extremely complex nature of these inhibitors makes their synthesis an extremely specialized process that makes them only of academic interest.

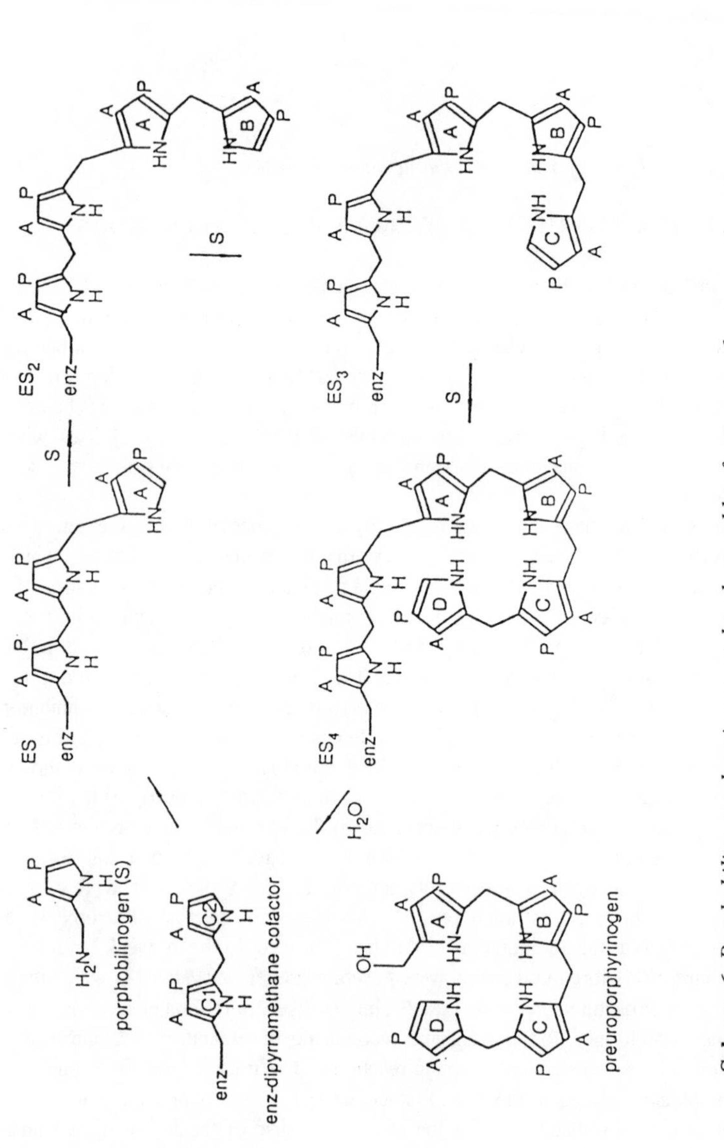

Scheme 6 *Porphobilinogen deaminase catalyzed assembly of preuroporphyrinogen from four molecules of porphobilinogen. The dipyrromethane cofactor acts as a primer for the reaction and remains permanently bound to the enzyme*

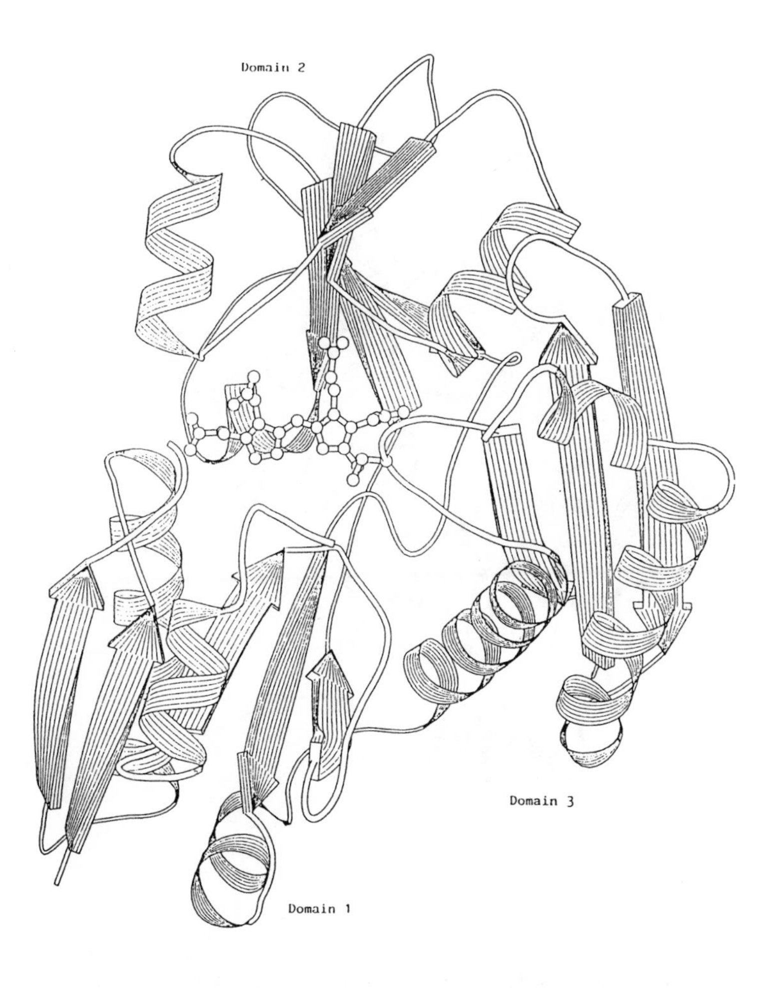

Figure 1 *Ribbon structure of E. coli porphobilinogen deaminase showing the 3 domain structure with the dipyrromethane cofactor located between domains 1 and 2*

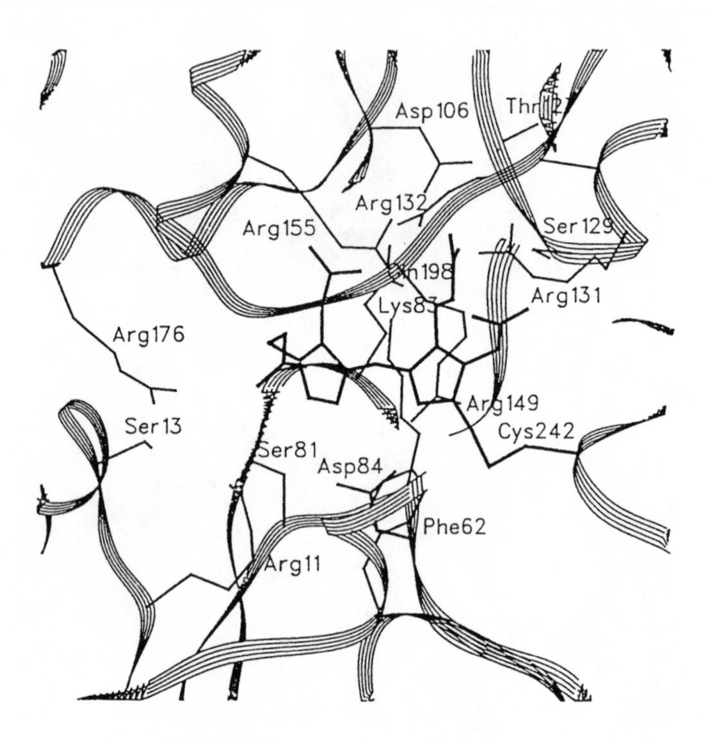

Figure 2 *Detail of the catalytic cleft of porphobilinogen deaminase showing the interactions between the protein and the dipyrromethane cofactor*

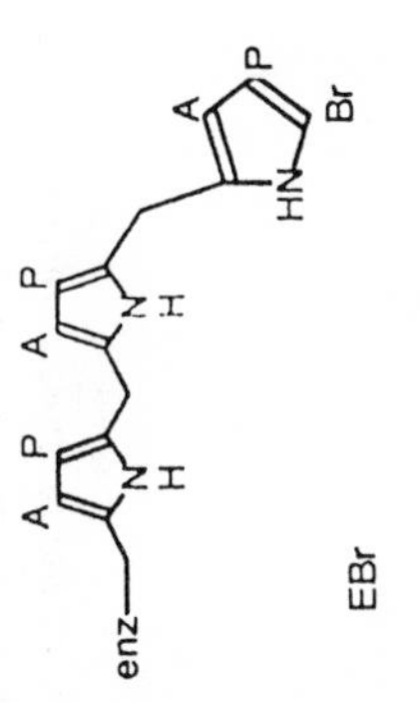

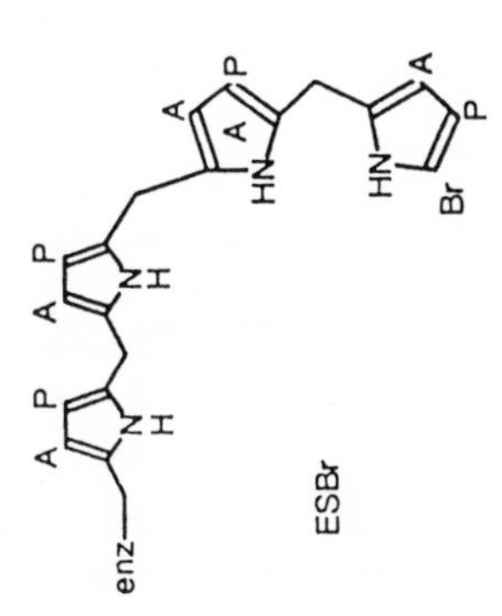

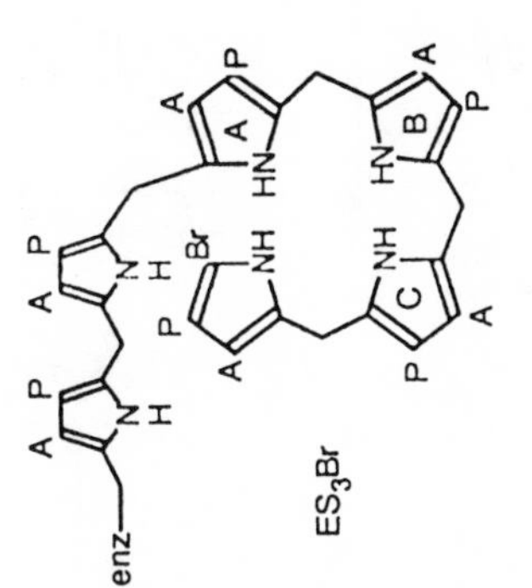

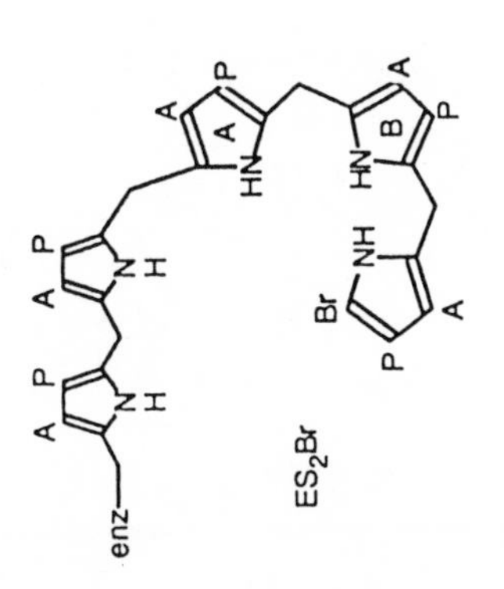

Scheme 7 *Inhibition of the porphobilinogen deaminase catalyzed reaction by the chain terminator, α-bromoporphobilinogen*

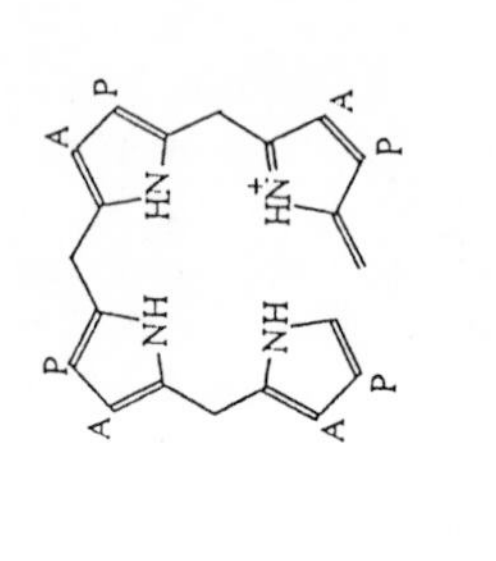
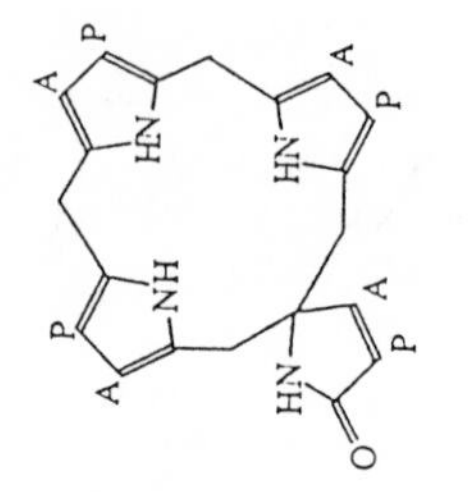
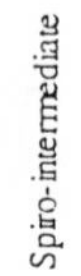
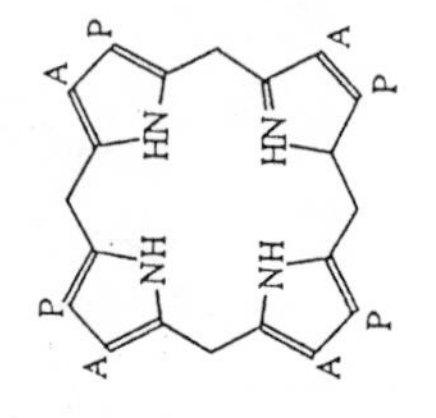

Scheme 8 *Structure of the proposed spiro-intermediate of uroporphyrinogen III synthase. Structural similarity of the spiro-lactam intermediate analogue*

6 CONCLUSIONS

The above discussion highlights the effects of several inhibitors aimed at enzymes catalyzing early reactions of the tetrapyrrole biosynthesis pathway involving the assembly of the central intermediate, uroporphyrinogen III. Some of the inhibitors rely on the fact that the biosynthesis of 5-aminolaevulinic acid occurs by two independent pathways. Other inhibitors exploit the differences in metal ion requirement for the enzyme 5-aminolaevulinic acid dehydratase. Where the reactions are common to all species, the search for specific inhibitors is more difficult and may require sophisticated modeling and design.

The presence of parallel reactions exist in other areas of tetrapyrrole biosynthesis, such as in the transformation of coproporphyrinogen III into protoporphyrinogen IX[3] and in the biosynthesis of vitamin B_{12}[2] where anaerobic and aerobic pathways exist. Many additional possibilities thus exist for the design of specific inhibitors for reactions in other parts of the tetrapyrrole biosynthesis pathways.

ACKNOWLEDGEMENTS

The work described in this account was supported by the BBSRC and Wellcome Trust.

REFERENCES

1. P. M. Jordan (1991) in *New Compr. Biochem.* Biosynthesis of Tetrapyrroles **19**, 1-66 (ed. P. M. Jordan) Elsevier, Amsterdam

2. D. J. Chadwick and K. Ackrill (eds) (1994) *Ciba Foudn. Symp.* **180**, The Biosynthesis of Tetrapyrrole Pigments. John Wiley

3. M. Akhtar (1991) in *New Compr. Biochem.* Biosynthesis of Tetrapyrroles **19**, 67-99 (ed. P. M. Jordan) Elsevier, Amsterdam

4. C. G. Kannangara, S. P. Gough, P. Bruyant, J. K. Hoober, A. Kahn and D. von Wettstein (1988) *Trends in Biochem. Sci.* **13**, 139-143

5. C. G. Kannangara, R. V. Andersen, B. Pontoppidan, R. Willows and D. von Wettstein (1994) in *Ciba Foundn. Symp.* **180**, The Biosynthesis of Tetrapyrrole Pigments. D. J. Cadwick and K. Ackrill (eds). John Wiley

6. R. A. John (1996) *Biochem. Biphys. Acta.* in press

7. P. M. Jordan and D. Shemin (1972) in *The Enzymes (3rd ed)*, **8**, 339-356

8. G. C. Ferreira and J. Gong (1995) *J. Bioenerg. Biomembr.* **27**, 151-159

9. E. K. Jaffe (1995) *J. Bioenerg. Biomembr.* **27**, 169-179

10. P. N. B. Gibbs, M. G. Gore and P. M. Jordan, *Biochem. J.* **225** 573-580

11. Q. F. Boese, A. J. Spano, J. M. Li and M. P. Timko (1991) *J. Biol. Chem.* **266**, 17060-17066

12. P. N. B. Gibbs and P. M. Jordan (1986) *Biochem. J.* **236** 447-451

13. G. V. Louie, P. D. Brownlie, R. Lambert, J. B. Cooper, T. L. Blundell, S. P. Wood, V. N. Malashkevich, A. Hädner, M. J. Warren and P. M. Shoolingin-Jordan (1996) *Proteins* **25**, 48-78

14. P. M. Jordan and M. J. Warren (1987) FEBS *Lett.* **225**, 87-92

15. P. M. Shoolingin-Jordan (1995) *J. Bioenerg. Biomembr.* **27** 181-195

16. M. J. Warren and P. M. Jordan (1988) *Biochemistry* **27**, 9020-9030

17. F. J. Leeper, in D. J. Chadwick and K. Ackrill (eds) (1994) *Ciba. Foundn. Symp.* **180**, 111-137 The Biosynthesis of Tetrapyrrole Pigments. John Wiley

The Industrial Utilisation of Natural Resources: How to Quantify Demand and Compensation

K. Beese

EUROPEAN COMMISSION, INSTITUTE FOR PROSPECTIVE TECHNOLOGICAL STUDIES, WORLD TRADE CENTER, ISLA DE LA CARTUJA S/N E-41092, SEVILLE, SPAIN

1 INTRODUCTION

1.1 The Institute

The Institute for Prospective Technological Studies (IPTS) is one of the eight institutes of the European Commission's Joint Research Centre. IPTS' mission is to observe technological change and assess its impact on society and economy. Several IPTS projects deal with European health care services and the competitiveness of the European pharmaceutical industry.

1.2 Why this Project ?

The member states of the European Union and the European Commission itself have signed the Convention on Biological Diversity. They are represented at the Conference of the Parties which negotiates the terms of implementation of the Convention. Article 15 of the Convention recognises the "sovereign rights" of nations over their genetic resources and hints at a new class of resource rights.

On the subject of payments, the Convention proposes in its article 15(7) “sharing in a fair and equitable way the results of research and development and the benefits arising from the commercial use and other utilization of genetic resources.....upon mutually agreed terms”. However, there is no attempt to identify appropriate payment approaches or a system for valuing germplasm for specific uses, as already observed by Lesser and Krattinger (1994). The Convention can be a means to promote world trade in biological resources, but must be based on clearer terms than in the past. While many papers on this subject deal with the particulars of the Merck/INBio agreement, most treat the financial aspects in generalities.

Models for benefit sharing in biodiversity prospecting can not be sustainable if they ignore the economic factors which determine the competitiveness of natural products. This competitiveness changes with increased costs of the access to genetic resources, the development of alternative sources of molecular diversity (combinatorial chemistry, etc.) and new approaches to treat diseases (rational drug design, gene therapy, etc.).

The EU delegations have to know about the relevance and value of biological resources for European pharmaceutical R&D. A study carried out by the Institute for Prospective Technological Studies is aimed at such an assessment.

2 NATURAL PRODUCTS

2.1 How Important Have Natural Products Been for Drug Development ?

Papers on the importance of natural products state that tens, if not hundreds of millions US$ are spent for the development of new drugs from natural products. Different estimations have been made to demonstrate the relative importance of natural products for drug production:

* tropical forests contain undiscovered drugs worth at least $147 billion, they should yield around 375 (potential) drugs (Kleiner 1995, without information on the calculation);

* 47 major drugs come from tropical plants (Kleiner 1995);

* 25% of modern prescription drugs contain at least one compound now or once derived or patterned after compounds derived from higher plants (15% of those produced synthetically or semisynthetically, Duke 1993),

* but probably only 10% of leading (excluding illicit) drugs contain phytochemicals still extracted directly from higher plants (Duke 1993);

* should such natural products as alcohol, vinegar, citric acid, resorcinol, be included, then well over 50%, perhaps 75% of medicines contain plant-derived phytochemicals (Duke 1993);

* of the roughly 120 pure chemical substances from higher plants used in medicine throughout the world, two thirds originate from temperate species (Pistorius & van Wijk, 1993).

All these figures have to be treated with caution and require interpretation. Duke (1993) illustrates the problems one encounters in trying to make a survey of this type. Depending on where one draws the line between natural and synthetic products one gets very different figures. Often the products are synthetic or semisynthetic but based on a naturally occurring molecule discovered in an earlier natural products programme.
In 1936, 45% of all therapeutic drugs were still derived from plant sources. This proportion decreased following the discovery of microbially-derived antibiotics and the increased industrial application of chemical synthesis (Joffe & Thomas 1989).

Since the launch of large screening programmes the demand of pharmaceutical industry for natural products has not been continuous:

1950 beginning of mass screening
1960 every company screens natural products
1970 decline of natural products screening
1980 reinforcement of screening programmes
1990 decline in the 1990s ?
2000 ???

The reasons for the shifts in interest in natural products seem to lie in the changes in the supply of molecular diversity for screening and simply fashion (in the 1960s natural products screening was considered as being old-fashioned). Recently, the supply of synthetic chemicals for screening has grown rapidly. This growth has come through the commercialisation of large chemical libraries and through the development of combinatorial synthesis techniques. Chemical diversity may no longer be in short supply. Currently direct comparisons are being made between the economics and overall success of chemical library and natural products screening.

Duke (1993, see Table 1) presents rather current data on the percentage of pharmaceuticals still containing natural products as *one of* the major ingredients. The average for major US based firms is less than 10%.

Table 1: Estimated Percentage of "Green" Drugs Produced by Major U.S. Firms, 1991

Company	Projected	%Natural
Bristol-Myers	11.2	<10
Merck	8.6	<5
Pfizer	7.2	<5
Smith Kline Beecham		<1
American Home Products	7.1	30
Eli Lilly	5.9	17
Warner Lambert	5.0	<5
Rhone Poulenc	3.7	15
Schering-Plough	3.6	<5
Upjohn	3.4	<2
Marion Merrel Dow	2.8	17
Syntex	1.8	<1
Johnson & Johnson		10

Source: Duke 1993

2.2 Design and Results of Natural Products Screening Programmes

2.2.1 Size of a Screening Programme:

From 10,000 up to 1 million assays per year can be achieved by a medium-size biotechnological or pharmaceutical company. The usual lifetime of a focused screen is 3-6 months, in exceptional cases up to one year (figures presented at the conference "Natural Products - Rapid Utilization of Sources for Drug Discovery and Development", San Francisco, May 15-16, 1995).

2.2.2 Costs of Natural Products Screening:

The primary assays are the cheapest part of the whole drug discovery process, ranging between US$ 0.10 and $1.00 per assay (according to presentations at the conference "Natural Products - Rapid Utilization of Sources for Drug Discovery and Development", San Francisco, May 15-16, 1995).

2.2.3 Chances of Detecting New Leads:

* Only 0.01%-0.05% of extracts contain a new drug candidate (Reid et al. 1993 and others), in the case of anti-cancer agents only one in 40,000 - 50,000 (0.0025 - 0.0020%, Cragg et al. 1995);

* 95% of active compounds detected in an assay usually have been found before, a fact which makes scrupulous and sophisticated chemical analysis of active extracts essential (presented at the conference "Natural Products - Rapid Utilization of Sources for Drug Discovery and Development", San Francisco, May 15-16, 1995).

* only one fourth of the chemicals reaching clinical trials will ever be approved as a new drug (Reid et al. 1993), one out of every 4,000-10,000 compounds discovered can be marketed commercially (Duke 1993).

* the hit rate can, in some cases of simple diseases, be improved if taxonomically and ecologically diverse sources and extracts used in traditional medicine are employed.

2.3 What Makes Natural Products so Interesting ?

The chemical diversity in living organisms has by far not yet been explored. Less than 1-5% of the 250,000 higher plant species on earth have been screened for pharmaceutical use. Now new scientific insights and innovation in human cell and microbe cultivation techniques make a far larger resource available for pharmaceutical assays. Natural products protagonists argue that nature provides an enormous variation of extremely complex molecules, infinitely more sophisticated than any molecule from any other source, e.g. combinatorial chemistry. The particular quality of natural products has been seen in:

* uniqueness: natural products have unique features which do not appear in synthetic chemicals;

* unlimited structural diversity, compared to molecules deriving from combinatorial chemistry;

* patentability due to distinct chemical identity, not always given in synthetic leads;

* the complex mixtures of compounds which often only in this composition lead to cures which cannot be achieved with single chemical compounds. (However, mixtures also have technical and other disadvantages);

* the high number of diverse chemical compounds in a single biological extract increases the likeness to find a new pharmaceutical lead, compared to homogenous synthetic samples (Reid et al. 1993).

3 COMPETITION FROM OTHER SOURCES OF CHEMICAL DIVERSITY

3.1 What Are the Competing Sources of Molecular Diversity?

In drug discovery two basically different approaches are applied:

* an active molecule with specific features is designed (rational drug design, protein engineering) or

* a broad range of molecules are more or less randomly tested for effectiveness against a specific disease. Sources of chemical diversity for these assays can either be synthetic processes (e.g. combinatorial chemistry) or the biochemical machinery of living organisms.

Combinatorial chemistry enables the rapid synthesis and screening of large compound libraries, comprising arrays of novel structures made by random or directed combination of simpler molecular building blocks. Depending on the technological approach used, a relative handful of scientists may be able to generate libraries containing thousands to millions of compounds and begin screening for promising lead structures (Stone 1993). By producing larger, more diverse compound libraries, companies increase the probability that they will find novel compounds of significant therapeutic and commercial value. Not surprisingly, the field has captured the attention of every major player in the pharmaceutical arena (Richon 1995). The growing string of acquisitions of early-stage, combinatorial-chemistry companies by pharmaceutical firms (Affymax / Glaxo, Selectide / Marion Merrell Dow, Sphinx Pharmaceuticals / Eli Lilly) affirms the potential of this technology.

As for natural products the process of finding novel, active compounds through combinatorial chemistry is akin to finding the proverbial needle in a haystack, using an array of technologically sophisticated tools but a smaller investment of time and resources than required by other methods including natural products screening (Richon 1995).

3.2 What is the Edge of Synthetics Over Natural Products ?

In many cases natural products screening is not looking good in economic terms. The higher costs of natural products screening are chiefly due to

* higher efforts to collect and describe the samples. The costs of acquisition are high (see further below) and would increase if developing countries get tougher on the financial compensation for access to natural resources.

* the more complex mixture and structure of natural compounds which makes it more difficult to identify and isolate a new lead.

New technical developments, in particular the whole field of combinatorial chemistry, are leading to an enormous supply of molecular diversity for screening by extraordinarily efficient procedures. These sources will increasingly compete with natural biodiversity prospecting. There is a trend for synthetics to replace natural compounds in prescription and over the counter pharmaceuticals. Today, ephedrine, salicylates, vitamins, and xanthines are mostly synthetic and steroids are often semisynthetic (Duke 1993).

3.3 Other Competing Medical Approaches

While biotechnology has improved the efficiency of screening for new leads, it also has led to new techniques which will increasingly compete with the drugs deriving from natural products in health care. Over a long term these developments will presumably make screening for biological activity of randomly chosen molecules an antiquated and inefficient approach.

This concerns, in particular, the rapidly developing field of clinical genetics. Human genes, their function and mutations are discovered with rapidly increasing frequency, at present one gene is discovered every day. At the same time diagnostic tools for genetic testing enter new dimensions of efficacy and cost effectiveness. Gene therapy is producing its first successes although technical difficulties still hinder a broad scale application. Within 10-15 years time these developments are likely to cause an enormous shift from medication towards genetic therapy. In particular cancer and HIV, which are widespread in rich countries and have a genetic cause, will be priority targets, eventually making natural products screening with the same therapeutic targets obsolete.

At the same time expectations to cure diseases with agents deriving from protein engineering and drug design are increasing with growing computer power and molecular knowledge and techniques.

3.4 Economical Comparison: Natural Products v. Synthetic Compounds

The pharmaceutical industry spends more on R&D than any other industry. In 1995 pharmaceutical companies in the USA will spend more than $14 billion on R&D (worldwide spending 1993: $22.7 bn, see Adkins 1994), compared to $600 million in 1970. 25-35% of these efforts are related to drug discovery. The average costs to develop one new drug increased in the US from $54 million in 1976 to $231 million in 1990 (Duke 1993).

At the same time the flow of new drugs has slowed. This is one of the reasons why drug companies have been under financial pressure. Another reason is the long time (average twelve years - see Reid et al. 1993) required between patenting a new chemical entity and its approval for marketing to the public. As patent protection lasts only 15-18 years, drug companies have just 5-6 years of exclusive marketing to recoup $200-300M R&D costs per drug (Duke 1993, Reid et al. 1993, Australian Academy of Science 1995). According to Kleiner (1995) a pharmaceutical company can expect to earn $94 million over the years its patent runs (this figure is far lower than other estimates).

The key reasons for decisions for or against biodiversity prospecting lies in the overall economic assessment of different drug development concepts which has not been assesed yet. The sample acquisition and screening costs alone do not provide sufficient information for this but

- incremental costs
- return on investment, and
- opportunity costs

have to be estimated.

Although the collection of samples and the screening contribute only little to the overall costs of the development of a new drug, these expenses can be crucial for management decisions. Studies on the economics of natural resources for new drug development do not exist, at least they are not known to a larger circle. It is obvious that without comprehensive economic assessments it is impossible to justify preference for the one or the other source.

The value of biodiversity usually is estimated by scientists chiefly concerned with biodiversity conservation (e.g. Pierce and Moran, 1994). The attitude of the pharmaceutical industry's management, weighing up the pros and cons of different approaches, might follow very different rules but is far more relevant for the estimation of the commercial value of natural products. High investment may lead pharmaceutical firms to prefer a proprietary synthetic or semisynthetic to a relatively less proprietary herbal or natural product. American pharmaceutical firms often seek the semisynthetic and avoid the natural compound, even if, at least in some cases, the natural compound might be best (Duke 1993).

4 COMPENSATION FOR ACCESS TO NATURAL RESOURCES

4.1 The Cost of Access to Biological Resources

During the last decades developing countries have become aware of the importance of genetic resources for pharmaceutical production although they usually overestimate the true market value. Most efforts to date which have attempted to identify a specific market value for genetic resources have been conceptionally very general or related to special cases.

The best known example of a contract on natural resources exploration is the Merck/INBio case which has raised more than $3.5 million during the years 1992-1994 through joint ventures. Under the agreement, INBio provides Merck with 1,000 biological samples, in return Merck has granted a two-year research and sampling budget of $1,135,000 ($1 million collecting fee and $135,000 in equipment) and will pay royalties on any resulting commercial product (Pistorius & van Wijk 1993).

Table 2, presented by Lesser and Krattiger (1994), summarises the different costs and returns in drug development based on natural products. Although one might criticise some of these figures they are valuable as a basis for an in depth discussion.

4.2 What Is the Price of a Biological Sample ?

Depending on how difficult a particular species is to collect, pharmaceutical plant sample providers receive payments of $50 to $200 per sample, which can be - as in

Table 2: Estimates of Variables for a Bioprospecting Agreement

	Per Sample Basis		
	Value Used in Calculations	**Range**	**Comment/ Reference**
Collection Fee	$50	$50 to $200	The figures are intended to cover actual costs (packaging, transport and related costs) but not return a profit (Laird 1993)
Royalty Payment	5%	1-5%	Royalty of gross sales (Nature 1993) indicates this range for basic materials.
Developing Country Interest	15%	10-25%	Discount rate used by developing countries with hard currency shortages; this is a likely minimum figure.
Corporate Interest Rate	7%	5-9%	Corporate interest rate charged to and by major corporations; lower than developing country rate because of better credit rating and more efficient credit markets in developing countries
Product Value	$500 million	$100 million to $1 billion+	Total worldwide sales once developed and over the life span of the product. Below an expected market of $100 million, returns generally do not cover development and regulatory costs.
Development Delay	10 years	10-12 years	Vagelos (1991)
Hit Rate	1:12,000	1:10,000 to 1:50,000	Frequency with which collected material will result in a marketable product; Sittenfeld and Gámez (1993)

Source: Lesser & Krattiger 1994 (changed)

the Merck/INBio case - paid in total, or in part, in services such as training or equipment. Merck pays to INBio many times more than $200. In exceptional cases up to $1,500 (including collecting, travel, and shipping costs) has been paid for particularly desirable samples. These costs would increase if developing countries get tougher on the financial compensation for access natural resources.

Payments for extracts can range between $200 and $250 per 25 gram sample. These payments are determined on a case-by-case basis, can include the costs of shipping, confirming initial field identifications, crating voucher specimens, processing data and conducting literature searches (Laird 1993).

4.3 What Royalties are Paid for Natural Products ?

The share of the parties involved in the pharmaceutical commercialisation of natural resources is determined on an individual basis, while details are kept confidential. This hinders a public international evaluation of the conditions under which biodiversity prospecting takes place. It appears that the handling of royalties differs widely. INBio will receive 2-6% of net sales, while the counterparts of Biotics get a half of that (Pistorius & van Wijk 1993, Thomas et al. 1994). This range of typical royalties applies also to new synthetic chemicals (Reid et al. 1993). The NCI follows a policy that royalty rates and other forms of compensation to the source country can only be discussed during the licensing process with the industry (G.M. Cragg, personal communication).
Paying royalties means that the firm has no interest costs, which are required if payments are made before the product is marketed and revenues flow. Since pharmaceuticals take up to 12 years to bring to the market, the interest cost can be considerable.

The results of a Harvard Business School study are presented in Table 3.

Table 3: Typical Royalty Rates by Product Category

Product Category Supplied	Percentage of Revenue to Supplier
Early Research	1-5%
Pre/Clinical Data	5-10%
Identified Product/Efficacy Data	10-15%

Source: Harvard Business School Case Study , 1992

5 CONCLUSIONS

The future relevance of natural products in drug development will depend on the efficiency and costs of access to chemically diverse extracts from living organisms, compared to new synthetic approaches like combinatorial chemistry. It appears that, at least for the next few years, natural products will retain their importance because true innovative products are only expected to come from nature. This situation is likely to change if

* the costs for access to natural resources exceed a certain limit,
* progress in synthetics increases the availability of cheap molecular diversity, or
* gene therapy and rational drug design replace random screening.

The Institute for Prospective Technological Studies will continue to work on this subject in order to get a more precise idea of these developments. IPTS would be grateful for offers of any company or academic organisation willing to collaborate with the institute in this challenging undertaking.

References

1. Adkins 1994. Pharmaceutical Equity Research.

2. Australian Academy of Science 1995. Submission to the Industry Commission. Inquiry into the Pharmaceutical Industry. WWW site: http://www.asap.unimelb.edu.au/aas/ policy/pharmsub.htm, last modified on August 28, 1995

3. G.M.Cragg *et al.* 1995. Drug Discovery and Development at the United States National Cancer Institute: International Collaboration in the Search for New Drugs from Natural Sources. Proceedings of the Eighth Asian Symposium on Medicinal Plants and Spices and other Natural Products, Melake, Malaysia, June 13-16, 1994

4. J.A.Duke, 1993. Medicinal Plants and the Pharmaceutical Industry. In: J. Janick and J.E. Simon (Eds.): Proceedings of the Second National Symposium. New Crops - Exploration, Research, and Commercialization, pp. 664-669

5. V. Glaser, 1995. Drug Companies Like Combinatorial Chemistry. Bio/Technology Vol. 13, April 1995, pp. 310-311

6. Harvard Business School Case Study 1992. A Case on the Merck/INBio JointVenture. Harvard Business School, Boston

7. S. Joffe and R.Thomas, 1989. Phytochemicals: A Renewable Global Resource. AgBiotech News and Information Vol. 1/5, 1989, pp. 697-700

8. K. Kleiner, 1995. Billion-Dollar Drugs Are Disappearing in the Forest. New Scientist, 8 July 1995, p. 5

9. A. Laird, 1993. Contracts for Biodiversity Prospecting. In: W.V. Reid *et al.* (eds.): Biodiversity Prospecting: Using Genetic Resources for Sustainable Development. Washington: World Resources Institute, pp. 99-130

10. W.H Lesser and F. Krattiger, 1994. The Complexities of Negotiating Terms for Germplasm Collection. Diversity Vol. 10, 3, 1994, pp. 6-10

11. D. Pearce and D Moran, 1994.The Economic Value of Biodiversity. Earthscan Publications, London.

12. R. Pistorius and J. van Wijk, 1993. Biodiversity Prospecting - Commercializing Genetic Resources for Export. Biotechnology and Development Monitor 15, June 1993, pp. 12-15

13. W.V. Reid *et al.,* 1993. A New Lease on Life. In: W.V. Reid *et al.* (eds.): Biodiversity Prospecting: Using Genetic Resources for Sustainable Development. Washington: World Resources Institute, pp. 1-52

14. R. Richon, 1995. Combinatorial Chemistry: A Strategy for the Future. Website http://edisto.awod.com/netsci/Issues/July95/feature2.html#plfig4, generated on 15 May 1995

15. A. Sittenfeld and R. Gámez, 1993. Biodiversity Prospecting by INBio. In: W.V. Reid *et al.* (eds.): Biodiversity Prospecting: Using Genetic Resources for Sustainable Development. Washington: World Resources Institute, pp. 69-97

16. D. Stone, 1993. The Hot New Field of Molecular Diversity. Bio/Technology Vol. 11, 11 December 1993, pp. 1508-1509

17. K. Ten Kate, 1995. Biopiracy or Green Petroleum ? - Expectations & Best Practice in Bioprospecting. Report for the U.K. Overseas Development Administration. London, October 1995

18. R. Thomas, A.G Brown and N.T Flaih, 1994. Proposed Establishment of Phytochemical Extraction Companies in Developing Countries. In: A.F. Krattinger *et al.* (eds.): Widening Perspectives on Biodiversity. IUCN 1994, pp.309-313

19. Presentations at the conference "Natural Products - Rapid Utilization of Sources for Drug Discovery and Development", San Francisco, May 15-16, 1995

International Instruments, Intellectual Property and the Collaborative Exploitation of Genetic Resources

Mark F. Cantley

BIOTECHNOLOGY UNIT, OECD, DIRECTORATE FOR SCIENCE, TECHNOLOGY AND INDUSTRY

1 INTRODUCTION

The elements assembled below are collected from a mixture of personal and bureaucratic experience over the past fifteen years, in the European Commission's Directorate-General for Science, Research and Development, and latterly in the Directorate for Science, Technology and Industry at OECD. Opinions expressed engage only the author.

The elements interact with one another, but may be represented as a tale of (at least) three strands: scientific, economic, and institutional-political. If we view the story as a kind of Greek drama, there should also be a Chorus; and indeed that has not been lacking. As to whether the play is a comedy, a tragedy, or some kind of mixture--a tragi-comedy--it may be too early to say--certainly there are elements of both.

Rather than reaching any crisp conclusions, we may just have to make do with posing some questions, the answers to which remain open, but are nonetheless important for that.

2 BACKGROUND: FAST "BIO-SOCIETY", BIOINFORMATICS AND PHYTOCHEMICALS

The European Commission in 1978 established a "Futures" group, "FAST": Forecasting and Assessment in Science and Technology. FAST's mandate was to assess what was going on, from a European perspective, and offer advice to the Commission on priorities for the Community's research and development programmes. The first such programme devoted a third of its time and attention to the life sciences and technologies, and the 1982 report concluded that in those fields, there was a lot happening, of great significance--though whether in the next year, or the next century, was not precisely spelt out.

In any case, FAST applauded the efforts of biology colleagues who since 1975, and from 1979 in parallel with the FAST programme, had been arguing for a European Community research programme in genetic engineering and enzymology--this was finally adopted in 1981, as "BEP", the "Biomolecular Engineering Programme", 1982-86, with the princely budget of 15 million ECUs, to spread over four years, nine countries, and the leading edges of molecular biology. Never mind--"great oaks from little acorns grow", as followed in this case.

The FAST programme had to commission and digest tens of consultant studies, workshops, and all the relevant literature, and present it to busy senior staff and politicians. At the time, futurists looking at information technology had been much impressed by the French report, "L'informatisation de la société" (Nora & Minc, 1978); and the French, who are rather good at this sort of thing, produced the following year the report by Gros, Jacob and Royer, "Sciences de la vie et société". So the chapters of the first FAST report, in the same style, had headings such as "The Information Society", and "The Bio-society".

To summarise the various analyses, scenarios and projections, particularly in the life sciences and technologies, we used four ponderous words:

-informatisation;

-dematerialisation;

-molecularisation;

-globalisation.

Each of these is good for a one-hour lecture. At the time, they led to operational consequences. The Commission set up a Task Force for Biotechnology Information in March 1982--bio-informatics was born, and we gave the first of many grants to the newly-born DNA sequence library which the European Molecular Biology Laboratory had established in Heidelberg the previous year. They were short of staff to read and annotate sequences, and needed to develop software, as some scientists were threatening to deposit sequence data in electronic form.

That first grant, of 50,000 ECUs, was not from the DG XII R&D programme; it came from a programme run by DG XIII, for stimulating the development of the European market for the provision of high quality information services. Many words; of which the key one is "service". And the culture and objectives of a service facility are quite different from those of a research laboratory. It was twelve years before EMBL acknowledged that to the extent of hiving off the European Bioinformatics Institute as a separate outstation.

Within the European Commission, the outcome of these scientific and bureaucratic developments was that the FAST report, while supporting the biology colleagues' demands for a larger programme, also argued--successfully--that the next programme must include two additional aspects:

i) provision for research infrastructure, such as data banks and culture collections, the genetic resources on which much of biology and practically all of modern biotechnology depend;

ii) initiatives to ensure that the efforts on the biotechnology research side should be accompanied by a coherent set of policy measures, "contextual" measures, ranging from patent law and safety guidelines, to collaborative work with developing countries.

Thus it was that the "BAP", Biotechnology Action Programme, 1985-89, with a budget initially of 55, later increased to 75, million ECUs, included a 1-MECU per year "Concertation Action"; and that Concertation Action, defined in much golden prose and a list of nine tasks, included among these:

"**Collaborative exploitation of phytochemical resources with developing countries**".

Shortly after that, we met Bob Thomas, and discovered that he knew what that phrase meant.

3 MODEL AGREEMENTS - 1: INFORMAL, SCIENTIFIC AND INDUSTRIAL

In this conference, there is little need to review in detail the rationale for natural product screening. Everyone here is aware that there are molecules out there which must be useful for something, by virtue of natural selection operating over 3,5 billion years. Survival, growth, defence against pests and predators, exploitation of the special features of exotic ecological niches, interesting secondary metabolites--all offer reasons for our interests, scientific, agricultural, pharmaceutical or whatever.

What was new in the 1980s were two relatively sudden developments:

i) the availability, in bulk and at affordable cost, thanks to gene cloning, of a growing number of the cell surface receptors, through which practically all biological functions are mediated; and

ii) the development of automated laboratory systems capable of handling the screening of several thousand samples per day.

Entrepreneurial companies developed to meet the corresponding needs and opportunities--companies such as Biotics, Pan Labs, Nova Pharmaceutical, Xenova, and others, and of course the internal departments with responsibilities for seeking new lead compounds, within large agrochemical, pharmaceutical and food companies.

There have been several significant efforts to develop equitable bases for such collaboration between "technology-rich" and "gene-rich" parties, some of them dating from long before 1992. There has also been much development of biotechnology based on organisms effectively in the global public domain, whether as pests, or as assets, or both--examples are easy to cite, of micro-organisms (vaccinia, yeast, E. Coli), insects (Drosophila), plants (Arabidopsis) or animals (mouse, rat).

Through Biotics, the European Commission financed the development of a model agreement, and its testing in the field; many thousands of plant samples have since been collected on this basis, from various developing countries, in Africa, Asia and Latin America, and screened in the laboratories of major companies or public research institutes. Many thousands of dollars have flowed back to the source countries, and the projects envisaged the progressive development of training and capacity-building in the developing countries. The published reports from Biotics (Thomas, 1988a, b; 1994) describe their approach, and there is a poster display at this conference giving further details. Similar activities were being conducted by the US National Cancer Institute--Gordon Cragg has spoken about the NCI experience.

A small--very small--number of the compounds have been identified as of potential interest for further investigation; at this conference several examples of the success stories are being presented.

What appears as common in all the screening agreements which have developed in scientific and industrial circles has been a commitment to equity in the terms of the screening agreements. Typically these agreements include a number of common features:

i) agreement on terms prior to the commencement of collecting activities;

ii) payment for the collection of samples;

iii) respect for local laws, regulations, and customary rights;
iv) defined exclusivity for the screening partner;
v) commitment to negotiate equitable royalties in the event of ultimate commercial exploitation attributable to the original collected sample.

Putterman (1996), who heads a Working Group on International Genetic Resource Issues with the Society for Biomolecular Screening, describes agreements, enshrining similar principles, which he has developed as basis for commercial screening agreements in a number of countries--his paper includes a policy model for implementing Article 15 of the Biodiversity Convention (see below) in a manner which defines source country rights without creating undue disincentives for private sector investment in natural products research.

Such model agreements did not spring into being without an earlier history. They typically reflect a heightened sensitivity to past and continuing criticism--I referred at the start to the role of the "Chorus". Under this pressure, there has been a progressive evolution from informal understanding to more explicit and formal agreements.

We cannot re-write the history of the last few centuries, in which the first nations to master gunpowder and navigation imposed their interests on other regions of the world, and made many disruptive transfers of germplasm with permanent effects. It is similarly futile to indict past generations for actions which appear indefensible in the light of later knowledge and democratic values. But the conditions of to-day oblige commercial and scientific agreements to respect considerations of equity rather than force, to take into account the long-term public interest (particularly regarding conservation), and to subordinate even the exercise of economic power to a framework of law based on democratic values.

There have therefore developed in recent decades, as developing countries acquired independence and the means to articulate their concerns, increasing arguments to make formal and transparent the conditions for access to and exploitation of genetic resources. The result has been formal international or inter-institutional agreements such as those discussed in the following section: between FAO and the IARCs; and the Convention on Biological Diversity.

The climate which led to these agreements, and now their formal terms, have altered the rules of the game for the "gene-hunters" of to-day. The behaviour of responsible organisations, such as those cited above, might be argued to render such agreements less necessary; but such political debates acquire a certain momentum of their own, and as with history, we must live with the consequences, adapt to what is inevitable, and focus on how we can improve the future.

In the early years, we must expect a "running-in" period, in which institutional bureaucracy will cause frustration and wasted opportunities. The long delays and frustrations over the development of Ancistrocladus korupensis is sadly instructive (Miller, 1993; Laird, 1995). The various players will have to shed some of their illusions and prejudices, and look to learn pragmatic lessons. Industry is supremely adaptable--it operates under Darwinian rules of selection--and the screening agreements being negotiated to-day reflect the new rules of the game. Industry is familiar with the man from the government, who says, "I'm here to help".

As examples, one might quote the publicly stated policies of Novo Nordisk (1995), a biotechnology company with a strong commitment to responsible corporate behaviour. They have an obvious need to maintain their competitiveness as a developer and manufacturer of enzymes, a sector totally dependent on the identification and exploitation of natural molecules. While many of the best known enzymes are widespread and familiar, the future development of industrial enzymology cannot ignore the capabilities of extremophiles, which may be located in unique localities and therefore clearly subject to national control and authorisation. Their stated policy (Novo Nordisk, 1995) is scrupulous with regard to national sovereignty, and the Biodiversity Convention. At the same time, they seek a pragmatic solution to bureaucratic obstacles, and make an intriguing suggestion to the European Commission: that, for the purposes of the Convention, the European Union act as a single entity. But why stop there?

The same logic is exemplified by Recombinant BioCatalysis Inc., a recent biotechnology start-up dedicated to tracking down and identifying the enzymatic potential of extremophiles (Gross, 1996). Their agreement with the Icelandic company GENIS has received the clear approval of the Icelandic government (RBI press release, 1996).

In scientific circles, although the commercial element may in many cases be absent, its potential presence, the changing climate referred to, and the potential for conflicts over scientific credit and access to materials, have obliged the scientific community similarly to evolve from unquestioned assumptions, through informal understandings with local collaborators, towards ever more formal codes of practice and legal agreements. The International Union of Pure and Applied Chemistry is currently considering an Australian proposal that they endorse the Manila Declaration and the Melaka Accord, concerning the ethical utilisation of Asian Biological Resources.

4 MODEL AGREEMENTS - 2: INTERNATIONAL: AGRICULTURE AND BIODIVERSITY

4.1 The CGIAR-FAO agreement on the ex situ collections

There was an extended and often heated debate through much of the 1980s, involving the conditions for collection, control and diffusion of the genetic resources assembled within the International Agricultural Research Centres (IARCs) funded through the Consultative Group for International Agricultural Research (CGIAR). This debate focused on the relationship between the (then) International Board for Plant Genetic Resources, and the Food and Agriculture Organization of the United Nations; IBPGR was located in the FAO premises in Rome.

It would be dangerous for a non-participant to summarise the conflict, but one of the underlying tensions seemed to be the same evolutionary process already referred to, between a long-established scientific tradition incorporating, almost implicitly and unquestioningly, two assumptions which gradually came to be intensely questioned, and then changed:

i) the free availability and exchange of germplasm, seen and referred to as the common patrimony of humanity (for a reasoned defence, see Jain, 1994);

ii) a tradition of governance of scientific research centres by a process of scientific peer review within a narrow community, rather than accountability to national authorities.

The CGIAR itself had been created in the 1970s, to bring a certain coherence into the private and public sector funding of the IARCs. It was (and is) chaired by the World Bank. It proved valuable in the 1980s, as common policy challenges were faced by almost all of the otherwise diverse IARCs. Several of these challenges appeared to be associated with rise of modern biotechnology: issues of biosafety, and the field release of genetically engineered organisms; issues of intellectual property, through patents, plant breeders' rights, or otherwise; and (closely related to the latter), relations with the private sector, whose finance would be essential to carry a research breakthrough to practical application and effective diffusion, but which demanded in return some exclusivity.

The CGIAR created in 1989 a CGIAR Task Force on Biotechnology (BIOTASK), "to raise awareness amongst all components of the CGIAR system of the issues involved in biotechnology" (Persley, 1990); and through this corporate mechanism, the IARCs were able finally to negotiate a satisfactory resolution with FAO of the status of their ex situ collections of germplasm.

FAO had through the same years gradually arrived at the establishment of their Commission on Plant genetic Resources, and the International Undertaking on Plant Genetic Resources, to which a growing number of countries agreed to adhere.

The agreements of October 1994 between the FAO and the twelve IARCs holding ex situ collections had to address not only the issues of governance and status of the collections, but also the questions of compatibility with the terms of the 1993 Convention on Biological Diversity, and in particular its central principle: the establishment of national sovereignty over genetic resources. However, the CBD did not cover these existing collections. This was noted as a problem at the May 1992 "Conference on the Adoption of the Agreed Text of the Convention on Biological Diversity", whose Resolution 3 addressed "The Interrelationship between the Convention on Biological Diversity and the Promotion of Sustainable Agriculture", and recognised:

> "the need to seek solutions to outstanding matters concerning plant genetic resources within the Global System for the Conservation and Sustainable Use of Plant Genetic Resources for Food and Sustainable Agriculture, in particular:
>
> (a) Access to ex situ collections not acquired in accordance with this Convention; and
>
> (b) The question of farmers' rights".

The FAO has sought to address this problem, in principle through the "Global System" referred to, within which the FAO-IARC agreements form one element. Their key terms are as follows (the example is that for IRRI, but the other agreements are substantially similar):

AGREEMENT BETWEEN THE INTERNATIONAL RICE RESEARCH INSTITUTE (IRRI) AND THE FOOD AND AGRICULTURE ORGANIZATION OF THE UNITED NATIONS (FAO) PLACING COLLECTIONS OF PLANT GERMPLASM UNDER THE AUSPICES OF THE FAO

ARTICLE 1
APPLICATION OF THIS AGREEMENT

This Agreement shall be construed and applied in a manner consistent with the provisions of the Convention on Biological Diversity and the International Undertaking on Plant Genetic Resources.

ARTICLE 2
BASIC UNDERTAKING

The Centre hereby places under the auspices of the FAO, as part of the international network of ex situ collections provided in Article 7 of the International Undertaking on Plant Genetic Resources, the collections of plant genetic resources listed in the Appendix hereto (hereinafter referred to as the "designated germplasm"),

ARTICLE 3
STATUS OF DESIGNATED GERMPLASM

(a) The Centre shall hold the designated germplasm in trust for the benefit of the international community, in particular the developing countries and in accordance with the International Undertaking on plant Genetic Resources and the terms and conditions set out in this Agreement.

(b) The Centre shall not claim legal ownership over the designated germplasm, nor shall it seek any intellectual property rights over that germplasm or related information.

For fuller history of these events, the reader is referred to the excellent publication, generously supported by the government of The Netherlands, "Biotechnology and Development Monitor". Mugabe and Ouko's "Control over genetic resources" summarises the issues (1994).

The germplasm samples collected in support of agricultural research have two special features: they are limited to the purposes of agriculture, broadly though the research scientist may wish to address his topic; and by the same reason, they do not attempt to address the totality of germplasm and biodiversity, in the sense that motivates conservationists or ecologists. Agriculture, at least as much as industry, has been charged with the widespread elimination of genetic diversity, both through direct effects of chemicals and wastes, and indirectly through short-term economic pressures driving farmers to focus on a limited range of plants, livestock and other germplasm giving the best near-term commercial returns. On this analysis, neither the agricultural research community, nor the agricultural producer interest--as very approximately symbolised by the CGIAR and the FAO--could logically be charged with the task of conservation.

4.2 The Convention on Biological Diversity (CBD)

The task of conserving biological diversity logically fell to the United Nations, and in particular has been addressed by the Convention on Biological Diversity (CBD), adopted at the "Earth Summit", the UN Conference on Environment and Development, held in Rio de Janeiro in June 1992.

The CBD addresses a complex combination of issues, of which some are relevant to the present paper. Parts of some key Articles are as follows:

Article 1. Objectives

"The objectives of this Convention ... are the conservation of biological diversity, the sustainable use of its components and the fair and equitable sharing of the benefits arising out of the utilization of genetic resources, including by appropriate access to genetic resources and by appropriate transfer of relevant technologies, taking into account all rights over those resources and to technologies, and by appropriate funding."

Article 15. Access to Genetic Resources

"1. Recognizing the sovereign rights of States over their natural resources, the authority to determine access to genetic resources rests with the national governments and is subject to national legislation."

Article 16. Access to and Transfer of Technology

"1. Each Contracting Party, recognizing that technology includes biotechnology, and that both access to and transfer of technology among Contracting Parties are essential elements for the attainment of the objectives of this Convention, undertakes subject to the provisions of this Article to provide and/or facilitate access for and transfer to other Contracting Parties of technologies that are relevant to the conservation and sustainable use of biological diversity or make use of genetic resources and do not cause significant damage to the environment.

2. Access to and transfer of technology referred to in paragraph 1 above to developing countries shall be provided and/or facilitated under fair and most favourable terms, including on concessional and preferential terms where mutually agreed, and, where necessary, in accordance with the financial mechanism established by Articles 20 and 21. In the case of technology subject to patents and other intellectual property rights, such access and transfer shall be provided on terms which recognize and are consistent with the adequate and effective protection of intellectual property rights. The application of this paragraph shall be consistent with paragraphs 3, 4 and 5 below.

3. Each Contracting Party shall take legislative, administrative or policy measures, as appropriate, with the aim that Contracting Parties, in particular those that are developing countries, which provide genetic resources are provided access to and transfer of technology which makes use of those resources, on mutually agreed terms, including technology protected by

patents and other intellectual property rights, where necessary, through the provisions of Articles 20 and 21 and in accordance with international law and consistent with paragraphs 4 and 5 below.

4. Each Contracting Party shall take legislative, administrative or policy measures, as appropriate, with the aim that the private sector facilitates access to, joint development and transfer of technology referred to in paragraph 1 above for the benefit of both governmental institutions and the private sector of developing countries and in this regard shall abide by the obligations included in paragraphs 1, 2 and 3 above.

5. The Contracting Parties, recognizing that patents and other intellectual property rights may have an influence on the implementation of this Convention, shall cooperate in this regard subject to national legislation and international law in order to ensure that such rights are supportive of and do not run counter to its objectives."

Article 20. Financial Resources

"4. The extent to which developing country Parties will effectively implement their commitments under this Convention will depend on the effective implementation by developed country Parties of their commitments under this Convention related to financial resources and transfer of technology and will take fully into account the fact that economic and social development and eradication of poverty are the first and overriding priorities of the developing country Parties."

The Articles quoted are central to some of the "bargains" enshrined in the CBD, and to some of the related controversy--including issues which have so far inhibited the United States from ratifying the Convention, in spite of their active participation in its preparation.

Defenders of the Convention stress that its main merit is that it exists, as an important political statement of intent; and assume that some of the more extreme interpretations of its language will remain theoretical. Critics express concern that developed country governments will "arm-twist" industry into giving access to their biotechnology patents, and the resulting products and services could then be used by their competitors. The language of Article 20.4 can be read as a sort of blackmail, that if the developed countries do not fund conservation of biodiversity, developing countries will allow social and economic development pressures to destroy it.

More technical criticisms are that those who drafted the Convention flattered biotechnology in the extent to which it is represented as, or implied to be, the key to unlocking the economic potential of biodiversity. Again, since most of the relevant techniques are in public domain, the key to access to biotechnology lies less in intellectual property, than in the usual matters of training, infrastructure, and supportive policies for research, investment and innovation.

Most importantly, the language and the existence of the Convention have tended to imply that the potential economic benefits derivable from biodiversity are comparable to the costs of its conservation. Work is in progress--in particular, at the OECD (1996)--to explore economic incentives for conservation (in line with Article 11 of the Convention, which states: "Each Contracting Party shall, as far as possible and as appropriate, adopt economically and socially sound measures that act as incentives for the conservation and sustainable use of components of biological diversity"). However, it is far from self-evident that the economic benefits will cover the costs; and if they do not, is one invited to conclude that conservation is not worth pursuing?

5 TOWARDS A KNOWLEDGE-BASED GLOBAL ECONOMY: IPR AND THE TRIPS AGREEMENT

5.1 Towards a knowledge and information intensive world economy

The terms "globalisation" and "knowledge-based economy" are almost clichés, but summarise major trends in the evolution of the world's economic system. The term "informatisation" similarly refers to a pervasive process of intensification of the knowledge and information content of almost all areas of human activity, including traditional economic sectors and long-established businesses, practices and institutions. Science is naturally global, but the above trends interact particularly with the life sciences and technologies, where "informatisation" has permeated all disciplines; especially when linked with "molecularisation". The result is an intensification of global competition in the corresponding sectors. E-mail shrinks the world and intensifies competition, just as the grain ships and refrigerated transport affected the cereal and meat sectors in the last century. The triumph of reductionism reaches its apotheosis in the structure and sequence of DNA; and the resulting illumination from the molecular level shines--or casts shadows--across the whole of the life sciences and technologies.

Operational consequences of these trends include a surge of interest in bio-informatics, genomics and related issues of intellectual property at the level of gene sequences. However, although genes are older than patent systems, intellectual property systems developed before Mendel and before molecular biology. Needs therefore arise, at least for interpretation, in the application of old concepts to new technological developments; beyond interpretation, there may be needs for adaptation.

For example, it was precisely because the innovative successes of plant breeders could not be described in such precise and reproducible terms as those of industrial inventors, that a separate body of law and practice was developed for plant breeders' rights. With the advent of molecular biotechnology, it is now evident that there are inventions, applied to plants, for the usual objectives of plant breeders (improved yield, pest resistance etc.) which satisfy the traditional criteria of patentability (inventive step, non-obviousness, utility, adequate description). It has therefore become necessary to clarify the relations between the two bodies of law (e.g. see Crespi, 1992).

5.2 The European Commission's proposed Directive on Biotech Patents

Such "boundary-crossing" problems prove contentious, where living materials are concerned, because they are bound up with culture and value-laden activities, and most of all where human genetic material is concerned. The history of the European Commission's proposed Directive for the Protection of Biotechnological Inventions illustrates the political passions, real and synthetic, which can be engendered. This proposal was initiated in the mid-1980s, following the publication of a report (Beier et al, 1985) indicating that "there is no other field of technology where national patent laws vary on so many points as they do in biotechnology"; and that Europe was disadvantaged relative to its main competitors in a number of aspects. The proposed Directive sought primarily to clarify certain matters of interpretation, without calling into question the European Patent Convention.

The proposal was carefully prepared, including extensive consultation with interested circles, and put forward in October 1988. Its progress was delayed because of a succession of interactions. There was a proposal in preparation for a Regulation to establish a Community-wide harmonised system of plant breeders' rights. The progress of this was delayed by the conference of UPOV, the Union for the Protection of Plant Varieties, which in 1991 made a number of changes to reduce the differences between European and US practice. In the 1990s, and with the rise of European Parliamentary interest in ethical aspects of biotechnology, there was a hot interaction with the questions surrounding the patentability of parts of the human body, such as cells or genes.

In the context of biodiversity and genetic resources, the linkages between these two apparently separate areas of debate were underlined by the public controversies over the patenting by US scientists or government agencies of cell lines from patients or from isolated populations, and over the "Human Genetic Diversity" project. At present, the proposed Directive remains unadopted, a carefully-worded compromise text having been rejected by the European Parliament on 1 March 1995. A new proposal is now under discussion.

5.3 The Marrakesh accords and the TRIPs agreement

The benefits of an open world trading system are widely acknowledged, not only within the Member countries of the OECD, but by the over 100 countries signatory to the GATT (General Agreement on Tariffs and Trade). The Uruguay Round was the latest such set of negotiations and led to the creation of the new World Trade Organization (WTO), effective from 1 January 1995, following signature of the several related agreements at Marrakesh in April 1994.

As a result, the protection of intellectual property became an integral part of the multilateral trading system. It is described by Otten and Wager (1996) as "one of the three pillars of the WTO, the other two being trade in goods (the area traditionally covered by the GATT) and the new agreement on trade in services." They continue, "The fact that the protection of intellectual property has thus moved to the centre stage of international economic relations is not surprising given its major and growing importance for the conditions of international competition in many areas of economic activity."

This applies with particular force to the life sciences and technologies, and their commercial applications, for the reasons indicated at the start of this section. So at the moment when agriculture, for instance, becomes for the first time subject to the GATT agreement, it is also in course of becoming an information-intensive sector, in which the related intellectual property will be of much increased significance--and such property will itself be subject to the TRIPs agreement.

The TRIPs agreement obliges signatories (i.e. all members of the WTO) to provide certain minimum standards of protection. However, it carries over certain special features concerning living materials, reflecting corresponding special treatment in the European Patent Convention, which amongst its exclusions from patentability includes:

"Plant or animal varieties or essentially biological processes for the production of plants or animals, except for microbiological processes or the products thereof." (The patentability of microbiological processes is a long-established practice, going back to Louis Pasteur, who patented a yeast in 1865).

A similar exception was carried into the wider international domain by the Patent Cooperation Treaty, which was established in 1970, and like the European Patent Convention came into force in 1978. It may be noted that these important Conventions, the fruit of heavy and extended negotiations, came into effect just before the advent to public notice of the techniques of modern biotechnology.

Countries are required under the TRIPs agreement "to make patent protection available for inventions in all areas of technology without discrimination--the only substantial sectoral exceptions being for plants and animals other than micro-biological processes. However, any country opting to exclude plant varieties from patent protection will be required to introduce an effective sui generis system, of protection." (Otten and Wager, op. cit.)

The continued special treatment of intellectual property in biotechnology has been vigorously attacked by the United States, for example by Bruce Lehman, Commissioner of Patents and Trademarks of the US PTO (Patent and Trademark Office), at the Pasteur Centennial colloquium held in Paris in 1995. He criticised the exclusions in the European Patent Convention and the Patent Cooperation Treaty, although acknowledging that at the time they were made in ignorance of their effects on the subsequent development of biotechnology. He continued:

> "There was little excuse, however, to perpetuate this unreasonable prejudice against biotechnology in the late 80s and early 90s, other than an unwillingness to review and revise earlier-held views despite overwhelming evidence of the benefits of biotechnology. Yet, the mistakes of earlier years were not only repeated in the TRIPs Agreement, they were compounded even further. In Article 27, paragraph 3, of that Agreement, biotechnology has not merely been forced to take a back seat it has virtually been forced out of the arena altogether.
>
> ... This development is an international giant step backward as far as biotechnology is concerned. ... Thankfully, Article 27 of the TRIPs Agreement contains a clause mandating its review in 1999. It is our fervent hope that our negotiating partners will finally be convinced by then that their previous actions regarding biotechnology were not in their best interests and that the patent system should not

be used as a tool to satisfy concerns that could easily be addressed through other means." (Gallochat, 1995)

It is thus clear that the ipr aspect of the world trade agreement remains contentious so far as concerns biotechnology and genetic resources. The possibility of conflict between obligations of the CBD and the TRIPs agreement was raised by some countries at the Second Conference of the Parties to the CBD, at Djakarta in December 1995.

6 PROBLEMS, DELUSIONS AND OPEN QUESTIONS

6.1 Combinatorials v. Screening

Companies trying to earn their living through screening and the sale of samples to pharmaceutical companies have had a difficult existence during the past three years. The rise of combinatorial chemistry (defined, according to Floyd et al, 1996, as "a method of increasing the size of the haystack in which to find your needle") has apparently given the companies the tools to generate an almost infinite diversity of oligonucleotides and oligopeptides at minimal cost, and they can then apply to this "synthetic" biodiversity the screening tools referred to earlier. In consequence, some of those responsible within companies for the search for lead compounds have for the moment lost interest in phytochemical and other natural product screening. One might contrast dramatically the $1m paid in the much-reported Merck deal with InBio, Costa Rica, with the $533m paid by Glaxo in acquiring Affymax--reportedly for their mastery of the technology of combinatorial chemistry and related developments.

It remains nonetheless clear that there are valuable and complex molecules in nature which are unlikely ever to be found by random synthesis. For example, there is an interesting snail toxin, fifty amino-acids long, and variants on the basic sequence reveal a range of significant and potentially useful biological activities (Atlas, 1996). But not even combinatorial chemistry and supercomputers can explore the phase space of all combinations of fifty amino acids, each being any of the twenty-plus such acids.

Nature has at least two advantages. The target molecules most often used in screens--the enzymes, antibodies, and cell surface receptors employ a relatively limited range of motifs. Nature is economical in reusing repeatedly an idea, embodied in a secondary structure and once found useful. The bundle of alpha-helices used in a trans-membrane tunnel, e.g. for an ion pump, is likely to have some structural commonalities, whether it be a pump for protons, calcium, sodium or potassium. The homologies can be identified at sequence level, albeit sometimes with difficulty.

Similarly, there are several success stories of natural isolates such as taxol, penicillin, cyclosporin, serving as starting point for a range of semi-synthetics. Clearly there is a place both for natural product screening, and the combinatorial approach to exploring variations on a theme. Where the boundary will be established will in some degree depend upon the ease of access to natural biodiversity, and it must be remembered that there is a vast repertoire of diversity not under exclusive national control.

6.2 The Value of Germplasm

President Nasser of Egypt once remarked that "the Arabs cannot drink their oil"--to realise its value, they are dependent upon the markets of the industrialised world. Similarly, there are many factors--some referred to--which can block or conversely facilitate economic interest in natural product screening. Mugabe and Ouko (op. cit.) cite with approval the partnership between Costa Rica and Merck, particularly because of the local added value:

> "The value of genetic raw materials can be enhanced if the material is put through a process of identification, collection and screening by the owners of the biological resources before presenting it to a potential recipient. Developing countries should establish a reputation as a reliable partner in the screening process. ... The Costa Rican example serves to show that a developing country can indeed succeed in this so-called "technology-intensive" venture with the right attitude and the right policies."

Where the balance will be struck between screening natural extracts and combinatorials remains an open question, whose answer will depend upon the interests and goals of the screener, but also upon the conditions of partnership with the source country. Reference has been made to the Cameroon experience (with Ancistrocladus K.). At the "Diversitas" meeting in Paris, September 1994, Mugabe spoke of the unrealistic assumptions which he had encountered at preparatory meetings of African countries in advance of the CBD first Conference of the Parties.

We mentioned earlier the "Chorus": the constant criticism and loaded language, with accusations of "bio-piracy", will certainly do little to help developing countries find partners with whom to add value to their resources. Di Castri (1995) has as President of the International Union of Biological Societies voiced the widely felt complaint of the scientists:

> "the niche of public visibility in these fields (ecology, biodiversity, hydrobiology, environmental biology, etc.) is taken over--much too often--by "instant scientists", with unknown university backgrounds, almost devoid of research experience, with no associated students, and not belonging to the so-called scientific community. These individuals, however, tend to answer to the whims and demands of the media, that is to say, sensationalism, catastrophicism, or a romantic attachment to a wilderness past."

The sharp riposte by Goldstein to Shand (both 1994) makes similar important points, about the impossibility of imputing either moral or financial credit to past generations. He disputes the equation of "resources" with "germplasm", stating, "If Ms Shand equates resources to germplasm, we should ask why after 500 years of Latin American history, the region still cannot achieve anything meaningful with its germplasm." Recalling the indifference or hostility with which Latin American countries had treated various discoveries of potential economic significance (demonstrated elsewhere), he says, "I do not see here traces of imperialism in action, but the destructive effects of the ignorance that characterizes underdevelopment".

6.3 Germplasm and Genomics

Goldstein goes on to make a sharp prediction:

"Ms Shand cares about germplasm, but fails to mention that germplasm soon will be not worth a penny. Strategic genes are identified in and isolated from model organisms and, suitably modified, used to construct the desired transgenics. Yet the South does not participate in the genome projects, and is marginalized from explicit technological research agendas aimed to the discovery of strategic genes."

He then makes a related point about human genes, noting that the gene for Huntington's disease was isolated in 1993 in the USA, using the DNA from an afflicted family in Venezuela--for whose scientific community, this disease "was a non-existing problem". This point may be linked with the warning by Sandy Thomas and colleagues in a recent issue of Nature: to the effect that "While Europe has over the past eight years debated the pros and cons of a harmonized directive on biotechnology intellectual property rights, Japan and the United States have between them secured rights to 70% of EPO patents for human gene sequences." (Thomas et al, 1996)

Goldstein's point about the drop in value of germplasm as a result of progress in genomics and model organisms, although qualified in his recent plea (1996) for North - South partnership in exploiting biodiversity, is both inherently plausible, and empirically confirmed, by items such as the following:

- the paper by Kurata et al (1994) on "Conservation of Genome Structure Between Rice and Wheat", as the authors point out, "may provide a basis for novel isolation strategies in wheat, one of the world's major crop species"; and inter-species homology will broaden the applicability of insights developed from any chosen starting-point;
- in February 1996, Pioneer Hi-Bred International signed a $16m comprehensive corn gene sequencing and discovery collaboration agreement with Human Genome Sciences, proprietors of TIGR, The Institute for Genome Research.

Res ipsa loquitur. Science has a habit of moving faster than institutional thinking and legislative processes. Scientific progress, in conjunction with political and bureaucratic delays and difficulties in establishing and enforcing property rights over germplasm, may indeed diminish the economic value of raw germplasm or "genetic resources." The case for conservation cannot rest on economic potential alone.

References

1. Atlas, R., personal communication, 1996.

2. Beier, F. K., Crespi, S., and Straus, J., 'Biotechnology and Patent Protection: An International Review', OECD, Paris, 1985.

3. Crespi, S., 'Patents and Plant Variety Rights: Is There an Interface problem?', *International Review of Industrial Property and Copyright Law*, 1992, **2**, 168.

4. di Castri, F., 'Matching Rigot with Openness in Biology', *Biology International*, July 1994, **29**.

5. Floyd, C.D. et al, 'More leads in the haystack', *Chemistry in Britain*, March 1996.

6. Gallochat, A., ed., 'The Protection of Biotechnology Inventions', Colloquium organised by Institut Pasteur, 28 September 1995.

7. Goldstein, D., 'A critique of the critics', *Biotechnology and Development Monitor*, September 1994, **20**.

8. Goldstein, D. J., 'Biotech Could Play a Strategic Role in Renegotiating Latin America's Debt', *Genetic Engineering News*, 1996, **35**, 1.

9. Gross, N., 'Exteme Enzymes: Science is commercializing nature's diehard proteins', *Business Week*, April 1, 1996.

10. Jain, H. K., 'The Biodiversity Convention: More losers than winners', *Biotechnology and Development Monitor*, December 1994, **21**.

11. Krattiger, A., ed., 'Widening Perspectives on Biodiversity', IUCN, the World Conservation Union, Gland, Switzerland, and the International Academy of the Environment, Geneva, 1994.

12. Kurata, N. et al, 'Conservation of Genome Structure Between Rice and Wheat', *Biotechnology*, March 1994, **12**,. 276.

13. Laird, S. A., Fair Deals in the Search for New Natural Products', Working Paper of the WWF International People and Plants Program, WWF, Panda House, Weyside Park, Godalming, Surrey GU7 1XR, UK, 1995.

14. Laird, S. A. and Cunningham, A. B., 'One in ten-thousand? The Case of Ancistrocladus korupensis', Paper prepared for the Rainforest's Natural Resources and Rights Program, 1995.

15. Miller, S. K., 'High hopes hanging on a "useless" vine', *New Scientist*, January 1993, **16**, 12.

16. Mugabe, J. and Ouko, E., 'Control over genetic resources', *Biotechnology and Development Monitor*, December 1994, **21**, 6.

17. Novo Nordisk, 'Position: Biodiversity on the international agenda,' Corporate Communications, Novo Nordisk, Bagsvaerd, Denmark, 1995.

18. OECD, 'Saving Biological Diversity: Incentive Measures', OECD, Paris, 1996.

19. Otten, A., and Wager, H., 'Compliance with TRIPS: The Emerging World View,', Paper prepared for Symposium on "Universal Intellectual Property Standards: The TRIPs Agreement", section on Intellectual Property Law, Annual Meeting of the American Association of Law Schools, San Antonio, Texas, January 1996.

20. Recombinant BioCatalysis Inc., 'Genis and RBI Collaboration,' Press Release, February 2, 1996.

21. Persley, G., 'Beyond Mendel's Garden: Biotechnology in the Service of World Agriculture,' CAB International for the World Bank, Wallingford, UK, 1990.

22. Putterman, D., 'Trade and the biodiversity convention', *Nature*, October 1994, **13**, 553.

23. Putterman, D. M., 'Model Material Transfer Agreements for Equitable Biodiversity Prospecting', *Colorado Journal of International Environmental Law and Policy*, 1996, **7** (1), 145.

24. Shand, H., '1993: A landmark year for biodiversity or bio-piracy?', *Biotechnology and Development Monitor*, December 1994, **17**.

25. Thomas, R., 'Collaborative Exploitation of Biological Resources in the African Region', Final report on work undertaken on behalf of the Commission of the European Communities by Biotics, Ltd., 1988a, Contract No. CUBI-0001-UK.

26. Thomas, R., 'Phytochemical Screening Projects in South East Asia: Training Programmes and Phytochemical Databases', Final report on work undertaken on behalf of the Commission of the European Communities by Biotics Ltd., 1988b, Contract No. CUBI-001-UK.

27. Thomas, R., Brown, A. G. and Flaih, N. T., 'Proposed Establishment of Phytochemical Extraction Companies in Developing Countries', in Krattiger, 309.

28. Thomas, S. M. et al, 'Ownership of the human genome', *Nature*, 4 April 1996, 387.

In Pursuit of Insecticidal Compounds from Plants

B. P. S. Khambay[1,*], D. G. Beddie[1], and M. S. J. Simmonds[2]

[1] BIOLOGICAL AND ECOLOGICAL CHEMISTRY, IACR-ROTHAMSTED, HARPENDEN, HERTFORDSHIRE AL5 2JQ, UK
[2] JODRELL LABORATORY, ROYAL BOTANIC GARDENS, KEW, SURREY TW9 3DS, UK

1 INTRODUCTION

Effective methods of pest control are needed to solve the urgent problems of producing enough food for the world's population. Control can involve the use of chemicals or biological control agents. Whilst biological methods of pest control are perceived to be more benign than chemical control, the use of insecticidal compounds will remain a major contribution for the foreseeable future[1,2].

The current array of insecticides is based on a narrow range of chemical classes[3]. As use of each class of compounds increases, the proportion of insects becoming resistant to them also increases. Furthermore, there have been changes in the target pest species as a consequence of changes in cropping systems and climatic changes[1]. Therefore there is an urgent need to discover classes of compounds with novel modes of action.

Natural products have played an important role in crop protection over thousands of years[4,5]. In the last two decades the industry has concentrated on evaluating chemicals from microorganisms and synthetic screening programmes. However, limited success has led to a revival of interest in plants which in the past have provided the agrochemical industry with important lead compounds[6]. Plants offer a wider source of chemical diversity than most synthetic based screens and this diversity warrants further study. Such a study should take into account the needs of both the market and the environment. Thus, it should seek compounds with:

1. activity against a range of target species e.g. mites, beetles, flies, caterpillars and ticks;
2. low toxicity against non-target species e.g. beneficial insects and mammals;
3. novel modes of action;
4. activity against target species resistant to other pesticides;
5. the ability to break down in the environment to non-toxic products.
6. cheap routes to production;
7. ease of use in the field;

These considerations in turn influence the type and design of both primary and secondary bioassay screens. Practical and financial limitations invariably limit the number of routine tests possible. In contrast to pharmaceutical screens, the design of bioassays for insecticidal activity requires additional consideration of factors such as:

1. life stages e.g. adults, larvae and eggs;
2. use of model insect species;
3. application routes e.g. topical, residual, vapour and systemic;
4. throughput;
5. *in vivo* or *in vitro* tests.

Insecticidal tests in our primary screen are based on contact activity eg. topical application or vial confinement. Although time consuming, they are more sensitive and repeatable compared to residual assays which are preferred in industry due to their higher throughput.

2 CHOICE OF PLANT SPECIES

Plants for screening can be gathered at random or selected using ethnobotanical information. The latter approach has been particularly effective in drug discovery[7] in recent years but this has not been the case for pesticides. However, the success rate can be improved by selecting plants by applying additional criteria such as:

1. selecting plants with minimal pest damage;
2. screening plants with known biological activity;
3. ability to survive in harsh environments.

Other general considerations in choosing plants include the type of plant (eg. trees, shrubs and ferns) and whether a species is endangered.

Chemotaxonomic directed studies on plant species and genera related to the active plant can often provide an array of related compounds for SAR investigations which may be difficult to synthesise and which would not have been chosen in a preliminary analogue synthesis programme. Selectivity to pests can often be achieved through structure-activity relationship (SAR) studies aimed at exploiting differences in metabolic processes.

Bearing in mind that insecticides are in general, a low cost commodity, industrial by-products have become a new area of investigation for cheap and readily available botanical insecticides[8]. For example, seeds from grapefruit contain limonene, a potent antifeedant to the Colorado potato beetle[9] .

3 FINDING ACTIVITY IN PLANTS

Once the choice of plant species to be studied has been made, the correct identification of the species and accurate record keeping during collection are vital. This is especially important when re-sampling becomes necessary. Production of secondary metabolites in plants is known to be influenced by many factors including the environment (eg. altitude, sunlight levels, nutrition, temperature and stress) and growth stages[10,11]. Major variation can often be found within species and strains of a given genus. For example[12], the yield of linalool can vary between 1% and 84% within various cultivars of *Ocimum bacilicum*. Therefore it is essential not to reject plants on the basis of negative results from a single sample.

4 COMMERCIAL CONSIDERATIONS

Although over 2,400 plant species have been reported[13] to possess pest control properties, only a few (Table 1) are sold commercially as botanical insecticides[8].

Table 1. Current commercial botanical insecticides.

Plant species	Insecticidal constituents
Tanacetum cinerariaefolium	Pyrethrins
Derris, Tephrosia and *Lonchocarpus spp.*	Rotenoids
Ryania speciosa	Ryanodine alkaloids
Schoenocaulon officinale	Veratrum alkaloids
Azadirachta indica	Limonoids

Reasons which account for this small number include :

1. narrow spectrum of activity;
2. photolability;
3. mammalian toxicity;
4. cost of registration which is as high as for synthetic compounds;
5. problems with supply of plant material on large scale;
6. difficulty of patent protection especially if ethnobotanical knowledge is involved;
7. optimised activity of synthetic analogues is often higher;
8. production cost of synthetic analogues is often lower than that of natural products.

Plants have provided lead structures for the development of carbamates and pyrethroids, two of the three main classes of established insecticides. It is interesting to note that whilst in the latter case[14], the plant (*Tanacetum cinerariaefolium*) containing the lead compound, pyrethrin I, was known through folklore, the discovery of physostigmine[15] and hence the carbamates, was based on the search for compounds which exhibited similar symptoms of poisoning to those observed with organophosphorus insecticides. In contrast, the structural resemblance of imidacloprid[16] (a recently introduced commercial insecticide) to nicotine appears to have been recognised after the development process (Figure 1).

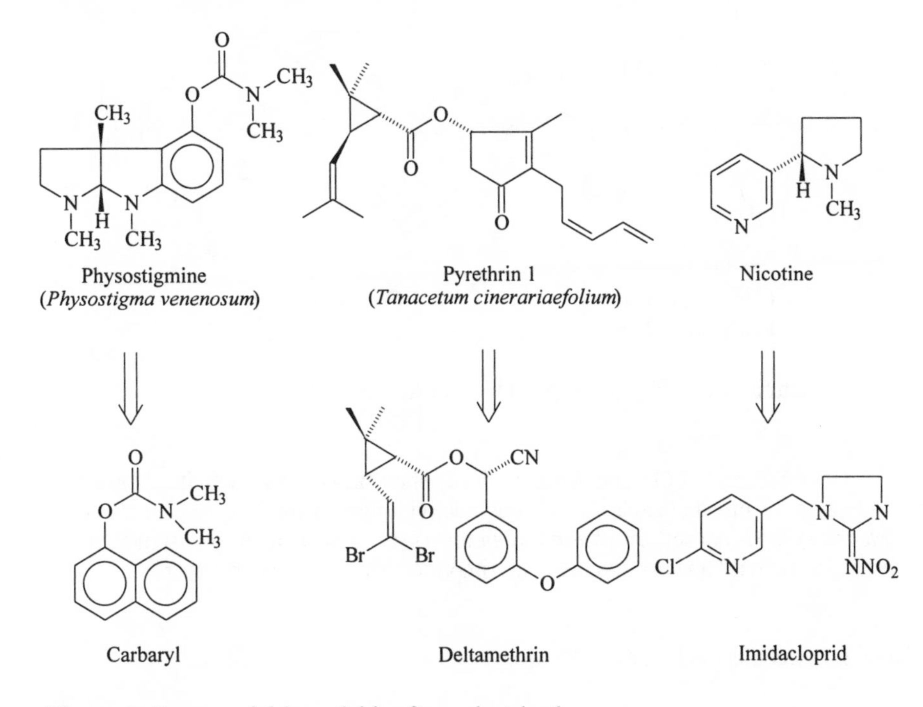

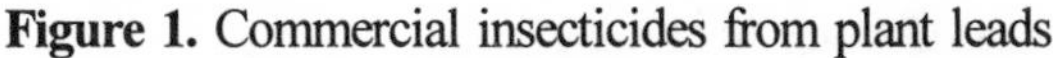
Figure 1. Commercial insecticides from plant leads

5 CASE STUDIES

The following two case studies are selected from in-house work carried out at Rothamsted in collaboration with the Jodrell Laboratory at Kew. Case study I is based on the random screening of plants supplied by Professor R. Thomas of Biotics Ltd. Case study II is based on Chilean plants selected and collected by Professor A. H. Niemeyer of University of Chile and initially screened for activity by Dr. M. Mead-Briggs of the Agrochemical Evaluation Unit, University of Southampton.

5.1 Case study I

The insecticidal components, Kunzein 0 (**1**) and Kunzein 1 (**2**) (Figure 2) were isolated from the hexane extract of *Kunzea sinclairii* (Myrtaceae) using an extensive chromatographic procedure. Attempts to reduce tailing effects during HPLC by the addition of 1% trifluoroacetic acid to the mobile phase resulted in cyclisation to form the non-insecticidal compounds (**3**) and (**4**). Compounds (**1**) and (**2**) were subsequently isolated from other related plant species.

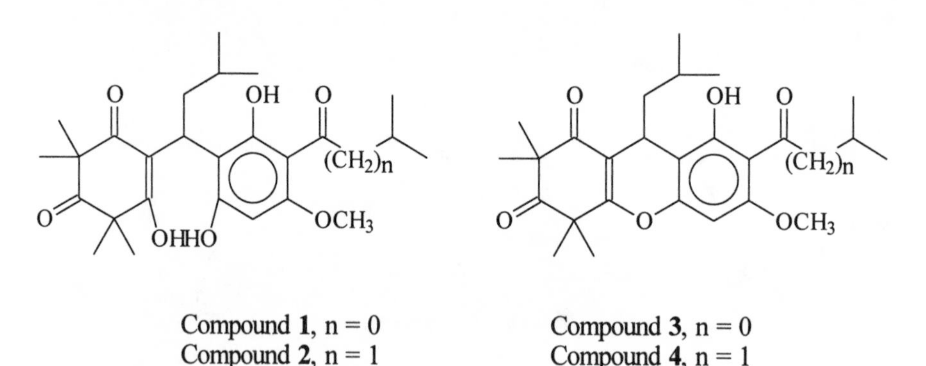

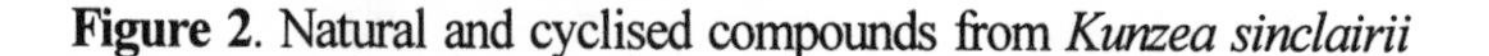

Figure 2. Natural and cyclised compounds from *Kunzea sinclairii*

Both Kunzein 0 (**1**) and Kunzein 1 (**2**) show insecticidal activities towards a narrow range of insects (Table 2). Although activity against mustard beetles (*Phaedon cochleariae*) and houseflies (*Musca domestica*) is greater than for some natural insecticides, activity against commercially important insects is much lower.

Table 2. Insecticidal activities of Kunzeins

	Insecticidal activity*				
Insect	Kunzein 0	Kunzein 1	Pyrethrins	Rotenone	Nicotine
Housefly	0.5	2.0	0.01	0.02	6.0
Mustard beetle	2.2	3.5	0.3	0.6	19.0

* LD50 in µg per insect

Over 150 related plant species, identified by taxonomic considerations, were tested and over 30% found to be active. To date, we have isolated (Figure 3) an additional five novel insecticidal compounds (**5-9**) from plants of related genera using bioassay guided fractionation. Of the seven *Kunzea* species studied, three contained a higher percentage (up to x10) of the Kunzeins (**1** and **2**) enabling purification to be simplified. The ratio of compounds **1** and **2** varied both among and within species.

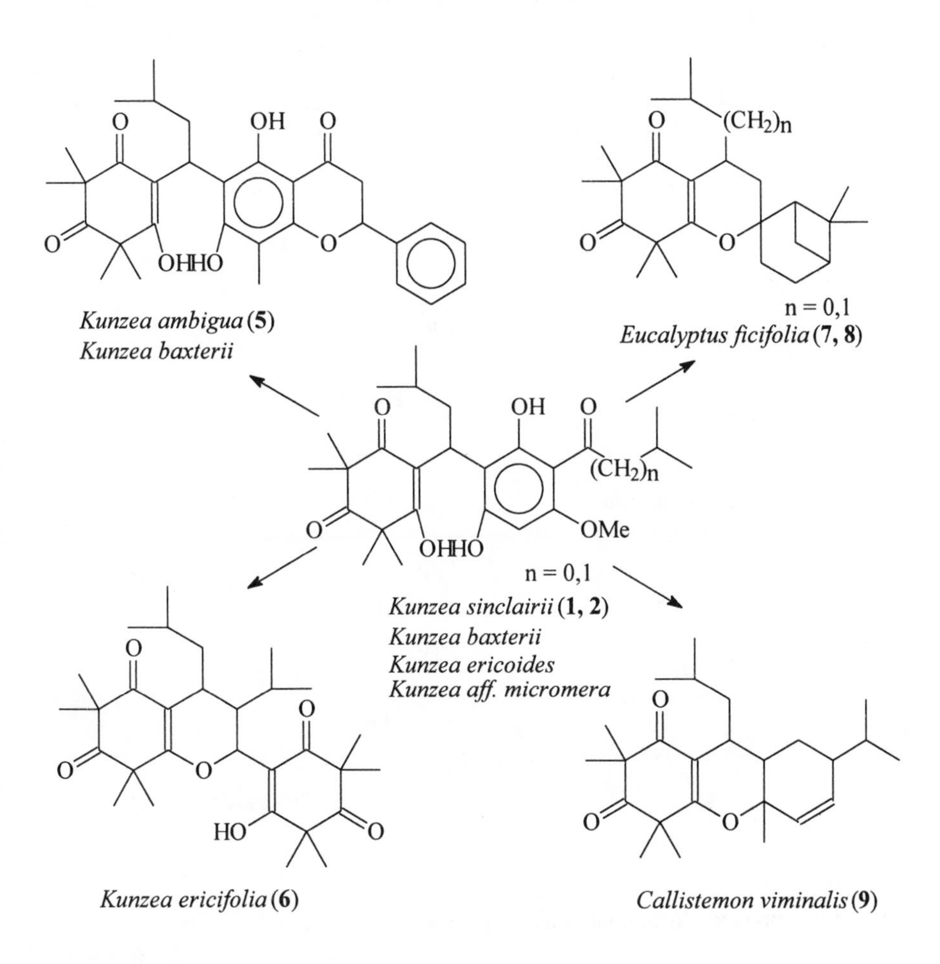

Figure 3. Compounds isolated from related plant species

All compounds isolated possess a common syncarpic acid moiety. Interestingly, the cyclised compounds (**3**) and (**4**) were non-insecticidal, whereas the compounds isolated from *Eucalyptus ficifolia* (**7**) and *Callistemon viminalis* (**8**), also with cyclic structures, have good insecticidal activity. A literature search[17] on related compounds revealed examples in which the syncarpic acid moiety was replaced by a phloroglucinol moiety (Figure 4).

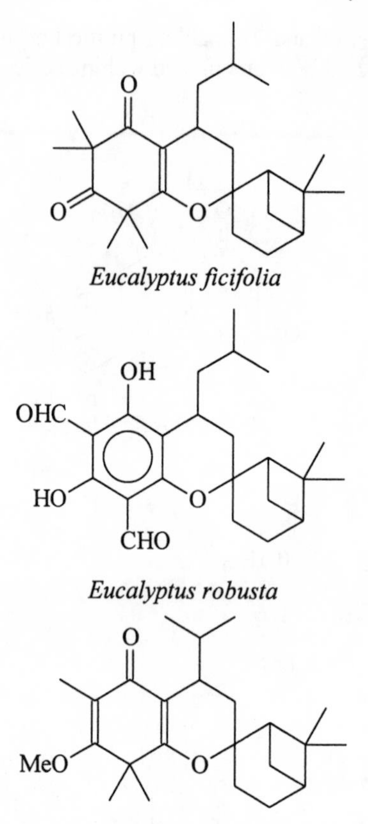

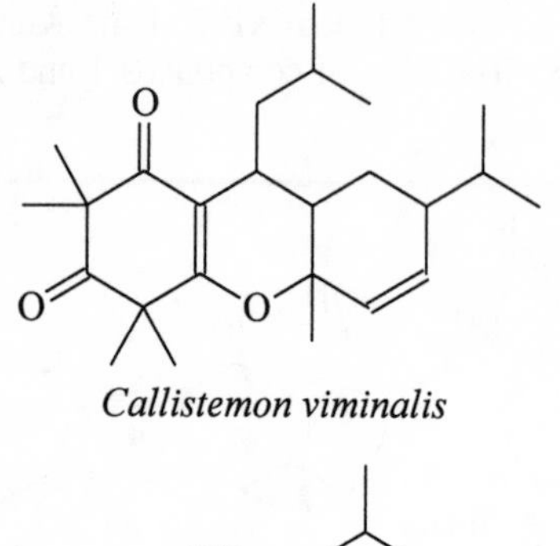

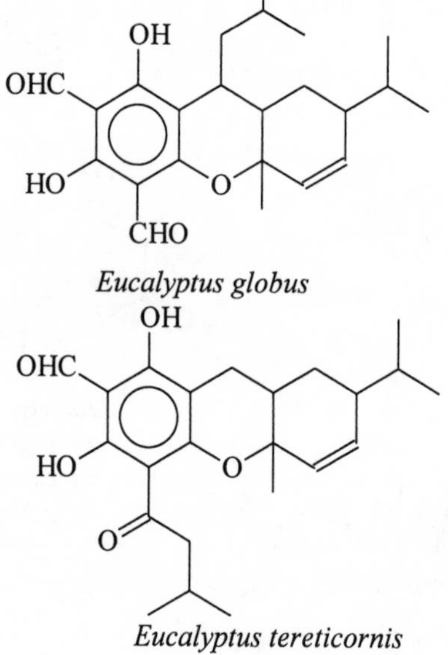

Figure 4. Compounds from related plant species

A range of compounds based on the kunzeins were synthesised (Figure 5) to examine SARs. These included variations of the central isobutyl group and the syncarpic acid moieties. Synthesis of analogues with variations in the phloroglucinol were limited by low reactivity of the intermediates. None of the synthetic compounds had activities superior to those of the natural products.

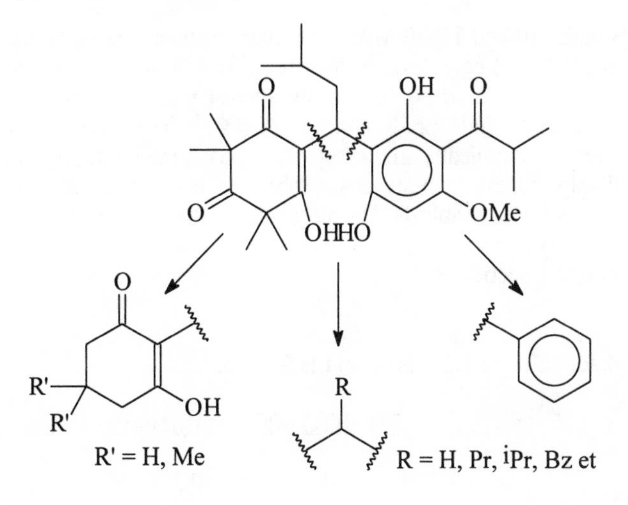

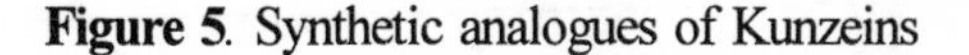

Figure 5. Synthetic analogues of Kunzeins

In summary, this study illustrates the value of taxonomy-based approaches for the provision of a broad range of structurally diverse compounds from plants to complement analogues from synthetic work. Some SARs are clearly discernible from the bioassay results. Although more active compounds with an increased spectrum of activity compared to Kunzein 0 and Kunzein 1 have been identified, none is, as yet, of commercial interest. Further work is in progress.

5.2 Case study II

Two naphthoquinones, BTG 504 (**10**) and 505 (**11**) (Figure 6), were isolated by silica gel column chromatography from the hexane extract of *Calceolaria andina* (Scrophulariaceae).

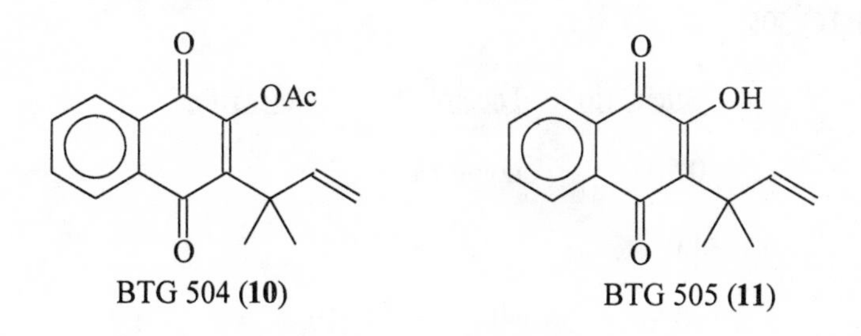

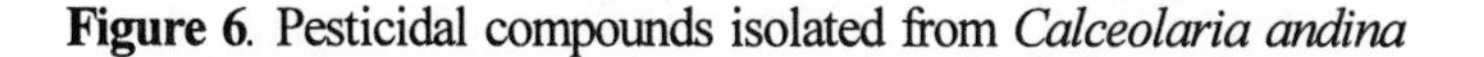

Figure 6. Pesticidal compounds isolated from *Calceolaria andina*

Both compounds (**10** and **11**) showed activities[18] against houseflies and mustard beetles (*Phaedon cochleariae*) equivalent to the kunzeins described in Case study I. In addition, these compounds showed potent activity against whiteflies (*Bemisia tabaci*), mites (*Tetranychus urticae*) and aphids (*Myzus persicae*). Their activity in contact bioassays against these economically important pests was greater than that of the two commercially available botanical pesticides (Table 3) and comparable to several established synthetic pesticides (Table 4). The most significant observation was that these compounds were equally effective against corresponding field strains that showed resistance to established insecticides.

Table 3. Pesticidal activities of BTG 504 and BTG 505.

Insect	BTG 504	BTG 505	Pyrethrins	Rotenone
Housefly*	3.0	3.4	0.3	0.6
Mustard beetle*	3.0	0.7	0.01	0.02
Diamondback moth*	NA	NA	c10	>10
Whitefly**	5.0	7.0	40	NA
Aphid**	250	60	-	-
Mite**	30	80	1400	NA

* LD50 in μg per insect
** LC50 in ppm

Table 4. Pesticidal activities of reference insecticidal agents for comparison with BTG 504 and BTG 505.

Insect	Bifenthrin	Dicofol	Chlorpyrifos
Whitefly*	0.5	non-toxic	3.0
Aphid*	1.0	non-toxic	c5
Mite*	50	60	10

* LC50 in ppm

However, in residual bioassays, involving pretreatmeant of leaves with formulated materials (10% active ingredient in a solution containing 2.5% each of the surfactants Atlox 4851 and Atlox 3400B (ICI Surfactants) in Solvesso (Exxon Chemical) and diluted with water to the required level), the activity of both BTG 504 and BTG 505 was significantly less than that expected in comparative tests with a range of established pesticides. Loss of activity (up to x8) was greater in the case of BTG 505 than BTG 504 (up to x6) against whiteflies and aphids whilst activity against mites was affected to a much smaller extent.

In the case of BTG 505, one possible explanation is that its concentration decreases through ionisation on the alkaline leaf surface (especially cotton and cabbage). This ionisation can be monitored visually as colour changes from pale yellow to purplish red of the ionised form. Penetration of the ionised form into the leaf itself is evident from reddening of the veins in the leaves. The loss of activity was also apparent when residual tests were performed on whole plants, as in commercial screens. The relative levels of activities also varied with the type of formulant used. The observation that activity against mites was less affected may be due differences in feeding habits (rasping vs sucking in the case of aphids). Work is in progress to further investigate the factor involved, especially to account for the decreased activity of BTG 504 on leaf surfaces, bearing in mind that it is does not contain an easily ionisable acidic proton as in the case of BTG 505.

Screening of related plant species is also underway to identify the most economical way of producing the natural products as botanical insecticides and also for the identification of other analogues which may contribute to SAR studies.

In summary, this study has highlighted the importance of carefully defining the range of pest species used in bioassay screens and the methods used for applying the test compound (or extract) to them. It has demonstrated the value of contact assays in identifying potent pesticides from plants which may not have been discovered if only residual, high throughput assays had been used.

Conclusions:

In this paper we have attempted to demonstrate that the single most important factor in identifying pesticidal compounds from plants is the quality of bioassay screens and have highlighted some of the considerations involved in designing them. From the studies reported here, we have concluded that contact assays are simpler and more likely to succeed than residual assays. By applying the test compound (or extract) directly to the pest, the number of variables (e.g. influence of plant surface and nature of formulations used in residual tests) is kept to the minimum thereby increasing the confidence in the bioassay results obtained.

Finally, our belief that plants continue to be an attractive source of new pesticidal compounds has been reinforced. It is also clear that given the rapidly depleting world reserve of plant species, it is essential that all investigations on them be carried out as thoroughly as possible.

Acknowledgements:

The work reviewed was supported (in part) by the United Kingdom Ministry of Agriculture, Fisheries and Food and the British Technology Group Limited. IACR receives grant-aided support from the Biotechnology and Biological Sciences Research Council of the United Kingdom.

References:

1. E-C. Oerke, H-W. Dehne, F. Schonbeck and A. Weber, 'Crop production and crop protection', Elsevier, New York, 1994.

2. J. Stetter, *Reg. Tox. Pharm.*, 1993, **17**, 346-370.

3. K. Naumann, *Nachr. Chem. Tech. Lab.*, 1994, **42**, 255-262.

4. J. B. Pillmoor, K. Wright and A. S. Terry, *Pestic. Sci.*, 1993, **39**, 131-140.

5. B. P. S. Khambay and N. O'Connor, 'Phytochemistry and agriculture', Clarendon Press, U.K., 1993.

6. J. P. Benner, *Pestic. Sci.*, 1993, **39**, 95-102.

7. G. T. Prance, D. J. Chadwick and J. March, 'Ethnobotany and the search for new drugs', Ciba Foundation Symposium 154, Wiley and Sons, New York, 1994.

8. M. B. Isman, *Rev. Pestic. Toxicol.*, 1995, **3**, 1 - 20.

9. R. Alford, J. A. Cullen, R. H. Storch and M. D. Bentley, *J. Econ. Ent.*, 1987, **80**, 575 - 578.

10. W. M. Blaney and M. S. J. Simmonds, 'Proc. 8th Insect-plant relationships', Kluwer Acad. Publ., Germany, 1992, p. 159-161.

11. M. S. J. Simmonds and W. M. Blaney, 'Advances in labiate-derived compounds on insect behaviour', 1992, Royal Botanic Gardens, Kew, pp. 375-392.

12. R. J. Grayer, G. C. Kite, F. J. Goldstone, S. E. Bryan, A. Paton and E. Putievsky, *Phytochemistry*, in-press.

13. M. Grainge and S. Ahmed, 'Handbook of plants with pest-control properties', Wiley and Sons, New York, 1988.

14. M. Elliott, *Pestic. Sci.*, 1989, **27**, 337-51.

15. J. A. Klocke, 'Allelochemicals: Role in agriculture and forestry' American Chemical Society, Symposium series 330, Washington DC, 1987, 396 - 415.

16. A. Elbert, H. Overbeck, K. Iwaya and S. Tsuboi, 'Proc. Brit. Crop Prot. Conf.- Pests and Diseases', 1990, **1**, 21.

17. E. L. Ghisalberti, *Phytochemistry*, 1996, **41**, 7-22.

18. B. P. S. Khambay, D. Batty and A. H. Niemeyer, 'Pesticidal naphthoquinone derivatives', 1995. UK Patent Application 2289463A.

Oncogene-Modulated Signal Transduction Inhibitors from Plants

C.-j. Chang[1], C. L. Ashendel[1], R. L. Geahlen[1], M.-c. Hung[3], J. L. McLaughlin[1] and D. L. Waters[2]

[1] DEPARTMENT OF MEDICINAL CHEMISTRY AND PHARMACOGNOSY AND
[2] DEPARTMENT OF VETERINARY CLINICAL SCIENCES, PURDUE UNIVERSITY, WEST LAFAYETTE, IN 47907, USA
[3] DEPARTMENT OF TUMOR BIOLOGY, THE UNIVERSITY OF TEXAS MD ANDERSON CANCER CENTER, HOUSTON, TX 77030, USA

1 INTRODUCTION

Plants have a long history of providing antitumor agents with novel structures and unique mechanisms[1-6]. Many of these plant-derived antitumor agents have also served as new leads for total synthesis and for further development of active analogs of clinical importance. These compounds have also provided useful molecular probes for elucidating novel mechanisms for tumor growth and control. Several Vinca alkaloids (vinblastine, vincristine, videsine and vinorelbine), podophyllotoxins (etoposide and teniposide), taxol, taxotere, 10-hydroxycamptothecin, irenotecan, homoharringtonine, indirubin and oridonin have currently been used in the treatment of various cancers. There are other antitumor compounds which are either in clinical or preclinical trial stages. The discovery of these natural anticancer drugs is mainly derived from the early antitumor screening system based on the growth inhibition or cytotoxicity against murine leukemia. Consequently, almost all of them are highly toxic and often display marginal therapeutic indexes. Therefore, alternative approaches must be envisioned on the basis of differential mechanisms of cancer cell biology.

2 ONCOGENE-MODULATED SIGNAL TRANSDUCTION

In recent years, the most significant breakthrough in our understanding of oncogenesis is the discovery of oncogenes[7-9]. Oncogenes are genes that can induce neoplastic transformation of normal cells. Most oncogenes were initially identified in acute transforming retroviruses. Their counterparts in normal cells, called proto-oncogenes, often encode various mitogenic products to control the signaling processes in the regulations of cellular proliferation and differentiation[10,11]. Structural or functional alterations of proto-oncogenes through mutation, deletion, amplification or rearrangement can activate the oncogenic signal transduction cascades, which eventually result in malignancy. Therefore, signal transduction pathway modulated by oncogenes becomes an attractive target for the mechanism-based discovery of anticancer drugs from plants to control the growth and differentiation of cancer cells[12-15].

The signal transduction pathways of cancer cells are highly intricate. In-depth understanding of these pathways at the atomic level is still very scanty. One of the better understood processes is based on the intracellular phosphorylation of the tyrosine or serine/threonine unit of key signalling proteins. We have therefore designed four kinase mechanism-based bioassay systems to inhibit (a) protein-tyrosine kinase, (b) protein kinase C, (c) p34*cdc-2* cell-cycle dependent kinase, and (d) *raf*-1 kinase for screening numerous extracts of diversified plants.

2.1 Protein-Tyrosine Kinase (PTK)[16-20]

The search for inhibitors of protein-tyrosine kinases requires the selection of target enzymes. For protein-tyrosine kinases, we concentrate our efforts on the identification of inhibitors for Src-family kinases because of their widespread involvement in signal transduction pathways[21,22]. The nine members of this family (c-Src, Lck, Fyn, Lyn, Blk, c-Fgr, Yes, Hck, Yrk) are 53-60 kDa, cytoplasmic enzymes that are structurally related to $pp60^{c\text{-}src}$ (v-Src)[21]. The receptor-mediated activation of a member of the Src-family is an important and obligatory step in the signal transduction pathways activated by most mitogens, ranging from growth factors to antigens, that have been described to date[21,22]. The binding of polypeptide mitogens such as PDGF (platelet-derived growth factor) to their respective receptors stimulates receptor autophosphorylation, which leads to the recruitment of Src-family kinases to the receptor where they bind and become activated, a process necessary for mitogenesis[23]. Activated Src-family kinases are found in tumor cells in which receptor tyrosine kinase genes are amplified[24]. Members of the Src-family of kinases also lie in signaling pathways immediately downstream from many receptors that lack intrinsic kinase activity. Thus, inhibitors of Src-family kinases may demonstrate widespread activity against tumors of multiple tissue types. To approach the development of Src-family kinase inhibitors, we will use Lck ($p56^{lck}$) as our initial target.

2.2 Protein Kinase C (PKC)[25-27]

PKC is a serine/threonine protein kinase involved in cellular signaling. It was discovered as a proteolytically activated protein kinase by Nishizuka in 1977, and was later found to be a phospholipid-dependent, calcium-activated protein kinase, whose sensitivity to calcium and phospholipid was enhanced by the intracellular second-messenger diacylglycerol. PKC was found to be activated by tumor promoting phorbol esters in a manner similar to diacylglycerol but with much higher potency[28]. PKC cDNA was found to be a family of genes. There have been described at least 10 members of the PKC family and a few non-PKC gene products with some sequence similarity to PKC that bind phorbol esters[29]. However, the relative roles of these individual gene products in mediating the effects of phorbol esters remains unclear. In spite of this, the working dogma is that most forms of PKC are activated by phorbol esters or diacylglycerol in the presence of membrane phospholipid and divalent cations. The classical PKCs (α, β, and γ) require calcium, while a second class of PKCs, typified by δ, ε, and eta, do not require calcium. Some PKCs (e.g., zeta) require only phospholipid.

PKC is not an oncogene product, but overexpression of PKC does stimulate signalling, and increases the sensitivity of some cells to the transforming effects of oncogenic p21Ras[30,31] and cooperates strongly with overexpression of normal c-raf-1 for transformation of rodent fibroblasts in culture[32]. PKC activation has been observed in many, but not all, experiments involving stimulation of a variety of cells with growth factors or expression of oncogenes. Because it is now known that growth factor receptors activation of p21Ras via Grb-2 and SOS occurs simultaneously with activation of PIP_2-PLC (and also PI-3 kinase), it is clear that production of diacylglycerol and PKC activation are not essential for activation of raf and the MAP kinase pathway. On the other hand, some effects of oncogenic *ras* require PKC[33], and PKC-mediated signaling may serve as backup pathway for the one involving *ras*. Taken together, these observations justify targeting PKC for anti-cancer drug discovery.

2.3 p34^{cdc-2} Cell-cycle Dependent Kinase

The cell cycle is regulated by a family of protein serine/threonine kinases (cyclin-dependent kinase or Cdk's) in mammalian cells[34-36]. Alterations in the expression of cyclins and endogenous Cdk inhibitors have been reported in a variety of tumors[37-39] including breast tumors, squamous cell carcinomas of the head and neck[37], esophageal cancer[40], and hepatocellular carcinoma[41]. In normal cells, Ckd-cyclin complexes contain two additional proteins: PCNA (proliferating cell nuclear antigen) and p21[42] to control the progression of cell cycle. However, both PCNA and p21 are lost in transformed cells[43]. The reduction or deletion of p21 in tumor cells may largely be attributed to the mutation of tumor suppressor gene product p53. Furthermore, transformed cells with mutant or deleted p53 genes are unable to undergo apoptosis[44]. We therefore propose that inhibition of p34^{cdc-2} kinase catalytic activity may lead to the isolation of natural Cdk inhibitors, which can be developed into a novel class of antitumor agents.

2.4 raf-1 Kinase

The evidence strongly indicates that signal transduction is uniquely altered by certain cellular oncogenes during the transformation of normal cells into malignant cells[7-11]. Two oncoproteins, p21^{ras} and p74^{raf} have an essential role in oncogene-modulated signal transduction. *Raf*-1 was discovered as a virally transduced oncogene (v-*raf*), which is a member of the serine/threonine family of protein kinase genes[8,45]. The activation of wild-type *raf*-1 protein kinase activity is induced by the interaction of the negatively regulating amino-terminal and p21^{ras}[46,47]. The only known endogenous substrate for *raf*-1 kinase is another protein kinase called MAP-kinase/ERK-kinase (MEK)[48,49]. The phosphorylated MEK can then phosphorylate members of MAP kinase family[50]. This "MAP kinase cascade" is the critical signal transduction pathway of *ras* oncogenesis. Therefore, the inhibition of the *raf*-1 kinase and the MAP kinase cascade are novel targets for the discovery of natural products for controlling the growth of tumor cells.

3 SIGNAL TRANSDUCTION INHIBITORS FROM PLANTS

3.1 Acquisition of Plants

The first prerequisite for a natural products drug discovery program is the plant selection and collection[51]. Different approaches have been taken to optimize (a) the probability of isolating active compounds, (b) structural novelty and diversity, and (c) cost-effectiveness. Information-based approach has increasingly been incorporated into the planning of plant acquisition because of the availability of high speed and high capacity of personal computer and the low cost access of large chemical and biological databases. This approach may minimize the probability of reisolation of known natural products. However, there is only limited information about the structural types and phytochemical taxonomy of protein kinase inhibitors. Therefore, other approaches may also be taken to enhance the "hit rate". On the basis of the early plant-derived anticancer drug discovery programs, screenings of traditional/folklore medicinal plants have led to higher probability of identifying active leads[52]. Thus, one of our focuses in plant acquisition is ethnobotanically-based plant collection. On the other hand, we should probably be cautious and not overinterpret the folklore therapeutic activities of medicinal plants since there is no direct pathological or histological correlation with the biochemical mechanism of signal transduction in the traditional folklore medicine. In order to enhance the structural diversity and novelty of active compounds, we may select different plants from a large variety of families and genera or from unscreened or endemic genera. In summary, at the current stage of our knowledge in kinase-based signal transduction inhibitors and plant chemotaxonomy, it would be premature to overemphasize any specific approach for plant acquisition. However, a more systematic, information-based approach may be conceived in the future as we diligently accumulate more information about plant-derived signal transduction inhibitors.

Approximately 50% of the plants collected in India and Taiwan are known Indian and Chinese medicinal plants. Three important classes of signal transduction inhibitors (anthraquinone, stilbene and polythiophene) were derived from two Chinese traditional medicinal plants, *Polygonum cuspidatum*[53] and *Eclipta prostrata*[54] with renown anti-inflammatory activity without toxicity. In this report, a bioassay-directed isolation of protein-tyrosine kinase inhibitors from the roots of *Polygonum cuspidatum* is used to exemplify our approach to the discovery of oncogene-modulated signal transduction inhibitors as potential antitumor agents from medicinal plants.

3.2 Extraction, Fractionation and Screening

The ground roots of *Polygonium cuspidatum* (4.8 kg) were initially extracted with EtOH, and followed by partition between methylene chloride and water. The methylene chloride was further partitioned between 90% methanol and hexane. This protein-tyrosine kinase (PTK, $p56^{lck}$) inhibitory activity-directed fractionation led to 50-fold enrichment of the activity in the methanol fraction (95 g). The methanol fraction was then separated by silica gel chromatography to yield a series of anthraquinones as a new class of PTK inhibitors[55].

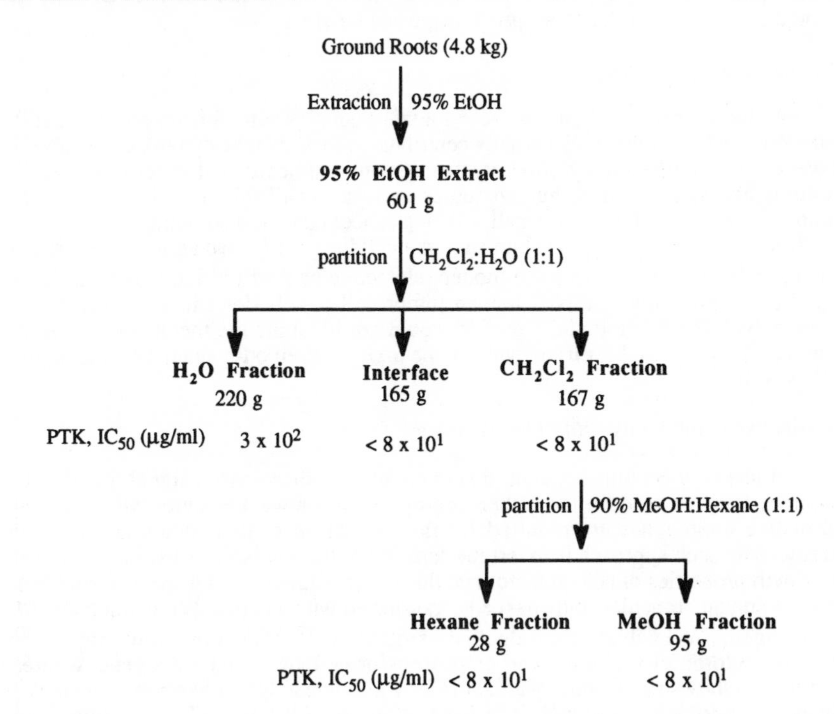

3.3 Inhibition of Protein-Tyrosine Kinase[20]

The strongest anthraquinoid PTK inhibitor isolated from *Polygonum cuspidatum* is emodin (3-methyl-1,6,8-trihydroxyanthraquinone (IC_{50}, 5 ug/ml)[55]. Deletion of the 6-hydroxyl group (aloe-emodin, 3-methyl-1,8-dihydroxyanthraquinone) or replacement of the 6-hydroxyl group with a methoxyl group (physcion, 3-methyl-6-methoxy-1,8-dihydroxy-anthraquinone) completely abolished the inhibitory activity ($IC_{50} > 8 \times 10^2$ ug/ml), indicative of the essential requirement for a free phenolic group.

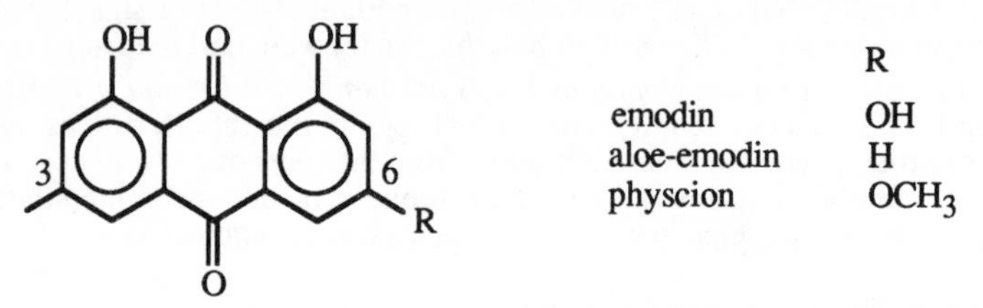

To determine if emodin could also inhibit the activity of Lck in intact cells, we turned to the lymphoma cell line LSTRA, which overexpresses Lck due to the insertion of a retroviral promoter upstream of the *lck* gene[56]. In this experiment, LSTRA cells were pretreated with the phosphotyrosine phosphatase inhibitor, vanadate, and then with emodin. Proteins were separated by SDS-PAGE, transferred to nitrocellulose and probed with anti-phosphotyrosine antibodies to detect tyrosine-phosphorylated proteins. Treatment of cells with emodin resulted in a dose-dependent inhibition of protein-tyrosine phosphorylation. These data indicate that emodin can enter intact cells and inhibit the activity of endogenous protein-tyrosine kinases.

3.4 Antitumor Cytotoxicity

Emodin appears slightly more cytotoxic than aloe-emodin, suggesting that the inhibition of PTK activity may partially contribute to their relative cytotoxicity (Table 1). However, the antitumor cytotoxicity results only indicate that emodin is weakly cytotoxic/cytostatic against human tumor cell panels (Table 1). The differential cytotoxicity against human non-small cell lung cancer panel is also unimpressive. Thus, it is difficult to select appropriate human tumor cell lines for *in vivo* antitumor evaluation although emodin was shown to be moderately active against murine leukemia[57]. We should also notice that the NCI human tumor cell panels don't include any normal human cells. These cytotoxicity profiles could not illustrate the therapeutic potential. Therefore, it is essential for us to evaluate the toxicity of emodin against normal human cells as well.

3.5 Selective cytotoxicity - *Ras* Oncogene

It has now become apparent that many of the kinase-based signal transduction inhibitors such as emodin are either non-cytotoxic or weakly cytotoxic/cytostatic. Alternative approaches are required for the evaluation of their potential antitumor activity. One such approach is to test the capacity of these inhibitors to alter selectively the growth properties of cells transformed due to the expression of a specific oncogene. We used human bronchial epithelial cells, transfected with a plasmid containing the *v-H-ras* oncogene, to evaluate the selective cytotoxicity[58]. Emodin shows significant selective cytotoxicity against the cells transformed by *v-ras* oncogene whereas doxorubicin shows no selectivity at all (Table 2). Interestingly, physcion also displays selective cytotoxicity, although it is inactive in the cell-free PTK inhibition assay. However, we have recently demonstrated that it inhibits tyrosine phosphorylation in LSTRA murine leukemia cells (unpublished results).

Table 1 *Antitumor Cytotoxicity Against Human Tumor Cells*

Cell Line	- log GI_{50}[a] (M)	
	Emodin[b]	*Aloe-emodin*[c]
A. Leukemia		
CCRF-CEM	4.42	4.50
HL-60	4.57	4.61
MOLT-4	4.49	4.39
RPMI-8226	4.61	4.50
SR	4.63	4.48
B. Lung Cancer		
A-549	4.67	4.40
HOP-18	5.40	< 4.00
HOP-62	5.09	< 4.00
HOP-92	4.94	< 4.00
NCI-H226	4.48	< 4.00
NCI-H23	4.84	< 4.00
NCI-H322M	4.45	4.31
NCI-H460	4.70	4.53
NCI-H522	4.84	4.64
C. Melanoma		
LOX	4.77	4.58
MALME-3M	4.93	4.68
M14	4.60	4.42
SK-MEL-2	4.69	4.04
SK-MEL-28	4.44	> 4.00
SK-MEL-5	5.58	4.80

a. GI_{50}: concentration for 50% growth inhibition
b. NSC No.: 408120, tested by the U.S. National Cancer Institute
c. NSC No.: 38628, tested by the U.S. National Cancer Institute

Table 2 *Selected Anti-oncogene Cytotoxicity of Emodin*

Compound	GI_{50} (μg/ml)[a]		*Selective Cytotoxicity Index*[b]
	TBE	*HBE*	
Emodin	4	> 100[c]	> 25
Physcion	< 2	> 10[c]	> 5
Doxorubicin	6	5	1
Retinoic acid	> 100	> 100	1

a. TBE: *ras*-transformed human bronchial epithelial cells.
HBE: normal human bronchial epithelial cells.
b. Selective cytotoxicity index = GI_{50} (HBE)/IC_{50} (TBE)
c. Dosage is limited by water solubility.

To explore the mechanism of action of emodin in these cells, we probed TBE and HBE cell lysates with anti-phosphotyrosine antibodies to estimate the relative levels of tyrosine-phosphorylated proteins present in each cell type. We found that TBE cells exhibited elevated levels of phosphotyrosine-containing proteins relative to those found in HBE cells, even though the transforming principal (activated *ras*) lacks protein-tyrosine kinase activity[58]. Treatment of intact TBE cells with emodin resulted in a marked decrease in the concentration of cellular protein-tyrosine phosphorylation[58].

3.6 Inhibition of Protein-Tyrosine Kinase and Growth Inhibition

We have demonstrated that the anti-*ras* and anti-*src(lck)* growth inhibitory effects of emodin are qualitatively related to its inhibition of protein-tyrosine kinase in TBE (*ras* activation) and LSTRA (*lck* overexpression) transformed cells. In order to evaluate semiquantitively if emodin could specifically inhibit a proto-oncogene encoded protein-tyrosine kinase in intact cells and its relationship with growth inhibition, we utilized a series of human breast cancer cell lines with differential degree of expression of *Her-2/neu* (c-*erb* B-2). *Her-2/neu* encodes a $p185^{neu}$ transmembrane tyrosine kinase growth factor receptor[59], which is related to epidermal growth factor receptor. Treatment with emodin inhibits hyperphosphorylation of $p185^{neu}$ in the *Her-2/neu* overexpressing human breast cancer cells, MDA-MB453, AU-565 and BT-483[60]. Furthermore, emodin demonstrated selective cytotoxicity against these three cell lines in reference to two human breast cancer cell lines (MCF-7 and MDA-MB231) and one normal human breast cell line (HBL-100) that express a base level of $p185^{neu}$[60]. These results strongly suggested that the selective cytotoxicity of emodin against cancer cells semiquantitively correlates with the overexpression of *Her-2/neu* oncogene.

In conclusion, systematic screenings of medicinal plant extracts using protein kinase assay systems can resuls in the discovery of unique oncogene-modulated signal transduction inhibitors. A strong protein-tyrosine kinase inhibitor, emodin, has displayed remarkably selective cytotoxicity against the *ras*-transformed and *Her-2/neu* overexpressing human cancer cells. The growth inhibitory activity may be attributed to the selective suppression of oncogene-modulated signal transduction pathways through the inhibition of protein kinase activity. Further *in vivo* antitumor efficacy tests are in progress. It is evident that oncogene-modulated signaling processes offer attractive targets for the discovery of novel mechanism-based antitumor agents from plants.

Acknowledgment: We gratefully acknowledge financial support from the National Cancer Institute (U01 CA50743) through the national cooperative natural products drug discovery groups.

References

1. M. Suffness, G.M. Cragg, M.R. Grever, F.J. Grifo, G. Johnson, J.A.R. Mead, S.A. Schepartz, J.M. Venditti and M. Wolpert, *Int. J. Pharmacog.*, 1995, **335**, 5.
2. F.A. Valeriote, T.H. Corbett and L.H. Baker, 'Anticancer Drug Discovery and Development: Natural Products and Molecular Models,' Kluwer Academic, Boston, 1994.
3. R.C. Donehower and E.K. Rowinsky, 'Cancer: Principles and Practice of Oncology,' V.T. DeVita, Jr., S. Hellman and S.A. Rosenberg, Eds., Lippincott, Philadelphia, 1993, Vol. 1, Chapter 18, p. 409.

4. J.L. McLaughlin, C.-j. Chang and D.L. Smith, *Am. Chem. Soc. Symp.*, 1993, **534**, 112.
5. J.M. Cassady, W.M. Baird and C.-j. Chang, *J. Nat. Prod.*, 1990, **53**, 23.
6. M. Suffness, D.J. Newman and K. Snader, *Bioorg. Marine Chem.*, 1989, **3**, 131.
7. R.A. Weinberg, 'Oncogenes and the Molecular Origins of Cancer,' Cold Spring Harbor Laboratory, Cold Spring Harbor, 1989.
8. R. Hesketh, 'The Oncogene Handbook,' Academic Press, New York, 1993.
9. E. Pimentel, 'Oncogenes,' 2nd Edition, CRC Press, Boca Raton, 1989.
10. J. Kurjan and B.L. Taylor, 'Signal Transduction,' Academic Press, New York, 1993.
11. L.C. Cantley, K.R. Auger, C. Carpenter, B. Buckworth, A. Graziani, R. Kapeller, S. Soltoff, *Cell*, 1991, **64**, 281.
12. D.J. Kerr and P. Workman, 'Molecular Targets for Cancer Chemotherapy,' CRC Press, Boca Raton, 1994.
13. G. Powis, R.T. Abraham, C.L. Ashendel, L.H. Zalkow, G.B. Brindey, C.J. Vlahos, R. Merriman and R. Bonjouklian, *Int. J. Pharmacog.*, 1995, **335**, 17.
14. C.-j. Chang, C.L. Ashendel, R.L. Geahlen and J.L. McLaughlin, "Anticancer Drug Discovery and Development: Natural Products and New Molecular Models,' F.A. Valeriote, T.H. Corbett and L.H. Baker, Eds., Kluwer Academic, Boston, 1994, Chapter 2, p. 27.
15. P. Workman, *Semin. Cancer Biol.*, 1992, **3**, 329.
16. S. Kellie, 'Tyrosine Kinase and Neoplastic Transformation,' R.G. Landes, Austin, 1994.
17. G. Hardie and S. Hanks, 'The Protein Kinase Facts Book-II: Protein-Tyrosine Kinase,' Academic Press, New York, 1995.
18. E.M. Dobrusin and D.W. Fry, *Ann. Repts. Med. Chem.*, 1993, **27**, 169.
19. W.J. Fantle, D.E. Johnson and L.E. Williams, *Annu. Rev. Biochem.*, 1993, **62**, 453.
20. C.-j. Chang and R.L. Geahlen, *J. Nat. Prod.*, 1992, **55**, 1529.
21. J.A. Cooper, 'Peptides and Protein Phosphorylation,' B.E. Kemp, Ed., CRC Press, Boca Raton, 1990, p. 85.
22. J.B. Bolen, P.A. Thompson, E. Eiseman and I.D. Horak, *Adv. Cancer Res.*, 1991, **57**, 103.
23. G.M. Twamley-Stein, R. Pepperkok, W. Ansorge and S.A. Courtneidge, *Proc. Natl. Acad. Sci. U.S.A.*, 1993, **90**, 7696.
24. S.K. Muthuswamy, P.M. Siegel, D.L. Dankort, M.A. Webster, W.A. Muller and W.J. Muller, *Mol. Cell. Biol.*, 1994, **14**, 735.
25. G. Hardie and S. Hanks, 'The Protein Kinase Facts Book-I: Protein Serine Kinase,' Academic Press, New York, 1995.
26. A. Basu, *Pharmac. Ther.*, 1993, **59**, 257.
27. A. Gescher, *Br. J. Cancer*, 1992, **66**, 10.
28. M. Castagna, T. Takai, K. Kaibuchi, K. Sano, U. Kikkawa and Y. Nishizuka, *J. Biol. Chem.*, 1982, **257**, 7847.
29. Y. Nishizuka, *Science*, 1992, **258**, 607.
30. W.-L. Hsiao, G.M. Housey, M.D. Johnson and I.B. Weinstein, *Mol. Cell. Biol.*, 1989, **9**, 2641.
31. R.S. Krauss, G. M. Housey, M.D. Johnson and I.B. Weinstein, *Oncogene*, 1989, **4**, 991.
32. W. Kolch, G. Heidecker, G. Kochs, R. Hummel, H. Vahidi, H. Mischak, G. Finkenzeller, D. Marme and U.R. Rapp, *Nature*, 1993, **364**, 249.
33. J.D.H. Morris, B. Price, A.C. Lloyd, A.J. Self, C.J. Marshall and A. Hall, *Oncogene*, 1988, **4**, 27.
34. C. Norbury and P. Nurse, *Annu. Rev. Biochem.*, 1992, **61**, 441.
35. L.H. Hartwell and T.A. Weinert, *Science*, 1989, **246**, 629.
36. T. Hunter, *Cell*, 1993, **75**, 839.

37. E. Schurring, E. Verhoeven, W.J. Mooi and R.J.A.M. Michalides, *Oncogene*, 1992, **7**, 335.
38. M.G. Buckley, K.J.E. Sweeney, J.A. Hamilton, R.L. Sini, D.L. Manning, R.I. Nicholson, A. DeFazio, C.K. W. Watts, E.A. Musgrove and R.L. Sutherland, *Oncogene*, 1993, **8**, 2127.
39. K. Keyomarsi and A.B. Pardee, *Proc. Nat. Acad. Sci. USA*, 1993, **90**, 1112.
40. W. Jiang, S.M. Kahn, N. Tomita, Y.-J. Zhang, S.H. Lu and I.B. Weinstein, *Cancer Res.*, 1992, **52**, 2980.
41. Y.-J. Zhang, W. Jiang, C.J. Chen, C.S. Lee, S.M. Kahn, R.M. Santella and I.B. Weinstein, *Biochem. Biophys. Res. Commun.*, 1993, **196**, 1010.
42. Y. Xiong, H. Zhang and D. Beach, *Cell*, 1992, **71**, 505.
43. Y. Xiong, H. Zhang and D. Beach, *Gene Dev.*, 1993, **7**, 1572.
44. D.E. Fisher, *Cell*, 1994, **78**, 538.
45. U.R. Rapp, G. Heidecker, J.L. Huleihel, J.L. Cleveland, W.C. Choi, T. Pawson, J.N. Ihle and W.B. Anderson, *Cold Spring Harbor Sym. Quant. Biol.*, 1988, **53**, 173.
46. S.G. MacDonald, C.M. Crews, L. Wu, J. Drilles, R. Clark, R. Erikson and F. McCormick, *Mol. Cell. Biol.*, 1993, **13**, 6615.
47. N.G. Williams, T.M. Roberts and P. Li, *Proc. Nat. Acad. Sci. USA*, 1992, **89**, 2922.
48. J.M. Kyriakis, H. App, X.-f. Zhang, P. Banerjee, D.L. Brautigan, U.R. Rapp and J. Avruch, *Nature*, 1992, **358**, 417.
49. C.-F. Zheng and K.-L. Guan, *EMBO J.*, 1994, **13**, 1123.
50. C.M. Crews, A.A. Alessandrini and R.L. Erikson, *Cell Growth and Differentiation*, 1992, **3**, 135.
51. G.A. Cordell, *Phytochemistry*, 1995, **40**, 1585.
52. J. Douros and M. Suffness, *J. Nat. Prod.*, 1982, **45**, 1.
53. S.-j. Lee, 'Peng Tsao Kong Mu,' Ming Dynasty, Vol. 26, 136.
54. S.-j. Lee, 'Peng Tsao Kong Mu,' Ming Dynasty, Vol. 16, 121.
55. H. Jayasuriya, N. Koonchanok, R.L. Geahlen, J.L. McLaughlin and C.-j. Chang, *J. Nat. Prod.*, 1992, **55**, 696.
56. J.D. Marth, R. Peet, E.G. Krebs and R.M. Perlmutter, *Cell*, 1985, **43**, 393.
57. S.M. Kupchan and A. Karim, *Lloydia*, 1976, **39**, 223.
58. T.C.K. Chan, C.-j. Chang, N.M. Koonchanok and R.L. Geahlen, *Biochem. Biophys. Res. Commun.*, 1993, **193**, 1152.
59. A.L. Schechter, D.F. Stern, L. Vridyanathan, S.J. Decker, J.A. Drebin, M.I. Greene and R.A. Weinberg, *Nature*, 1984, **312**, 513.
60. L. Zhang, C.-j. Chang, S.S. Bacus and M.-c. Hung, *Cancer Res.*, 1995, **55**, 3890.

Harnessing Phytochemical Diversity for Drug Discovery: The Phytera Approach

A. M. Stafford[1] and C. J. Pazoles[2]

[1] PHYTERA LIMITED, REGENT COURT, REGENT STREET, SHEFFIELD S1 4DA, UK
[2] PHYTERA INC, 377 PLANTATION STREET, WORCESTER, MA 01605, USA

1 INTRODUCTION

1.1 Biosynthetic potential of plant cell cultures

In most societies plants play a central role both in traditional (eg. ethnomedical) and in "modern" systems of medicine. The unique properties of certain plant products such as morphine, digoxin and vinblastine has ensured their enduring therapeutic role, and it was the continuing importance of such drugs and limitations to their production from whole plants which helped drive the development of alternative production strategies, including large-scale plant cell culture technology, particularly during the 1980's. More recently, the NCI and Bristol-Myers Squibb have promoted the development of *Taxus* plant cell culture systems as a potential route to commercial-scale taxol supplies.

In the course of the 1980's, substantial advances were made in plant cell culture technology, notably in two areas; in large-scale "fermentation" technology, and in the understanding of some of the factors influencing levels of known secondary metabolites in plant culture systems. Pioneering work by Zenk[1] was followed by a plethora of publications describing attempts to increase metabolic yields of specific products in diverse species, and more limited descriptions of large-scale processes. Despite substantial efforts, by the end of the decade few systems had achieved even a near-commercial stage (shikonin production from *Lithospermum erythrorhizon* being one notable exception[2]).

However, although few cell culture systems have been shown to be economically viable, most important plant-derived drugs can be produced in culture. A literature analysis of the list of Farnsworth[3] comprising c.125 phytochemicals (derived from only 90 species) used as drugs in various parts of the world showed that where plant cell culture had been attempted and some analysis performed, the target drug was detected in over 90% of cases.[4] Those compounds detected in culture exhibit high structural diversity (Table 1). Yields are usually lower than those typical of the whole plant, although occasionally cultures have been found to accumulate higher secondary metabolite levels than the plant as with rosmarinic acid, sanguinarine, and acridones from *Coleus* sp., *Papaver somniferum* and *Ruta* sp. respectively.[5]

Table 1 *Examples of Known Plant-Derived Drugs Produced in Plant Cell Suspension Cultures*

Compound Name	*Chemical Class*	*Plant Species Cultured*
ajmalicine	indole alkaloid	*Catharanthus roseus*
anabasine	pyridine alkaloid	*Nicotiana tabacum*
artemisinin	sesquiterpene lactone	*Artemisia annua*
berberine	isoquinoline alkaloid	*Coptis japonica*
colchicine	tropolone alkaloid	*Colchicum autumnale*
diosgenin	steroid saponin	*Dioscorea deltoidea*
emetine	alkaloid	*Cephaelis ipecacuanha*
podophyllotoxin	lignan	*Podophyllum hexandrum*
sennosides A & B	anthraquinones	*Rheum palmatum*
taxol	diterpene	*Taxus brevifolia*
valepotriates	iridoid	*Valeriana officinalis*
yohimbine	indole alkaloid	*Catharanthus roseus*

1.2 Plant Cell Cultures Produce Novel Structures

The phytochemical diversity generated by plant cell cultures is also illustrated by their production of novel structures, mostly discovered during the past 10 years by chemists performing systematic analyses of known drug-producing species. The 85 novel structures reported by Ruyter & Stockigt[6] in 1989 comprise mainly alkaloids, terpenes, quinones and phenylpropanoids; while most of these are modifications of previously known structures, entirely new chemical skeletons are also described. The biological activity of those structures has rarely been tested; exceptions are the anti-inflammatory compound dehydrodiconiferyl alcohol β-D-glucoside reported by Arens *et al*[7] and discovered during a screening programme of plant cell cultures by Nattermann,[8] and more recently 4 new taxoids promoting tubulin assembly from cell cultures of *Taxus baccata*.[9] It should be noted that in the majority of cases, the culture systems yielding this novel chemistry have not been optimised or engineered for production of secondary metabolites. The potential for novel structure discovery in cultures is therefore likely to be far higher than that observed to date.

2 THE PHYTERA APPROACH TO DRUG/LEAD DISCOVERY - ExPAND™

Of more than 250,000 plant species worldwide only a small proportion have been assessed for biological activity using modern screening methods. This, combined with the known structural diversity of phytochemicals, and the discovery each year of more than 1500 new chemical structures,[10] indicates the enormous potential of the plant kingdom for yielding new drugs and drug leads. The continuing development of new sensitive biological assays and improved chemical separation and analytical methods justifies the pharmaceutical industry's renewed interest in the plant kingdom.

From its foundation in 1992, Phytera has chosen to exploit plant cell cultures in its lead discovery effort. The generation of culture material is currently based in Sheffield, U.K., and to supplement the culture effort a new Danish subsidiary has recently been formed in Copenhagen, Denmark. Culture extraction, biological screening, natural products chemistry and synthetic chemistry are based at the company's US headquarters in Worcester, MA.

Phytera's ExPAND™ technology (Expanded Phytochemistry Aimed at Novel Discovery) lies at the core of the company's strategy. Comprehensive plant targeting and sourcing tactics are combined with the advantages of plant cell culture and the potential of culture manipulation strategies to yield small molecule-rich, renewable culture samples. The small molecule fractions of the cultures are then concentrated and enriched during extraction before screening is performed by Phytera or its corporate partners (Figure 1).

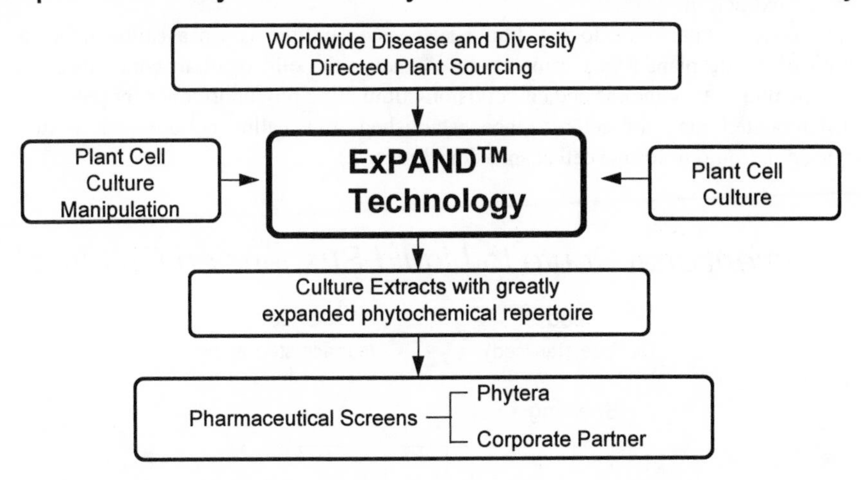

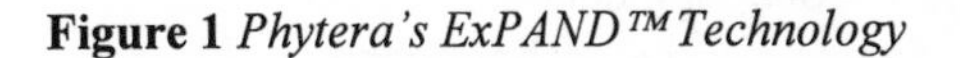

Figure 1 *Phytera's ExPAND™ Technology*

Some aspects of this strategy will now be described in more detail.

2.1 Maximising Phytochemical Diversity - Plant Sourcing and Culture

Phytera has a global plant sourcing strategy which integrates disease-targeting and diversity-targeting with the means of obtaining the desired species or genera. Ethnobotanical and pharmacological data contribute to a targeting rationale which places major emphasis on botanical and chemical diversity. Phytera's access to targeted material is facilitated by the minimal tissue needs of plant cell culture, as single, one-time supplies of only small amounts of tissue (often just a few seeds) are required. Phytera's "bioresponsible" approach reflects the tenets of the 1992 U.N. Convention on Biological Diversity and this further encourages the interest and cooperation of sourcing partners. The tactics of plant cell culture and manipulation are then applied to allow the expression of maximal chemical diversity from each accessed species.

The basic principles of plant cell culture, particularly with respect to the design of culture initiation procedures and growth conditions have been developed over the last 40 years and are the subject of numerous books and reviews.[11] Phytera's aim to apply such principles to the entire higher plant kingdom has led to the construction of an easily searchable database of in-house culture data which also guides daily culture workflow. The value of this database continually increases with the initiation of thousands of species into culture per year fuelling the knowledge base. In practice, this data resource has allowed the development of successful culture protocols, applicable across very wide species diversity.

In overview, the Phytera "standard" culture protocol requires a small section of native plant, commonly stem or leaf, or a germinating seedling; to initiate and establish a culture, the starting material must be sterile, this usually being effected using chemical disinfectants such as hypochlorite. The end product of the Phytera culture development process is a liquid suspension culture, usually but not always in an undifferentiated state. It is in this form that each accessed species can be most readily manipulated prior to chemical extraction.

The conventional route to any liquid suspension culture is via a callus stage; a cut section of sterile plant tissue responds to contact with solid medium containing sugars, inorganic nutrients, vitamins and cell division-stimulating phytohormones by producing an undifferentiated mass of cells. Once established, this callus culture can in turn be dispersed in liquid to form a cell suspension (Figure 2).

Conventional Route to Liquid Suspension Cultures:

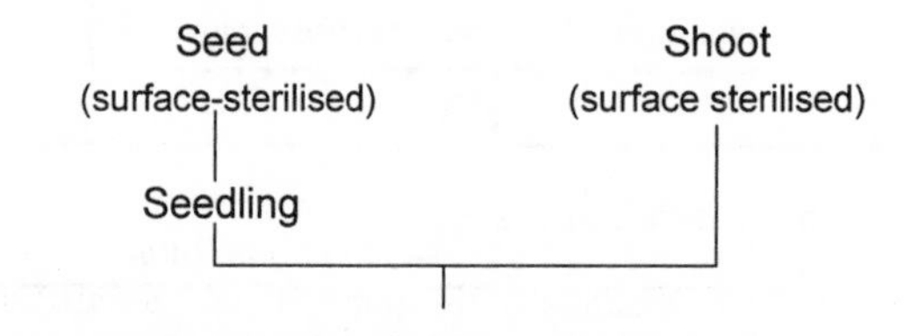

Callus initiation on range of agar media
containing inorganic nutrients, vitamins and phytohormones

Dispersion of callus in liquid and
establishment as a suspension culture

ExPAND™ Manipulations

Figure 2 *Plant Cell Culture Tactics*

The outcome of the sourcing and culture strategy to this point is a culture library which now covers c.80% of all orders of the flowering plants.[12] Needless to say, this culture library is also phytochemically highly rich and varied, with representation from, for example, the relatively widespread indole alkaloid-producing families and those producing diverse polyketide structures, to the gymnosperm groups producing complex diterpenoids.

Furthermore, this chemistry can be readily reaccessed. Compared with recollection of native plants (where botanical and/or chemical reaccess is often problematic), phytochemical resupply via cell culture regrowth and scale-up is simple and straightforward.

2.2 Maximising Phytochemical Diversity Using ExPAND™ Manipulation

In some respects, because of the large surface area available for uptake and rapid response to effector treatments, plant cell suspensions resemble microbial cultures. Some plant cultures will accumulate secondary metabolites, particularly in a growth-dissociated fashion, even under conditions designed to support cell division.[5] However, some metabolic pathways seem to require the design of specialised media or else demand stresses emulating those experienced by the native plant (eg. pathogen attack) in order for their expression to occur.[13]

Phytera's strategy requires the design of treatments which not only allow the accumulation of products to a level compatible with pharmaceutical screening, bioactivity-guided fractionation and structure elucidation, but which also increase the range of small molecule production beyond that normally found in any given specimen of the native plant. To this end, batteries of treatments which include both published and proprietary strategies are imposed on all cell suspension cultures in the Phytera library (Table 2).

Phenylpropanoid and sesquiterpene phytoalexins can be induced in field-grown plants using biotic elicitors - either whole pathogenic microorganisms or various fractions thereof (including enzyme preparations such as pectinase, or single chemical moieties such as arachidonic acid), and via wounding or abiotic elicitor treatment, eg. with heavy metals. Substantially similar effects are observed in plant cell cultures following such treatments. In some model systems this has led to the elucidation of previously poorly known secondary biosynthetic pathways in terms of their enzymology and regulation at the molecular level.[14] However, in cultures it is apparent that some of the initial recognition events lose specificity such that non-pathogenic and pathogenic microorganism preparations sometimes promote essentially the same effect. Significantly, elicitation of plant cultures has been observed to promote the accumulation of compounds not usually considered to be phytoalexins (Table 3). A recent publication describes the elicitation of a new class of flavonoid phytoalexin (an aurone) in cell suspension cultures of the cactus *Cephalocereus senilis*; comparison with flavonoids of the stem tissue of this species and non-elicited cultures indicate that elicitation induces a novel biosynthetic pathway in these cultures.[15] At Phytera, entirely novel compounds with significant biological activities have also been isolated only from elicited cultures; careful analysis of whole plants and non-elicited cultures failed to identify the same compounds (Figure 3, as an example).

Table 2 *Plant Cell Culture Manipulation to Increase Phytochemical Diversity*

Examples of ExPAND™ Techniques
♦ Infection/Elicitation
♦ Gene derepression
♦ Hormones
♦ Substrates
♦ Modified precursors

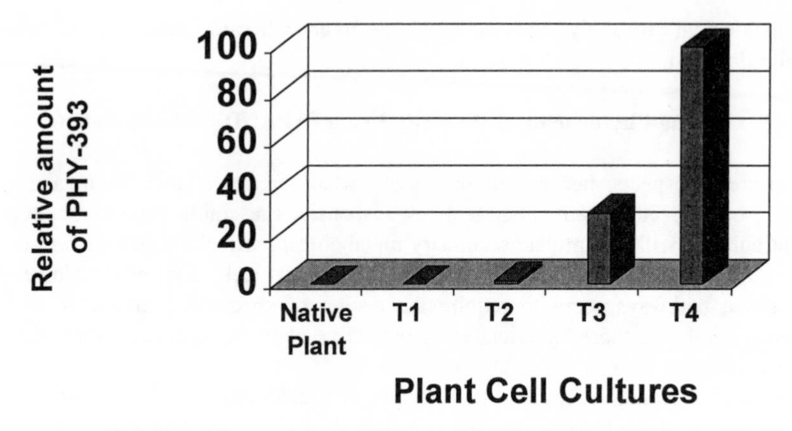

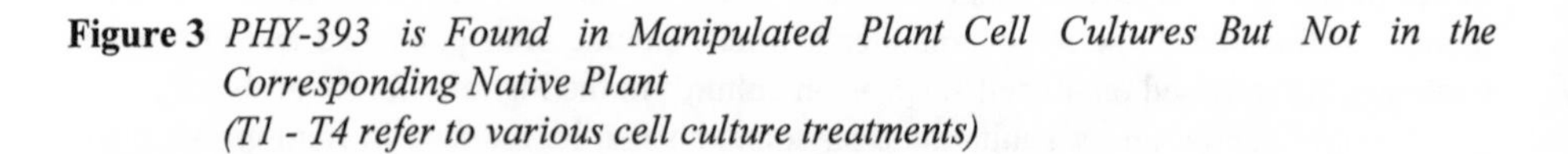
Figure 3 *PHY-393 is Found in Manipulated Plant Cell Cultures But Not in the Corresponding Native Plant*
(T1 - T4 refer to various cell culture treatments)

As an alternative approach, proprietary techniques designed to derepress gene expression in plant cell cultures have proven successful in expanding chemical complexity and pharmaceutical screening hit rate beyond that found in other treatments, and these continue to form an essential part of Phytera's ExPAND™ strategy.

Table 3 *Elicitation of Diverse Chemical Types in Plant Cell Cultures*[13,16]

Species	*Elicitor*	*Product*
Papaver somniferum	*Botrytis* sp.	sanguinarine (benzophenanthridine alkaloid)
Catharanthus roseus	*Pythium vexans*	catharanthine, ajmalicine (indole alkaloids)
Ruta graveolens	various fungal preparations	acridone alkaloids
Cinchona ledgeriana	*Phytophthora cinnamonii*	anthraquinones
Bidens pilosa	*Pythium aphanidesmatum*	aromatic polyalkynes
Glycine max	*Phytophthora megasperma*	glyceollin (pterocarpan phenolic)
Ailanthus altissima	*Saccharomyces cerevisiae* *Phytophthora megasperma*	canthin-l-one (indole alkaloid)

The removal or addition of "standard" medium components can significantly affect secondary product levels. In the late 1970's Zenk reported a culture medium specially designed to promote indole alkaloid production in *Catharanthus roseus*[1]. One of the main factors modified in Zenk's production medium was the hormone composition; the synthetic auxin 2,4-D (2,4-dichlorophenoxyacetic acid) was removed and replaced by other auxins. It is now known that 2,4-D, included in a number of cell culture growth media because of its cell division-promoting properties, tends to suppress secondary product formation in many species. Other medium constituents which affect secondary metabolite level either positively or negatively include the inorganic salt composition (probably often reflecting a major factor, the nitrate : ammonium ion ratio), phosphate level, and sugar level and composition.

With prior knowledge of the classes of compound most likely to be produced by a given species, substrate or precursor-feeding is a valid approach to increasing flux through a target biosynthetic pathway. This is exemplified by a recent publication[17] in which indole alkaloid accumulation in cell suspension of *Tabernaemontana divaricata* was apparently limited by availability of terpenoid precursors. The feeding of loganin overcame this limitation and increased the level of alkaloids by more than x100 from "barely detectable" to c.150 μmol/L culture. A related approach, but one with less literature precedent, is the use of metabolic inhibitors to reduce substrate flux to unwanted metabolites, thereby potentially increasing flux to the biosynthetic end point of interest. For instance, various sterol inhibitors developed as fungicides for application in agriculture or human healthcare, or as hypolipodoemic agents, have sites of action affecting different branches of terpenoid biosynthesis. One can selectively inhibit sterol 14-demethylase, squalene epoxidase, or HMG CoA reductase, thereby altering the balance of terpenoid products towards triterpenoids, cardenolides/sterols, or lower terpenoids.

Polyphenolics are widespread in the plant kingdom and these frequently fall into the category of "nuisance" compounds, often possessing low level, non-selective biological activity and masking the detection of more interesting but lower abundance compounds. While pre-treatments such as polyamide have been developed to remove polyphenols from plant extracts,[18,19] solid-phase steps in the isolation procedure are not always desirable, and the concept of specifically inhibiting the phenolic pathway in plant cell cultures as part of a manipulation battery is highly attractive. Inhibitors such as AOA (amino oxyacetate) are commercially available, and while not entirely specific, these can be very effectively applied to reduce phenolic yield.[20]

An analysis of the effects of running a battery of manipulations on a random set of cultures illustrates the value of the strategy (Figure 4).

2.3 Enrichment of Phytochemical Classes More Likely to Include Potential Drugs/ Leads

Manipulated cultures are harvested after a pre-determined incubation period and then harvested to yield cells and medium. The liquid medium is subjected to solid-phase extraction while the freeze-dried biomass is processed into relatively polar and relatively non-polar fractions using a combination of solvent extraction, solid-phase and molecular weight cut-off filtration, selectively discarding "nuisance" and "inert" compounds such as polyphenolics and sugars en route. Only then are the extracts introduced to the screening programme and lodged in the extract library.

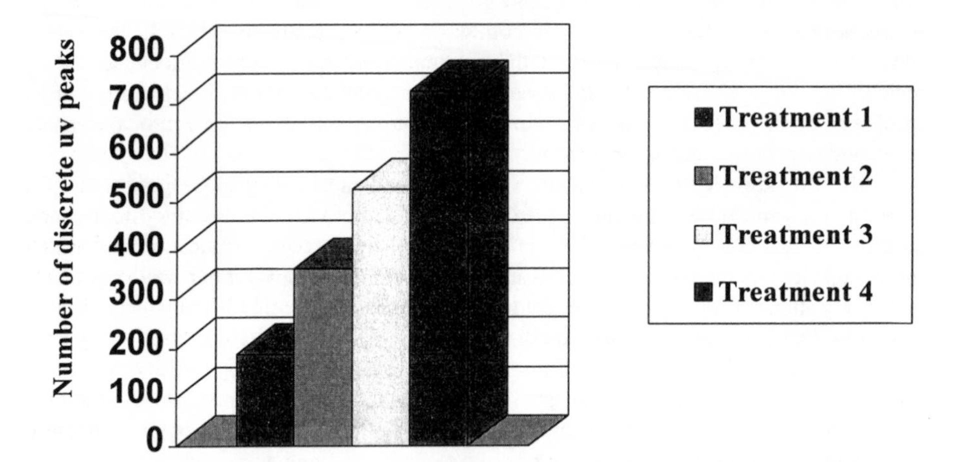

Figure 4 *Expanded Phytochemistry After Culture Manipulation as assessed by HPLC/UV (280 nm) analysis - bars represent the cumulative value of increasing numbers of manipulations imposed on the culture*

2.4 Phytera's Discovery Programme

The present emphasis is on anti-infectives, a therapeutic area which at the company's outset was agreed to be particularly relevant to its focus on plant materials. However, given the precedent for the discovery of extremely valuable plant-derived drugs in many other therapeutic areas (CNS, antihypertensives, anti-cancer, and so on) there is no reason to limit the scope of plant cell culture-based screening efforts. Phytera is therefore expanding its screening efforts beyond infectious diseases to other areas, as well as identifying collaborative partners with interests in further therapeutic targets.

In the context of the anti-infectives discovery programme, Phytera is concentrating on antifungals, antivirals (against hepatitis, herpes, CMV, influenza) and antibacterials, employing a combination of cell and molecular targets (in the latter case including as an example some novel enzyme targets required for hepatitis C replication). The screens are run in 96-well microtitre plate format and with the help of robotics automation, can handle large numbers of extracts. Acceptable rates of biological activity across these screens have been achieved (<1% primary screening hit rate). These efforts have already led to the identification of numerous phytochemicals with novel biological properties and the discovery of a series of chemically unique preclinical candidates in the antifungal area. Importantly, the vast majority (>75%) of interesting biological activity has been seen with extracts of manipulated (as opposed to unmanipulated) plant cell cultures.

2.5 Proving the Concepts

As previously stated, the core concept underlying the Phytera approach is that manipulated plant cell cultures provide access to phytochemistry that is not expressed in the corresponding native plant sample. Our experience to date supports and confirms this concept. With a considerable amount of data now accumulated we find that more often than not, the manipulated cultures of any given plant species exhibit greater chemical complexity than the native plant from which the culture was derived. At the simple level of HPLC separation and UV detection, the native plant profile is consistently less complex than that of the plant cell culture extract. In cases when biologically active lead compounds have been isolated from cultures, the same active compound is sometimes present in the source native plant, but frequently at a much lower level; to illustrate this point, the known triterpenoid jacoumaric acid was identified in culture extracts at levels ranging from x0.5 → x100 that found in the native plant, depending upon the type of manipulation employed. Furthermore, the novel lead compound PHY-393 could not be detected in the native plant at all, even when treatments similar to the relevant culture manipulations were imposed (as far as possible) on the plant (see Figure 3). Numerous similar cases have been observed.

To summarise, Phytera's experience supports and reinforces its approach to exploiting phytochemical diversity via the culture route, conceptually illustrated in Figure 5.

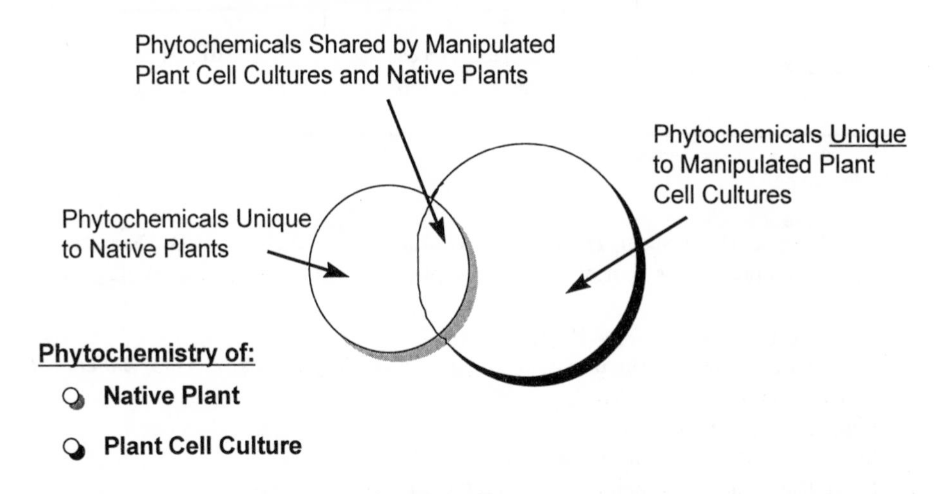

Figure 5 *ExPAND™ Technology Increases Access to Phytochemical Diversity*

Only time and wider experience will reveal the true relative sizes and extent of overlap of the chemical repertoires of plant and manipulated plant cell cultures. In the meantime, certain practical advantages over the whole native plant-based drug discovery paradigm are clear (Table 4). The expanding culture efforts and drug discovery programme ensure continued growth of the extract library and a stored culture library, the latter not only being a vehicle for lead molecule re-supply, but also a potential source of small molecules or enzymes to be discovered via future screening efforts. Furthermore, at a time when many plant species are rapidly disappearing from the earth's surface, Phytera's culture library ensures preservation of their biosynthetic machinery. This ensures that, beyond our knowledge or capabilities today, tomorrow's technologies can continue to exploit this rich resource.

Table 4 *Advantages of Phytera's ExPAND™ Technology*

TECHNOLOGICAL ADVANTAGE	PHYTERA'S ExPAND™ TECHNOLOGY	TRADITIONAL NATIVE PLANT-BASED DRUG DISCOVERY
Expansion of Phytochemical Diversity	Major Emphasis via plant cell culture manipulations	No capacity
Material Required	Small Quantities leaf, seed, root	Large Quantities Whole Plant
Material for Preclinical Development	Via Plant Cell Culture Scale Up	Can be a Major Limiting Issue
Endangered Species	Reasonable Access Facile Follow-up	Restricted Access Problematic Follow-up

References

1. M. H. Zenk, H. El-Shagi, H. Arens, J. Stockigt, E. W. Weiler and B. Deus, 'Plant Tissue Culture and its Biotechnological Application', Springer-Verlag, Berlin, 1977, p. 17.
2. M. E. Curtin, *Biotechnology*, 1983, **1**, 649.
3. N. R. Farnsworth, O. Akerele, A. S. Bingel, D. D. Soejarto and Z. Guo, *Bulletin of WHO*, 1985, **63**, 965.
4. Phytera internal report, 1996
5. D. V. Banthorpe, *Natural Product Reports*, 1994, 303.
6. C. M. Ruyter and J. Stockigt, *GIT Fachz.Lab.*, 1989, **4**, 283
7. A. Arens *et al*, *Planta Medica*, 1982, **46**, 210.
8. B. Ulbrich, H. Osthoff and W. Wiesner, 'Plant Cell Biotechnology', Springer-Verlag, Berlin, 1988, NATO ASI Series H : Cell Biology Vol. 18, p. 461.
9. W. Ma, G. L. Park, G. A. Gomez, M. H. Nieder, T. L. Adams, J. S. Aynsley, O. P. Sahai, R. J. Smith, R. W. Stahlhut and P. J. Hylands, *J. Nat. Prod.*, 1994, **57**, 116.
10. N. R. Farnsworth, 'Bioactive Compounds from Plants', CIBA Foundation, London, 1990, CIBA Foundation Symposium 154, p. 2.
11. R. A. Dixon, 'Plant Cell Culture, a Practical Approach', IRL Press, Oxford, 1985.
12. A. Cronquist, 'An Integrated System of Classification of Flowering Plants', Columbia University Press, New York, 1981.

13. M. W. Fowler, A. M. Stafford, 'Plant Biotechnology', Pergamon Press, Oxford, 1992, Comprehensive Biotechnology Second Supplement, p. 79.
14. C. J. Lamb, M. A. Lawton, M. Dron and R. A. Dixon, *Cell*, 1989, **56**, 215.
15. Q. Liu, M. S. Boness. M. Liu, E. Seradge, R. A. Dixon and T. J. Mabry, *Archives of Biochemistry & Biophysics*, 1995 **321**, 397.
16. G. Krauss, G-J. Krauss, R. Baumbach and D. Groger, *Z. Naturforsch.*, 1989, **44c**, 712.
17. D. Dagnino, J. Schripsema and R. Verpoorte, *Phytochemistry*, 1995, **39**, 341.
18. M. E. Wall, H. Taylor, L. Ambrosio and K. Davis, *J. Pharmaceutical Sciences*, 1969, **58**, 839.
19. G. T. Tan, J. M. Pezzuto, A. D. Kinghorn and S. H. Hughes, *J. Nat. Prod.*, 1991, **54**, 143.
20. U. Mackenbrock, W. Gunia and W. Barz, *J. Plant Physiol.*, 1993, **142**, 385.

Biotransformations Applied to Plant Chemicals: A Natural Combination

John T. Sime

ZYLEPSIS LTD, 6 HIGHPOINT, HENWOOD BUSINESS ESTATE, ASHFORD, KENT TN24 8DH, UK

1 INTRODUCTION

In a socioeconomic environment where consumer demands are directing products more towards perceived environmentally acceptable formulations, plant derived raw materials present an important commercial opportunity. The market pull means that products and ingredients which can be described as natural usually command a premium, and functional molecules derived from a natural, renewable resource and which have been produced by methods which are "non-chemical" or which are environmentally friendly are also in demand.

Plants represent an important source of a wide diversity of molecular structures, many of which are in common usage. When in combination with biotransformation technology a much more extensive range of products can be obtained, often with commercial advantage.

2 BIOTRANSFORMATIONS

Use of the term biotransformation in the context of this paper is defined as the use of enzyme systems to convert a molecular entity to another compound by functional group transformation. This, therefore, excludes fermentation for the production of secondary metabolites. The enzyme systems used can be whole cells or isolated, purified or semi-purified proteins. The present work concentrates on microbial systems as whole cell catalysts.

The accepted advantages of enzyme catalysed reactions over non-biochemical reaction systems include;

- selectivity; stereo-selectivity, regio-selectivity, chemo-selectivity
- mild reaction conditions; avoiding harsh reaction conditions
- environmentally friendly; avoiding extremes of pH and metal catalysts
- specificity; reaction on particular molecules
- wide range of catalytic activities available

These properties highlight some of the attributes which promote the perception that enzyme catalysed reactions give products which are "green", are naturally produced and hence have a high level of consumer acceptance.

3 THE COMBINATION

Materials from renewable plant sources help fulfil the need for an exceptionally wide range of functional chemicals which have an equally wide range of molecular structures. When there is a need for purified materials of this type they are often obtained by extraction technology. There is also the potential to extend the range of available materials by converting the plant substances into other, related or derivative, materials. The combination of plant chemicals with the power of biocatalysis potentially opens the door to an extended range of chemical entities. When the products of biotransformation are in themselves natural products there is a multiplier applied to the value generated as in many cases the product can be marketed under a natural label.

The selling price of truly natural functional chemicals can often be greater than that of a synthetic equivalent by a factor of ten or more. This presents a driver for the generation of materials which can carry a natural label, especially when they can be generated from lower cost natural starting materials.

Two examples are presented here which demonstrate the application of Zylepsis technology to phytochemicals as source materials and the generation of desirable products of use in the food/feed/cosmetics industries. One of these is a representative of a relatively small tonnage specialist flavour material from which unexpected advantages have been derived as a result of Zylepsis investigations. The other demonstrates a divergent use of plant material as a starting point for the generation of a number of products.

4 NOOTKATONE

Nootkatone is the bicyclic ketone responsible for the distinctive flavour of grapefruit. It is present in the fruit at low concentrations and even in high grade grapefruit essential oil the concentrations of nootkatone are typically about 0.3% and it is therefore a speciality product which is high value , low volume. For a number of reasons the demand of grapefruit flavour outstrips supply and so nootkatone retails at a significant price.

Valencene, on the other hand is a sesquiterpene present in orange oil and sells at a much lower price. The conversion of valencene to nootkatone involves an oxidation at a specific position of the sesquiterpene and provides a commercially desirable product. Chemical conversion gives a product which cannot be described as natural, although it is nature identical, and which sells at a fraction of the price of natural nootkatone. There is, therefore, a drive to find a biotransformation of valencene to nootkatone.

There are literature reports[1,2] of attempts at this transformation but these have suffered from low conversions, irrespective of whether microorganisms or plant callus cultures had been used. Some of the publications report high yields but the concentrations are so low as to make the reaction non-viable as an industrial process.

Scientists at Zylepsis discovered a number of organisms, bacteria and fungi, which were capable of carrying out the necessary double oxidation of valencene and some of these gave significantly higher levels of conversion than had previously been reported. It did appear, however, that in combination with these organisms a reactor design which was capable of removing the reaction products as they were formed would give an advantage in this bioreaction.

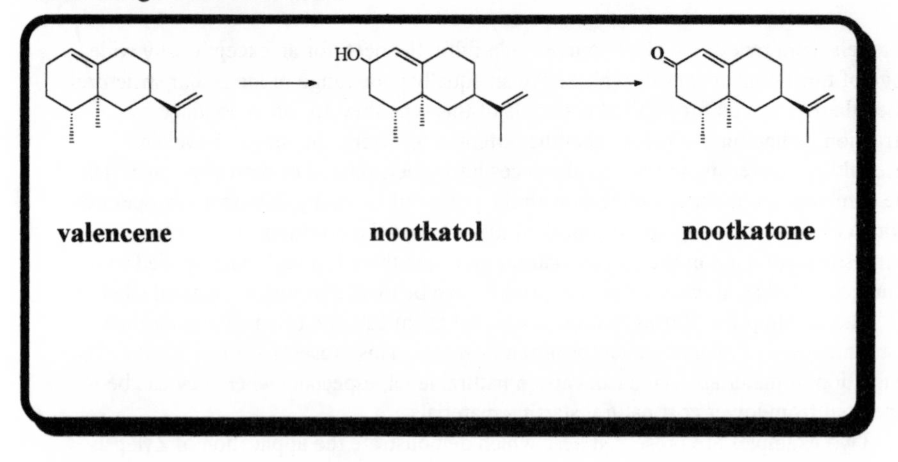

As part of the investigation into optimising this process it was necessary to find a way of separating the product from the substrate as the reaction progressed. The selective removal of the oxidised sesquiterpene from the valencene, an extremely similar molecule with a similar solubility profile, whilst all the substances were in an aqueous medium was to present a challenging *in situ* product removal problem. A configuration was, however, defined which would give selective removal of the nootkatone from the valencene in an extremely selective manner - even when the concentration of valencene was vastly greater than that of the nootkatone. No trace of valencene was observed in the isolated nootkatone.

It would appear that the nootkatone is a good inhibitor of certain cytochrome P450 enzymes. This observation is of interest in light of the medical reports[3] concerning the advantages of grapefruit juice when used in conjunction with certain medication regimens. In particular, coadministering grapefruit juice with certain chemical families of hypotensive agents[4] and also with cyclosporin, used as an immunosuppressant, can provide some clinical advantage.[5]

Research into the reasons for this pointed to the involvement of a cytochrome P450 inhibitor activity in the grapefruit juice preventing some of the first pass metabolism of the drug thereby leading to improved plasma levels and improved clinical effect. Most of the research on this inhibition has targeted the flavonoid component in grapefruit juice as being responsible, and naringin appeared to be a likely candidate as a causitive agent. Clinical studies have, however, failed to fully comfirm this hypothesis As a minor component present in grapefruit and which is shown to be a source of cytochrome P450 inhibition, nootkatone now presents itself as a candidate molecule for this application. Preliminary experimental work has confirmed the inhibitory activity of nootkatone against a range of cytochrome P450 enzymes responsible for drug inactivation in the gut.

It is therefore the case that the investigation of the production of a value flavour component of fruit has led Zylepsis to the discovery of a range of offshoot applications as well as the original target of the project.

5 OIL SEED

The commercial importance of plant derived oils is widely known and appreciated. The materials left over after extraction of the oils can also provide an opportunity for commercialisation when combined with enzymatic reaction.

Zylepsis identified a market gap in the growing sunscreen market. There is an absence of potential UV absorbing molecules which can be included in cosmetic formulations and which are natural. There are structures in plant materials which have the typical functionalities found in UV absorbing molecules, although most of these are substructures of larger molecular entities. It has now been discovered that there is an enzyme activity which is capable of releasing these substructures and so provides the ability to gain access to large quantities of high value molecules for an important expanding market sector.

Whilst developing the technology to produce such a sunscreen functional active from natural sources by use of an enzyme, it was discovered that it is also possible to degrade undesirable molecules present in plant materials. Chlorogenic acid is naturally present in much plant material presently used as animal feedstuff. The family of chlorogenic acids [one of which is shown] are present in some oil seed residues which are currently used in feedstuffs for animals and are known to have an antinutritional effect as they bind to the proteins in the feed and thus render it unavailable for absorbtion. Attempts have been made, over the years, to devise a method of removing these materials from plant material but because of the solubility profile of the undesirable elements in the feed any method which is capable of removing them also removes many of the desirable substances and so reduces further the nutritional value of the feed.

The enzyme activity which has been discovered has the ability to hydrolyse chlorogenic acids into quinic acid and caffeic acid so that the resulting undesirable moiety can be more easily removed to leave a higher value feedstuff. This enzyme activity is unusual in that the enzymic ester hydrolysis shown is not facile, possibly because of steric factors in the substrate. Certainly the range of first choice

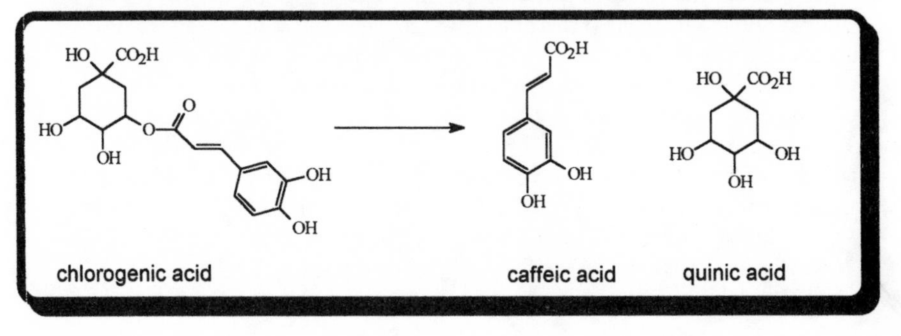

commercially available enzymes failed to show significant activity against this substrate. It has been shown that it is possilble to design a process which is capable of removing the antinutritional components without adversely affecting the protein content of the material.

The quinic acid generated also has functionality in its own right. Caffeic acid as a fine chemical presently commands a significant price.

6 CONCLUSION

The combination of phytochemicals with the rich variety and diversity of chemical structures, enzyme catalysed transformations and the advantages of a natural, renewable resource can lead to a great many scientific and commercial opportunities.

ACKNOWLEDGEMENTS

The work presented here arises from intellectual and practical contributions from Zylepsis staff including Drs Peter Cheetham, Michelle Gradley, Nigel Banister and Mr Steve Myers.

REFERENCES

1 H. Willershausen and H. Graf, *Chemiker-Zeitung*, 1991, **115**, 358.
2 Kao, 1993, JP 06303-967.
3 G. Yee, D. Stanley, L. Pessa *et al.*, *The Lancet,* 1995, **345**, 955
4 P. Soons, B. Vogels, M. Roosemalen, *et al*, *Clin. Pharmacol. Ther.*, 1991, **50**, 395
5 M. Ducharme, R. Provenzano, M. Dehoorne-Smith and D. Edwards, *Br. J. Clin. Pharmac.*, 1993, **36**, 457.

Partial and Semi-Synthesis with Biological Raw Materials

J. H. P. Tyman

DEPARTMENT OF CHEMISTRY, BRUNEL UNIVERSITY, UXBRIDGE, MIDDLESEX UB8 3PH, UK

1 INTRODUCTION

The terms partial and semi-synthesis may be conceived of as different, the former relating to total synthesis of a natural product and the latter to the formation of an unnatural industrial material although both commence with a naturally-occurring substance or intermediate. The designation 'semi-synthesis' implies the construction of an organic molecule by the modification of a natural product generally of replenishable, biological rather than of fossil fuel origin.The term total synthesis conveys no information about the nature of the starting material and indeed many different intermediates have frequently led to the same target intermediate or final product. Semi-synthesis can include a wide range of approaches, the construction of a larger compound and the degradation or breakdown of a larger to a smaller molecule by chemical means or perhaps by biotransformations. At the most rudimentary level it can involve the modification of a natural product by minor functional group or structural changes and more ambitiously by using the intrinsic stereochemistry of the starting material as the major feature of the strategy. Thus while in total synthesis the correct stereochemistry of the final product is the objective, in semi-synthesis the stereochemistry of the starting material is vital and in partial synthesis both are essential features. Of course the minor modification of unnatural synthetic molecules has for many years been the pathway adopted in classical organic pharmaceutical research enlightened by biochemical and pharmacological information and has been a rewarding method giving landmark achievements such as for example the range of 'M and B' sulphonamide drugs.[1] Although modified natural products could well have proved worthy models, the vast majority of synthesised products seem to have generally involved the use of petrochemical and coal-derived intermediates and only comparatively recently have biological raw materials begun to be employed.

The procedures of semi-synthesis and partial synthesis have been in fact in progress for many decades in both academic and industrial laboratories and might be deemed in time to be the forerunners of the 'chiron approach',[2] the alternative methodology being asymmetric synthesis.[3] A general account has appeared[4] covering several of these aspects and a review written some years ago.[5] Although partial synthesis might be construed as a judicious and ingenious approach to total synthesis and semi-synthesis as the more dramatic area,where industrial developments have captured the imagination the distinction between the two methodologies has become blurred with time . It could perhaps be said that semi-synthesis with a naturally-occurring intermediate is a step in the partial synthesis of another natural product and that both in this sense are two of the many pathways to total synthesis. Although the possibilities in the latter have been enormouslyadvanced in the last two decades through increased methodology and computer-aided retrosynthetic analysis, the intuitive procedures in partial and semi-synthesis have been in existence for many years.

Despite the abundance of fossil-fuel intermediates biological raw materials have begun to attract interest. Thus for example in a compilation of thirty five syntheses[6] of major natural products those of ascorbic acid, bradykinin, β-carotene, cephalosporin, cycloartenol, patchouli alcohol and penicillin commenced with a natural product. Increasingly in recent years in the pharmaceutical, food and flavour industries chirality has become an important consideration[7] and from this aspect the employment of natural product intermediates, for example in semi-synthesis, could be significant. The organic chemical industry frequently has faced environmental problems[8], sometimes stemming from multistage syntheses and conceivably the conciseness of semi-synthesis and partial synthesis can offer some simplification.

Although progress in organic chemical methodology has had in recent decades an unprecedented influence and led to remarkable success in total synthesis as seen, for example by that of vitamin B_{12} [9] and of palytoxin[10], a large number of compounds useful to humanity in pharmaceutical and in other technical ways are obtained from replenishable natural resources in comparatively few steps. In such manufacture partial or semi-synthesis is the only step involved. The objective of this account is to recall some of the achievements in these approaches and to emphasise that natural organic resources are varied and widespread in the vast majority of countries in the world. Classical examples of semi-synthesis in the derivation of C_{10} alcohols of interest in the olfactory terpenoid industries, in penicillin and cephalosporin chemistry, vitamins, the transformation of morphine alkaloids, steroids, and in lipid reactions are described.

2 MONOTERPENOID ALCOHOLS

The acyclic C_{10} alcohols linalol, geraniol and nerol, citronellol and the cyclic terpineols are valuable in perfumery while (-)-menthol in the cyclic series has wide use in the flavour industry. The traditional source for these and their derivatives is from essential oils, for example, lavender, lavendin, petitgrain, geranium, citronella, peppermint oil and others. The C_{10} alcohols isolated from such natural sources possess a characteristic odour associated with trace but intensely odorous minor components, as for example 'rose oxide' from both Bulgarian rose oil and geranium oil[11]. In the fifties and early sixties the greatly increased use of compounded perfumes in consumer products and other factors led to the chemical development of readily available natural resources such as pine oil (*Pinus sylvestris*) a rich source of α- and β-pinenes. Thus after separation by fractional distillation, β-pinene (1) was pyrolysed to give myrcene (2) and some pseudo-myrcene. Reaction of (2) with hydrogen chloride in the presence of cuprous chloride gave a complex mixture of halides which with sodium acetate afforded racemic linalyl acetate (3), together with geranyl (4) and neryl acetates[12] as depicted in Scheme 1. Separatory difficulties and stringent olfactory requirements led to more selective procedures and process simplifictions such as avoidance of the initial α/β pinene fractionation by hydrogenation of the mixture to *cis*-pinane, autoxidation to give pinanol (6) and pyrolysis of this compound to give linalol (Scheme 2) and thence linalyl acetate (3).

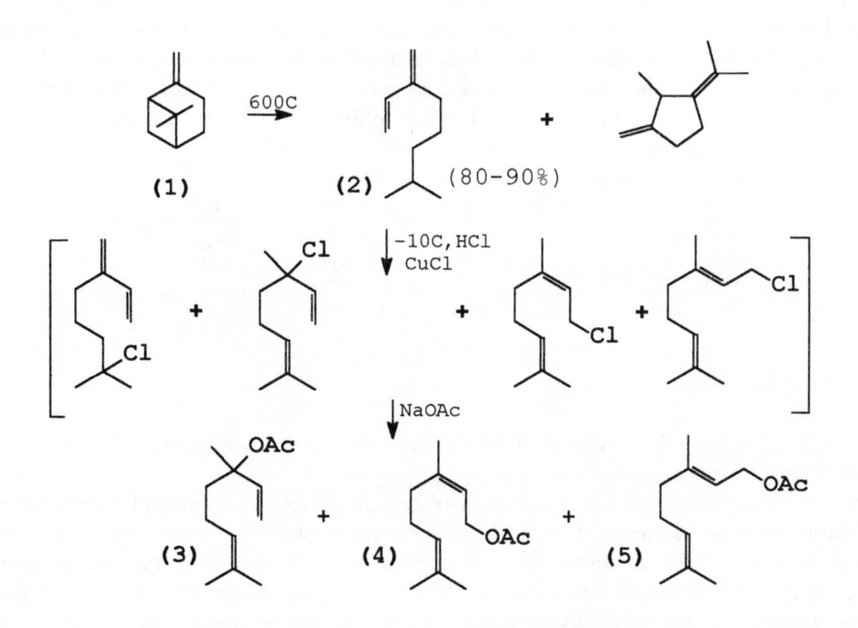

Scheme 1 Synthesis of C_{10} Alcohols from β-Pinene

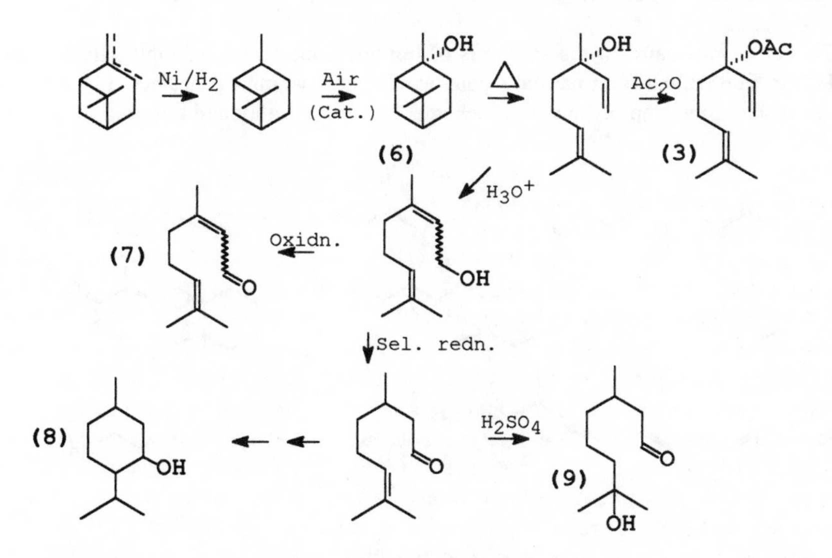

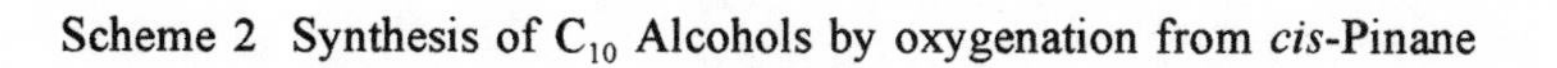

Scheme 2 Synthesis of C_{10} Alcohols by oxygenation from *cis*-Pinane

Mild acidic rearrangement of linalol afforded geraniol/nerol and thence by oxidation geranial/neral (7). Selective reduction of the α,β double bond gave citronellal which while of no perfumery value, gave via isopulegol, the highly useful menthol (8) after equilibration with sodium menthoxide (giving the product primarily in the all-equatorial form) or hydroxycitronellal (9) by classical methods as shown in Scheme 2. R-(+)-Citronellol (10) was obtained from *cis*-pinane more directly by pyrolysis to citronellene and thence Ziegler reaction with aluminium hydride followed by oxidation (Scheme3).

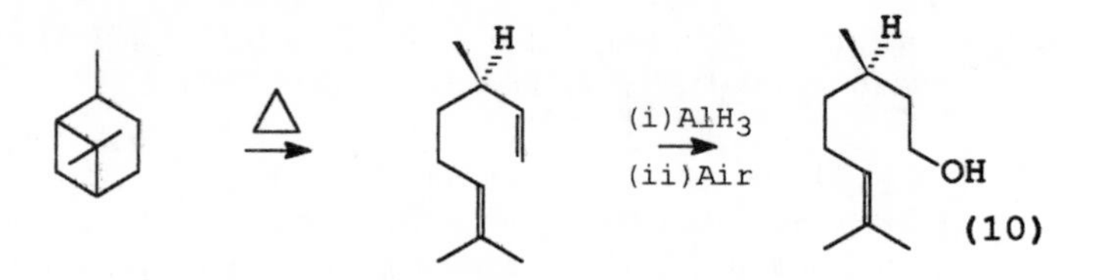

Scheme 3 Synthesis of R-(+)-Citronellol from *cis*-Pinane

Other technical developments have taken place and details are frequently shrouded in industrial secrecy. Concurrent with the use of replenishable raw materials, synthetic advances in acetylenic chemistry by Hoffmann La Roche have led to important technical procedures for obtaining C_{10} alcohols and the pure products obtained by both avenues have been regarded by perfumers as new materials since they no longer contain the trace substances which typify C_{10} alcohols isolated from essential oils.

3 THE LIPIDIC VITAMINS

The long side-chains in the synthesis of lipidic vitamins, for example vitamins A, E and K_1 can be obtained from natural resources. Thus for vitamin A acetate (11), citral can be used at the initial step as shown in Scheme 4. The C_6 compound required for step (v)

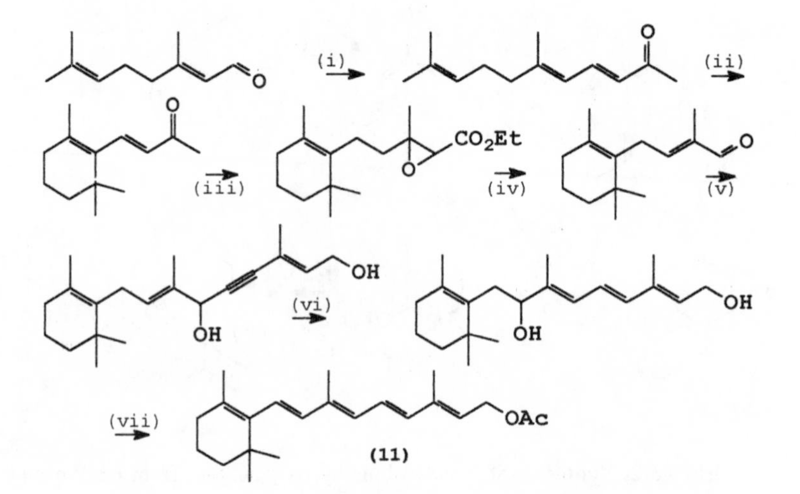

is obtained (route a) from butenone.

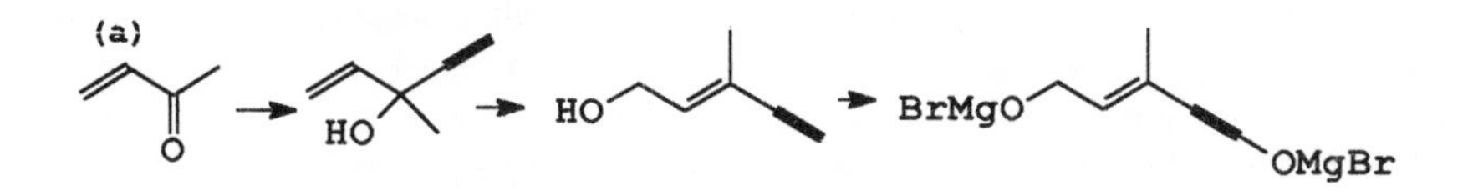

Scheme 4 A Synthesis of Vitamin A acetate from Citral

Reagents: See Appendix* (i) Me_2CO, HO^-, (ii) H^+,(iii) $ClCH_2CO_2Et$, NaOEt, (iv) HO^-, -CO_2 ,(v)diMgBr compd. from (a), H_3O^+ , (vi) Pd-C,H_2, Lindlar; Ac_2O; I_2

Phytol can be employed for the synthesis of phenolic vitamins[13], E and K_1 although synthetic isophytol prepared from acetylenic routes is the preferred intermediate as indeed is synthetic pseudoionone in the current syntheses of both vitamin A and β-carotene.[5,14] The extensive acetylenic chemistry developed for C_{10}, C_{20} and C_{40} by Hoffmann la Roche permits a large range of natural products to be synthesised on a commercial scale.
Vitamin D_3 (cholecalciferol) the important anti-rachitic lipidic compound is obtained industrially by a well-established route[15] from cholesterol through formation of 7-dehydrocholesterol with N-bromosuccinimide and dehydrobromination with collidine followed by ultraviolet irradiation.

4 THE STEROIDAL HORMONES

Cholesterol and the sapogenins are well-known intermediates for the industrial partial and semi-synthesis of steroidal hormones such as the gestogens, estrogens, androgens and corticosteroids. The early successful use of diosgenin (12) from a Mexican dioscorea species for the synthesis of progesterone[16] (13) (Scheme 5) was a landmark in the production of steroidal hormones.

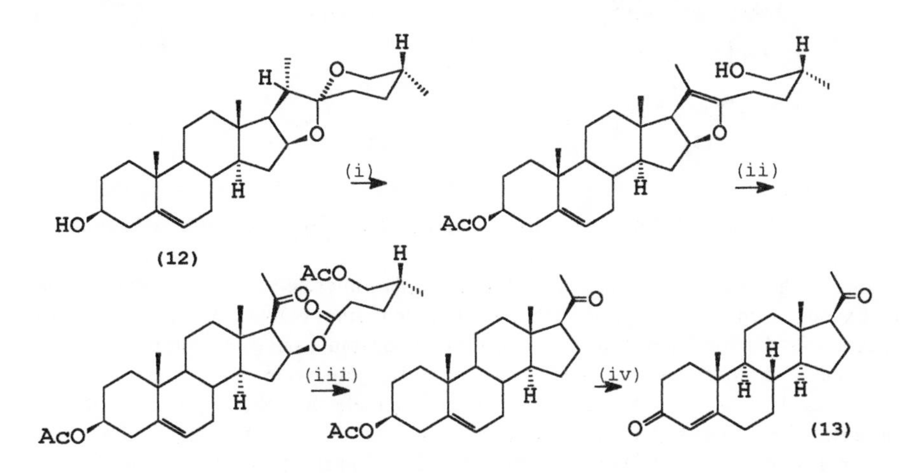

Scheme 5 Synthesis of Progesterone from Diosgenin

Reagents: Ac_2O, H_3O^+, (ii) CrO_3, AcOH, (iii) AcOH,Pd-C,H_2 , (iv) HO^-; Oppenauer.

Stigmasterol (14) from soya bean oil after conversion to stigmastadienone by Oppenauer oxidation was found subsequently to be convertible in three steps in higher overall yield (Scheme 6) to progesterone via the oxidation of an intermediate enamine and this proved to be important at that time in the use of the latter for the derivation of corticosteroids.

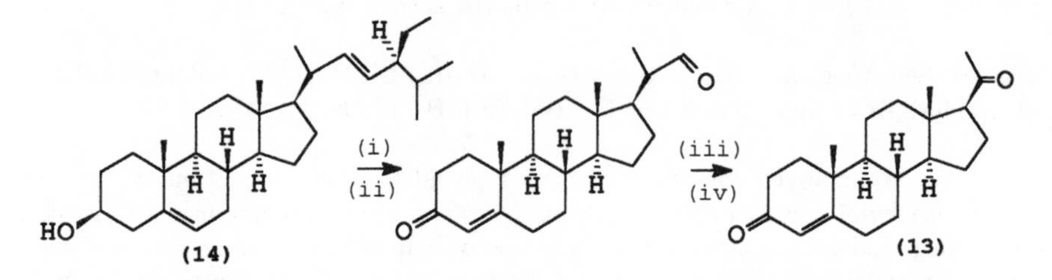

Scheme 6 Synthesis of Progesterone from Stigmasterol

Reagents: (i) Oppenauer, (ii) O_3 , CH_2Cl_2 ,(iii) morpholine, Δ , (iv)$Na_2Cr_2O_7$, AcOH, PhH.

Although diosgenin after aromatisation to a phenolic intermediate was then converted to estrone (14) by oximation, Beckmann rearrangement and hydrolysis of the α,β-unsaturated acetylamino product[17], a route from cholesterol subsequently was the preferred way to this hormone. Microbiological degradation of cholesterol to androst-1,4-diene-3,17-dione and two chemical steps [18] was one approach while in another 19-hydroxycholesterol (15) obtained chemically in three steps[19] from cholesteryl acetate was converted microbially with the organism CSD-10 to estrone in 72% yield in one step[20] (shown in Scheme 7).

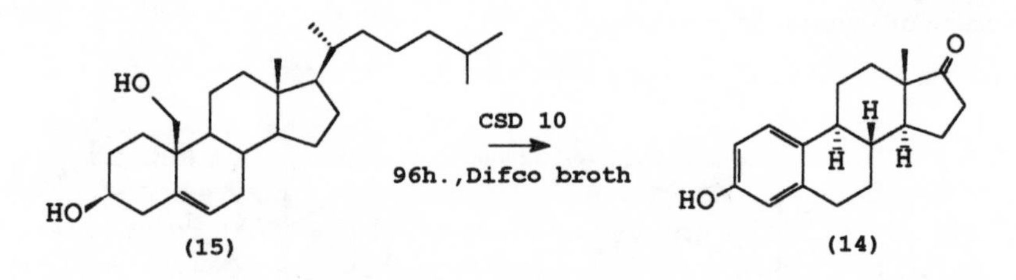

Scheme 7 Microbial Conversion of 19-hydroxycholesterol to Estrone

Estrone methyl ether in turn has proved an important intermediate for the semi-synthetic compounds norhisterone and norethynodrel which are obtained essentially by Birch reduction as the first step. Progesterone, from sapogenin sources and from stigmasterol rather than from cholesterol or the lengthy route from methyldesoxycholate used in early work, has likewise proved an essential intermediate for the generation of many corticosteroids. Thus a dramatic development in this direction was the finding by the Upjohn group[21] that selective hydroxylation at the 11-α position in progesterone took place in 80-90% yield by the action of the microorganism, *Rhizopus nigricans*. A further six steps led to a synthesis of cortisone acetate. Parallel with this remarkable finding was the development by another group of the sapogenin source, hecogenin (16), from the East African sisal plant which led to a five stage synthesis of cortisone[22] (17). The method involves the generation of an hydroxyl group in ring C of hecogenin, Marker degradation and further reactions upon the C_{17} group as depicted in Scheme 8.

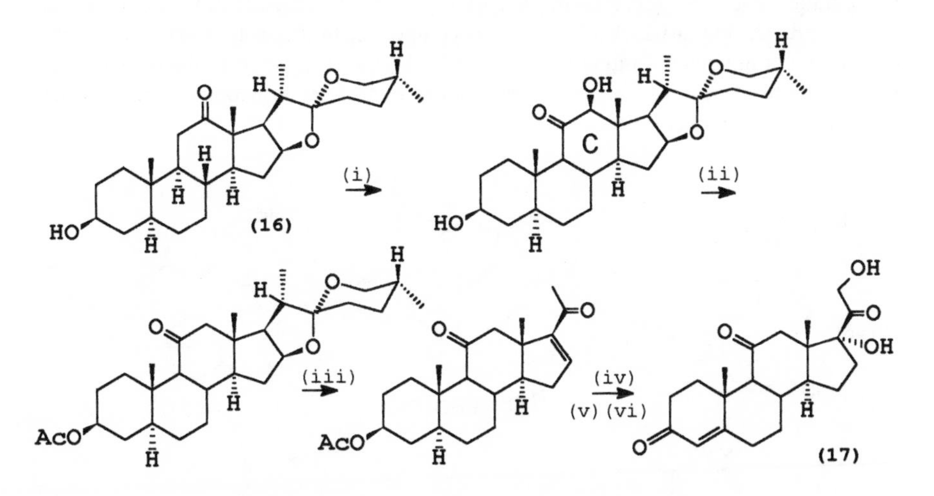

Scheme 8 Synthesis of Cortisone from Hecogenin

Reagents: (i) Ac_2O; Br_2 ; HO^- , (ii) Zn, H_3O^+ ;Ac_2O; Ca/NH_3(liq.), (iii) Marker degradation, (iv) $PhCO_3H$, (for 16,17-epoxide), (v) HBr; Br_2 ,(vi) KI; NaOAc; Ni/EtOH, H_2.

Corticosteroids have themselves been used as intermediates. Thus, corticosterone acetate has been converted by the 'nitrite' method into the highly active compound aldosterone[23].

5 THE MORPHINE ALKALOIDS

Steroidal intermediates have been employed for the synthesis of the medicinally useful conessine[24] and semi-synthesis has been employed for the preparation of anti-depressant derivatives of reserpine,but perhaps a main interest has resided in the morphine group of alkaloids[25]. The opium alkaloids contain morphine (18) the most powerful analgesic and addictive narcotic member of the group, codeine (19), a mild highly popular analgesic and an anti-tussive and thebaine (20) which is non-analgesic. The majority (90%) of the

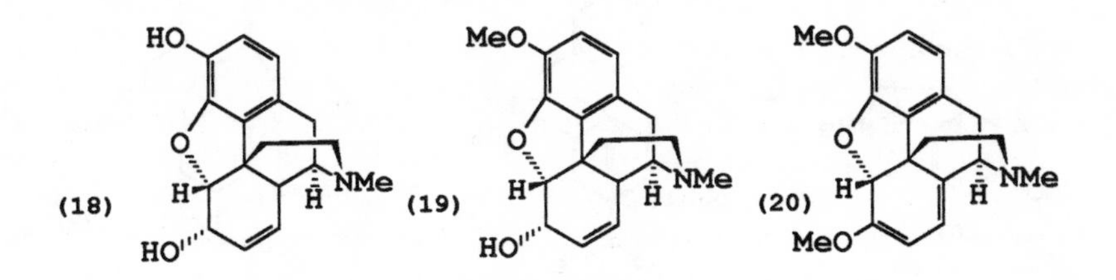

morphine from opium, which comprises generally morphine (20%) and minor amounts of codeine (0.5%) and thebaine, is converted commercially to codeine by methylation. Due to the potential diversion of morphine from opium (*Papaver somniferum*) to illicit heroin production with a consequent effect on the the legitimate availability of medicinal codeine, consideration has been given to the use of thebaine from the alternative natural source *Papaver bracteatum* in which it is the main component. In this way it was conceived that the narcotic drug situation might be monitored. The transformation of thebaine to codeine in 74% has been achieved[26] by oxymercuration, hydrolysis to neopine, isomerisation to codeinone and reduction (Scheme 9).

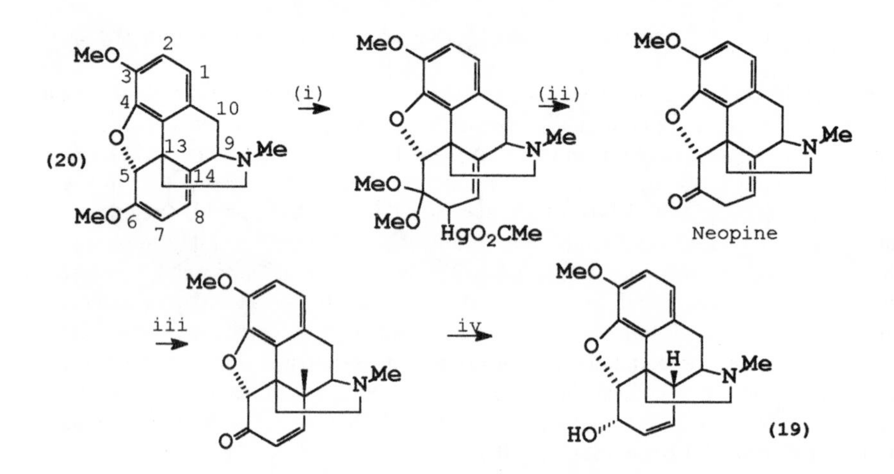

Scheme 9 Synthesis of Codeine from Thebaine (ex *Papaver bracteatum*)

Reagents: (i) $Hg(OAc)_2$, MeOH, (ii) AcOH or $NaBH_4$; 3M HCO_2H ,(iii) 3M HCO_2H, HCl, CH_2Cl_2 , (iv) $NaBH_4$, MeOH.

Apart from strategies to avoid the use of *Papaver somniferum* by for example obtaining morphine through the demethylation of codeine[27] derived from thebaine, a vast amount of research over many years mainly by semi-synthesis has been carried out to retain the powerful analgesic properties of morphine but without its addictive effect and certain less desirable characteristics.

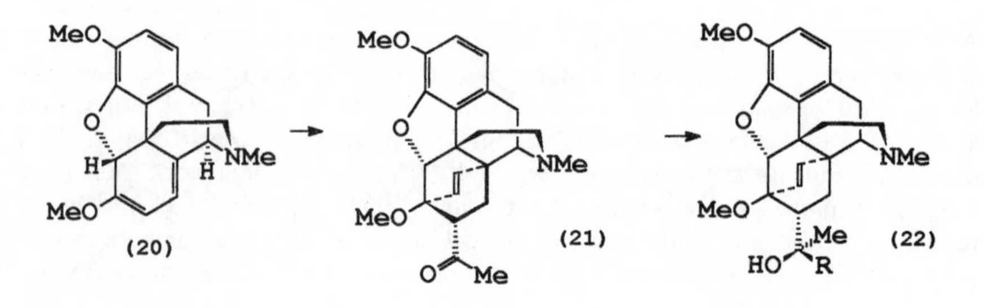

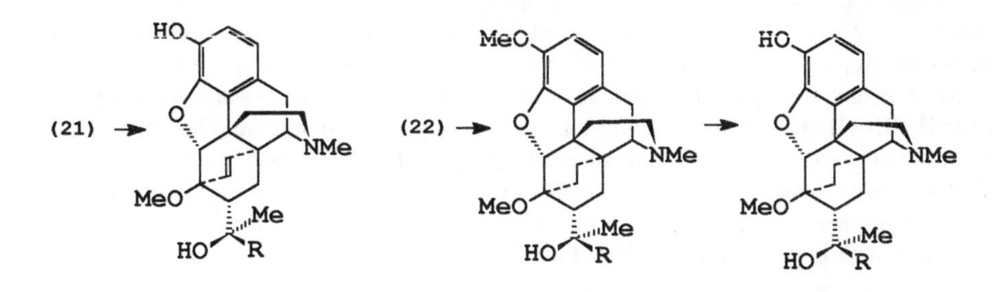

Scheme 10 Synthesis of Diels-Alder adducts and Derivatives from Thebaine

In most of the classical compounds such as dilaudid (7,8-dihydro-6-oxomorphine), metopon (5-methyldilaudid) and numorphan (14-hydroxydilaudid), improvements resulted. However, in a second generation of compounds derived by Diels-Alder addition reaction of thebaine with butenone to afford 6,14-endoethenotetrahydrothebaine (21) followed by Grignard formation with RMgBr of a tertiary alcohol (22) (Scheme 10), greatly increased analgesic activity has been found[28] both in the etheno, hydrogenated ethano products and their demethylated analogues. In this group the phenolic base, butoprenine prepared from thebaine has a 3-hydroxyl group, an ethano group bridging the 6,14-positions, no substituent at the 7-position and a N-cyclopropylmethyl group in place of the usual N-methyl. Although the alkaloids may be regarded as heterocyclic compounds spectacular achievements in semi-synthesis have been made in the field of heterocyclic antibiotics.

6 THE ANTIBIOTICS

1.5.1 Penicillins and Semi-synthetic Penicillins. The chemistry of the penicillins, (23), in *Penicillium notatum*, namely penicillin V (R = $PhOCH_2$), G (R = $PhCH_2$), F (R = n-C_5H_{11}), X (R = 4-HOC_6H_4) and K (R = n-C_7H_{15}), their heterogeneity, structure determination and synthesis has become legendary[29] involving a massive effort of manpower over many years.

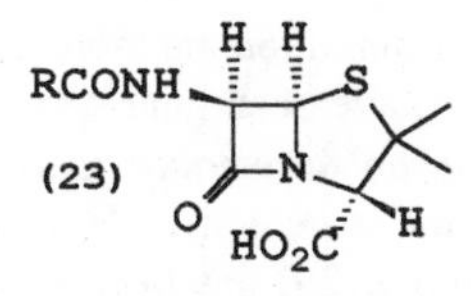

Although the curative power of the fermentation-produced penicillins for example V and G made medical history, it was apparent by the middle of the fifties that they were no longer effective against several important pathogenic bacteria. This was attributed to the ability of these bacteria to produce an enzyme, β-lactamase which cleaved the four-membered ring. Beecham scientists, who were prominent in this work, examining the inactivation proposed that a bulky R substituent could sterically block this hydrolytic reaction. To effect this notion required the availability of the parent amino compound of (23), 6-aminopenicillanic acid (25, 6-APA) a new compound which remarkably had not been isolated previously degradatively. By the use of precursor-free broths, the enzymatic hydrolysis of natural penicillin and novel separatory techniques, this compound was recovered. Acylation by a variety of methods with new acylating groups led to the semi-synthetic penicillins (26) such as ampicillin [26, R^1 = PhCH(NH_2)], amoxycillin [R^1 = $HOC_6H_4CH(NH_2)$] and broxil [R^1 = $(MeO)_2C_6H_4$]. These compounds were low in toxicity, orally administerable and had reasonable stability towards β-lactamase. For the production of these compounds, the use of *Penicillium chrysogenum* has been favoured and following special fermentation, iso-penicillin N (24) isolated, cleaved enzymatically and reacylated with the appropriate side-chain (Scheme 11).

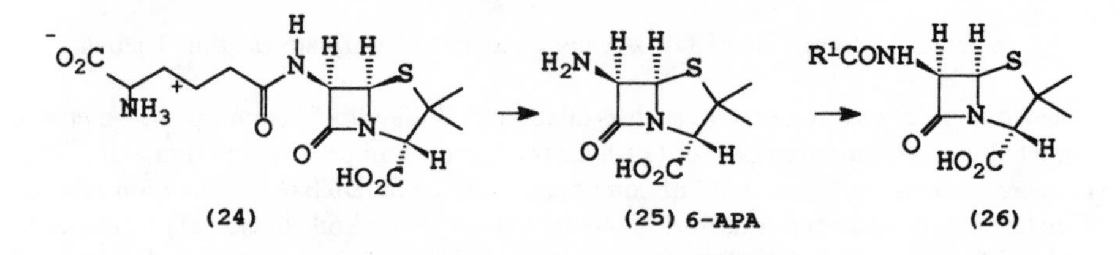

Scheme 11 Semi-synthesis of Unnatural penicillins

1.5.2 Cephalosporins and Semi-synthetic Compounds. Cephalosporin C (27) was isolated in 1956 and was found to possess greater stability to acidic conditions and to β-lactamase than the natural penicillins although a weaker anti-bacterial action. This was enhanced however in certain acyl derivatives of 7-aminocephalosporanic acid (28). Since

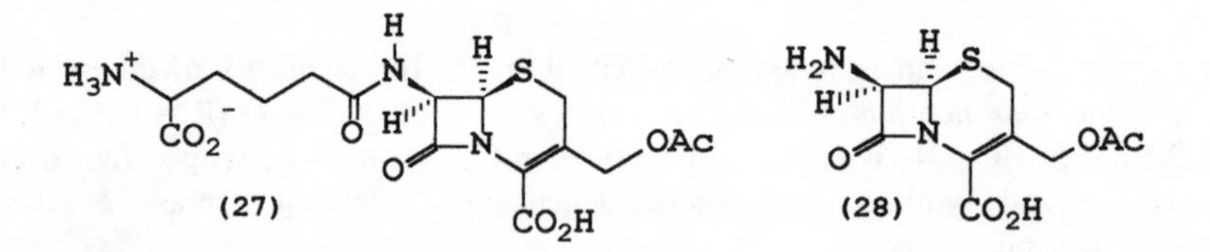

no straightforward enzymatic or fermentation method is accessible as in the penicillin series, semi-synthetic cephalosporins have been generally obtained by acidic hydrolysis of Cephalosporin C from *Cephalosporium acremonium*, to afford 7-aminocephalosporanic acid, separation by precipitation at the iso-electric pH, and acylation. Thus the powerful broad spectrum compound ceftaziidime (29) has been obtained by semi-synthesis.

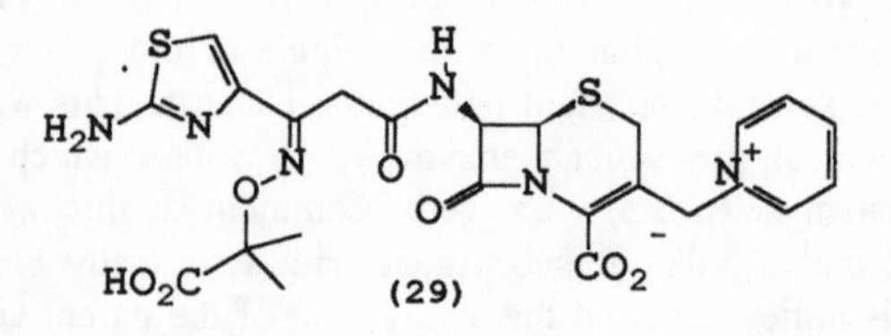

For compounds lacking the 3-acetoxy group it has proved possible to transform penicillins to cephalosporins[30]. Thus, from penicillin V, cephalexin (30) has been derived (Scheme 12), through protection of the carboxyl group, sulphoxide formation, Pummerer rearrangement, acylation and removal of the two protective groups.

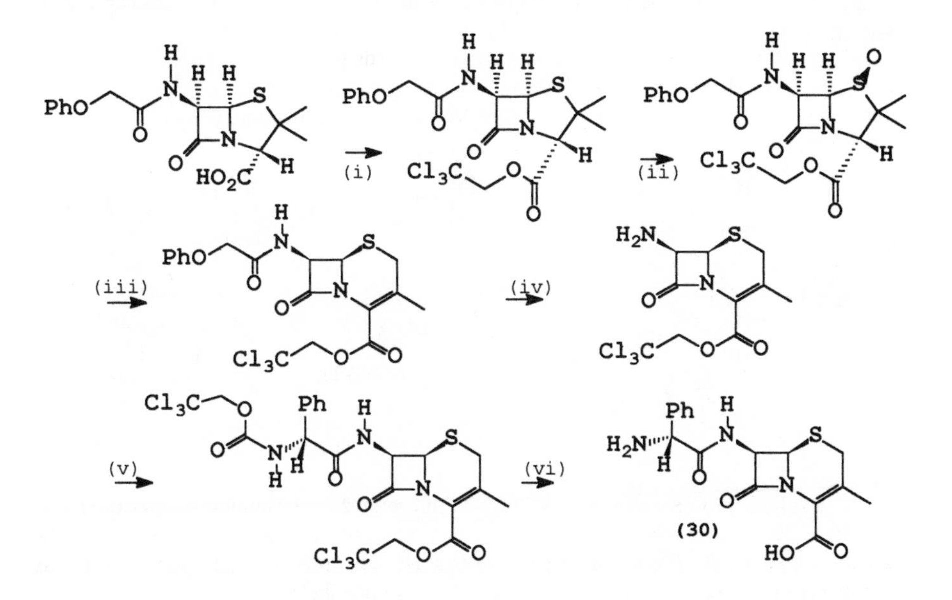

Scheme 12 Synthesis of Cephalexin from Penicillin V

Reagents: (i) Cl_3CCH_2OH, DCC,py., (ii) MCPBA, (iii) DMF, Δ, (iv)PhH, PCl_5 ; MeOH; H_2O, (v) $Cl_3CCH_2O_2CNHCH(Ph)CO_2H$, DCC, (vi) PG removal, $-CO_2$.

Cephalosporins have broad spectrum activity often against organisms which had become resistant to penicillins through lactam ring cleavage, non-toxicity and acid stability all of which probably account for their 52% share of the market (in 1989) compared with the penicillins (14%) and tetracyclines (6%).

1.5.3 Tetracyclines. Semi-synthesis has been widely used in the tetracycline class[31] for the interconversion of the principal members such as the reductive single-stage conversion of chlorotetracycline to the parent compound, tetracycline, and the transformation of oxytetracycline to doxycycline in three steps[32].

7 CARBOHYDRATES

Carbohydrate intermediates have been widely used as stereochemical templates in many syntheses[2,33]. The industrial synthesis of Vitamin C, (ascorbic acid) from D-glucose and from D-galactose by chemical stages are classic examples predating the chiron approach. More recently in a two-step microbiological route[34] D-glucose has been converted to L-2-oxogulonic acid and thence by acidic treatment to ascorbic acid. An intermediate L-gulonolactone, obtained by the hydrogenation of Vitamin C has been employed[35] in a synthesis of a polyhydroxypiperidine, deoxymannojirimycin (31), one of three related compounds which are powerful inhibitors of glycosidases, substances capable of the selective hydrolysis of sugars. The method (Scheme 13) was dependent on selective hydroxyl group protection, azide formation at the free hydroxyl group, reductive rearrangement to connect C1 and C5 giving a pyridone, followed by reduction to the piperidine (31).

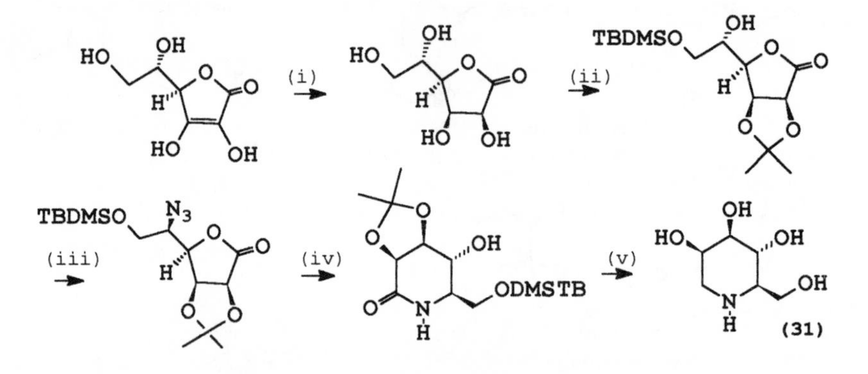

Scheme 13 Synthesis of a 3,4,5-Trihydroxy-2-hydroxymethylpiperidine

Reagents: (i) Pd-C, H_2 ,(ii) Me_2Bu^tSiCl ; Me_2CO, H^+ , (iii) $(CF_3SO_3)_2O$, NaN_3 , (iv) Pd-C, H_2 , (v) BH_3Me_2S, H^+.

8 LIPIDS

Carbohydrate intermediates have been employed especially for the synthesis of prostaglandins, leukotrienes and thromboxanes. Thus α-methyl-D-glucose has been the starting point for a seven step partial synthesis[36] of thromboxane B_2 , while triacetyl-D-glucal has afforded leukotriene B_5 in twelve stages[37].
The lipidic vitamins were considered earlier and in this final section the use of semi-synthesis in glyceride, phospholipid, pheromonic and phenolic lipid chemistry is described.

1.7.1 Glyceride-derived Products. This class of substances represents the largest volume of natural products in which semi-synthetic transformations are universally involved . Thus, the preparations of fatty acids, alcohols, amides, amines, mono- and diglycerides and their surfactant derivatives almost entirely devolve on the nucleophilic substitution reactions of glycerides[38]. Likewise the synthesis[39] of platelet activation factor (PAF), 1-O-alkyl-2-acetyl-sn-glycero-3-phosphocholine is dependent on a glyceride intermediate.
Methyl linoleate (32) derived by interesterification from linolein has been converted to a pheromone bouquet comprising ZZZ-3,6,9-eicosane and ZZZ-3,6,9-heneicosane by the reactions shown (Scheme 14)

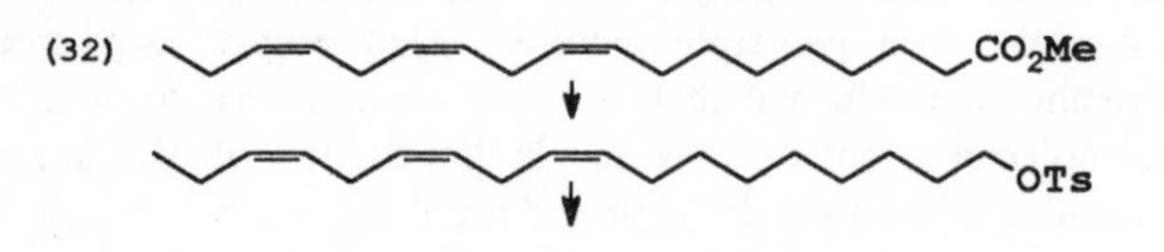

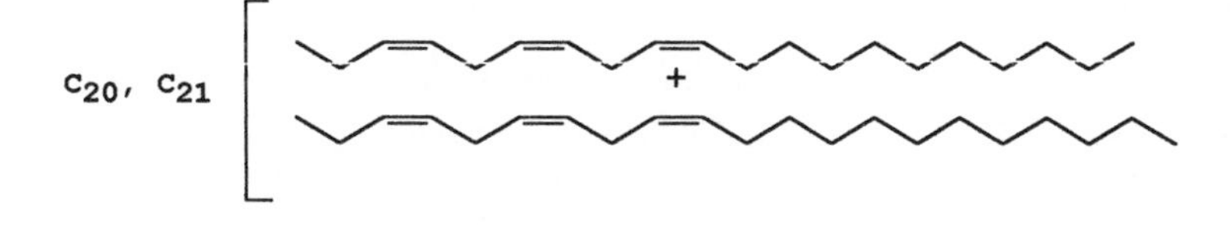

Scheme 14 Synthesis of C_{20} and C_{21} Alkatriene Pheromones

Reagents: (i) LAH, Et_2O ; 4-TSCl, Py. ,(ii) Li(n-Pr)$_2$Cu + LiEt$_2$Cu ; NH_4Cl

1.7.2 Phenolic Lipid-derived Products. The phenolic lipids comprise a comparatively little known group of alkyl, alkenyl, alkadienyl and alkatrienylphenols, resorcinols and salicylic acids[5,13] derived from *Anacardium occidentale* by hot industrial processing, which results in decarboxylation of the the principal natural component, anacardic acid (34), the recovery of the cashew kernel and formation of the by-product, technical cashew nut-shell liquid. Cardanol (33) is the principal component phenol in this complex mixture (R = $C_{15}H_{31-n}$, n = 0, 2, 4, 6) and one of its primary uses after mildly acidic side-chain polymerisation and formaldehyde polymerisation (Scheme 15) is in friction dusts with the polyimide 'Kevlar' for the automobile industry.

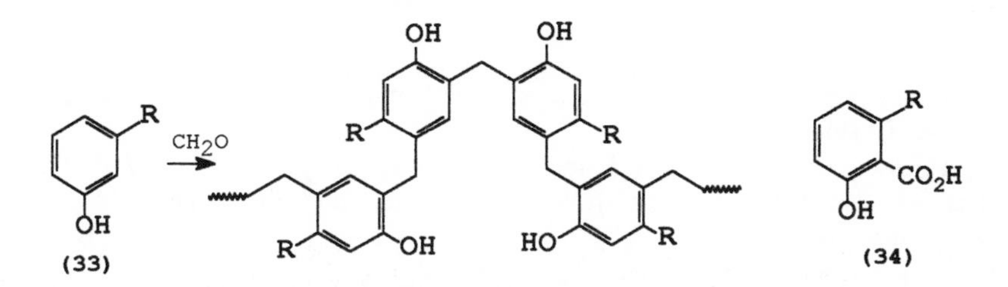

Scheme 15 Simplified Representation of Cardanol/Formaldehyde Reaction

It can be used for many of the same purposes as t-nonylphenol. Thus polyethoxylation[40] affords a non-ionic detergent which is substantially more biodegradable than its t-nonyl analogue. Many other potential applications have been found, for example as an aldoxime for the extraction of copper, a diol for the recovery of borates, a Mannich base for the curing of epoxy resins, an antioxidant by t-butylation, and as a thiobisphenol. Anacardic acid (34) the main component of natural cashew nut-shell liquid also has experimental uses, for example as a water soluble component for the fabrication of particle boards. In conclusion it is clear that there are many other instances of semi-synthesis as for example in the extensive chemistry of cellulose, in the peptides such as the dipeptide sweetener 'aspartame' and in other fields. Recent examples in semi-synthesis are found in important anti-cancer drugs, the 9-amino derivative of camptothecin and in taxol chemistry. Myrcene continues to attract interest as shown by its usage for a technical synthesis of one of the components of San Jose scale's pheromone, 3,7-dimethyleneoctyl propionate.[41]

9 CONCLUSIONS

Total synthesis, utilising raw materials from a wide variety of sources, is probably the main avenue through which progress in organic synthetic methodology is made, an activity of wide general value. Industrial implementation directed to producing useful medicinal and technical compounds has inevitably to adopt alternative more compact strategies, one of which can be partial synthesis from a natural product intermediate. In this account some historical and topical examples of the partial synthesis approach to obtaining a variety of natural products have been instanced in Schemes 1 to 9 and 13. In semi-synthesis the opportunity to either improve on nature or to devise an unnatural substance from a natural intermediate source, exists and has been used as depicted in the examples of Schemes 10, 11, 12, 14 and 15. These are only a few of many instances which have employed these two methodologies.

Semi-synthesis/partial synthesis, asymmetric synthesis, enzymatic/biochemical experimentation[42] and microbiological/genetic techniques probably are all likely to have or will have a future influence at some stage in the preparation on an industrial scale of organic compounds of value to humanity. The accessibility and cheapness of the natural product source is probably a major factor in the first named procedures of partial and semi-synthesis and for the other more recent techniques, cost /benefit considerations may well be a dominant factor.

References

1. P. G. Sammes, in 'Comprehensive Medicinal Chemistry', Pergamon, Oxford, 1990, Vol. 2, Chapter 7.1, p. 255.
2. S. Hanessian, J. Franco and B. Laroche, *Pure and App. Chem.*, 1990, **62**, 1887; S. Hanessian, in 'Total Synthesis of Natural Products; The Chiron Approach', Pergamon, Oxford, 1983.
3. C. H. Heathcock, in 'Asymmetric Synthesis', Ed. J. D. Morrison, Academic Press, New York, 1984, Vol. 3, p. 111; A. Koskinen, in 'Asymmetric Synthesis of Natural Products', Wiley, New York, 1993.
4. A. N. Collins, G. N. Sheldrake and J. Crosby, Eds., 'Chirality in Industry', J. Wiley, 1992, Chichester.
5. J. H. P. Tyman, in 'Studies in Natural Products Chemistry', Ed. Atta-ur-Rahman, Elsevier, 1995, Amsterdam, Vol. 17, p. 601.
6. I. Fleming, in 'Selected Organic Syntheses', J. Wiley, London, 1973.
7 E. J. Ariens, *Med. Res. Rev.*, 1988, **8**, 309.
8 R. A. Sheldon, *Chem. and Ind.*, (London),1992, 903.
9 R. B. Woodward, *Pure and App. Chem.*, 1968, **17**, 519.
10 Y. Kishi, *Chem. Scripta*, 1987, **27**, 573.
11 C.F. Seidel and M. Stoll, *Helv. Chim. Acta*, 1959, **42**, 1830.
; J. H. P. Tyman and B. J. Willis, *Tetrahedron Letters*, 1970, 4507.
12 B. D. Sully, Chem. and Ind., 1964, 263; M.F. Carroll, Rep. Progr. Appl. Chem., 1961, 500; USP 2871271, (Glidden Co.).
13 J.H.P. Tyman, 'Synthetic and Natural Phenols', Ch. 13, Elsevier, 1996, (in the press).
14 O. Isler and P. Schudel, *Advan. Org. Chem.*, Eds. R.A. Raphael and E.C. Taylor, Interscience 1963, **4**, 1.
15 L.F. Fieser and M. Fieser, 'Topics in Organic Chemistry', Reinhold, New York, 1963.

16 R.E. Marker and J. Krueger, *J. Am. Chem. Soc.*, 1940, **62**, 3349; R.E. Marker, T. Tsukamoto and D.L. Turner, ibid., 2525.
17 G. Rosenkranz, O. Mancera, F. Sondheimer and C. Djerassi, *J. Org. Chem.*, 1956, **21**, 520.
18 H.L. Dryden, G.M. Webber and J. Wieczorek, *J. Am. Chem. Soc.*, 1964, **86**, 742.
19 J. Kalvoda, K. Heusler, H. Uebenwasser, G. Anner and A. Wettstein, *Helv. Chim. Acta*, 1963, **46**, 1361.
21 D.H. Peterson and H.C. Murray, *J.Am. Chem. Soc.*, 1952, **74**, 1871.
22 J.H. Chapman, J. Elks, F.H. Phillips, T. Walker and L.J. Wyman, *J. Chem. Soc.*, 1956, 4344.
23 D.H.R. Barton, G.M. Beaton, G.M. Geller and M.M. Pechet, *J. Am. Chem. Soc.*, 1961, **77**, 2400.
24 E.J. Corey and W.R. Hertler, *J. Am. Chem. Soc.*, 1958, **80**, 2903.
25 W.F. Michne, 'Analgesics', in Kirk-Othmer Encyclopdia of Chemical Tecnology, Vol.2, p.274, 3rd. Edn., Wiley-Interscience, 1978.
26 R.B. Barber and H. Rapaport, *J. Med. Chem.*, 1976, **19**, 1175.
27 K.C. Rice, J. Med. Chem., 1977, 20, 164; J.A. Lawson and J.I. DeGraw, ibid., 165.
28 K.W. Bentley and D.G. Hardy, *J. Am. Chem. Soc.*, 1967, **89**, 3281.
29 H.T. Clarke, J.R. Johnson and R. Robinson, Eds., 'The Chemistry of Penicillins', Princeton Univ. Press, 1949; M.S. Manhas and A.K. Bose, 'Synthesis of Penicillins, Cephalosporins and their analogues', M. Dekker, New York, 1969; J.E. Baldwin, 'On the Biosynthesis of Penicillin and Cephalosporin', 16th IUPAC Symp. Chem. Nat. Prod., Kyoto, Japan, 1988.
30 R.M. Morin and M. Gorman, Eds., 'Chemistryand Biology of β-lactam Antibiotics', Vol.1, 'Penicillins and Cephalosporins',Acadeic Press, New York, 1982.
31 J.J. Hlavka, G.A. Ellestad and I. Chopra, 'Tetracyclines', Kirk-Othmer Encycl. Chem. Tech., 4th Edn., Vol. 2, p.331.
32 R.K. Blackwood, J.J. Beereboom, H.H. Rennard, M. Schach von Wittenau and C.R. Stephens, *J.Am. Chem. Soc.*, 1963, **85**, 3943.
33 B.J. Fitzsimmons and B. Fraser-Reid, *Tetrahedron*, 1984, **40**, 1279.
34 A.J. Pratt, *Chem. in Brit.*, 1989, **5**, 282.
35 G.W.J. Fleet, *Chem. in Brit.*, 1989, **25**, 287.
36 E.J. Corey, M. Shibasaki and J. Knolle, *Tetrahedron Letters*, 1977, 1625.
37 E.J. Corey, S.G. Pyne and W.-G. Su, *ibid*, 1983, **24**, 4883.
38 K. Coupland, in 'Surfactants in Lipid Chemistry', Ed. J.H.P. Tyman, Royal Society of Chemistry, Cambridge, 1992, p1.
39 H. Eibl, *Angew. Chem. Internat. Ed. Engl.*, 1984, **23**, 257.
40 J.H.P. Tyman, ref. 38., p.159.
41 H.-S. Zhang, J.-W.Chen and Z.-H. Xu, *Yuoji huaxse*, 1993, **13**, 504.
42 M.J. Warren, C.A. Roessner, S.-L. Ozaki, N.J. Stolowich, P.J. Santander and A.I. Scott, *Biochemistry*, 1992, **31**, 603.

Appendix* (The Hoffmann la Roche route uses pseudoionone (A)prepared from 'methylbutynol' to methylheptenone with diketene (-CO_2), to dehydrolinalol and repeat).

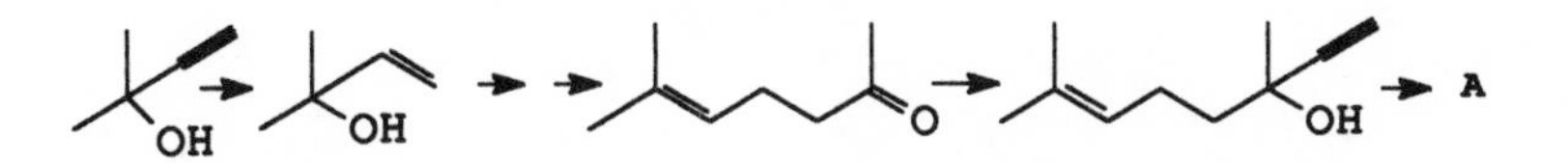

Biosynthetically-Patterned Terpenoid Biotransformations

J. R. Hanson

SCHOOL OF CHEMISTRY AND MOLECULAR SCIENCES, UNIVERSITY OF SUSSEX, BRIGHTON, SUSSEX BN1 9QJ, UK

1 INTRODUCTION

The biotransformation of natural products can significantly enhance the value of phytochemicals. The bioconversion of plant sterols for steroid hormone synthesis is a well-known example. Microbiological hydroxylations are mild, environmentally friendly reactions that often possess a regio-specificity which complements that of chemical transformations. The underlying carbon skeleton of a molecule is rarely altered, a feature which is of value when biotransformations are used for structural correlations.

Microbiological hydroxylations can be divided into two broad areas. On the one hand there are xenobiotic transformations in which the substrate, for example a steroid, is alien to the transforming organism. In contrast there are biosynthetically-directed transformations in which the substrate possesses a formal relationship to a natural biosynthetic intermediate on one of the biosynthetic pathways found in the organism. Apart from their obvious potential in yielding structural analogues of biologically active fungal metabolites, these transformations can yield useful information on the stereoelectronic requirements and constraints of biosynthetic pathways.

The biosynthesis of the diterpenoid fungal metabolite, aphidicolin (2) has been examined [1-6] in the fungus, *Cephalosporium aphidicola.* This tetracyclic diterpenoid has attracted interest as an anti-viral agent and as a specific inhibitor of DNA polymerase α.[7] The late stages in the biosynthesis were shown to involve a sequential series of hydroxylations of aphidicolan-16β-ol (1) at C-18, C-3α and then C-17. Aphidicolin is then metabolized further by the fungus by hydroxylation at C-6β and C-11β.

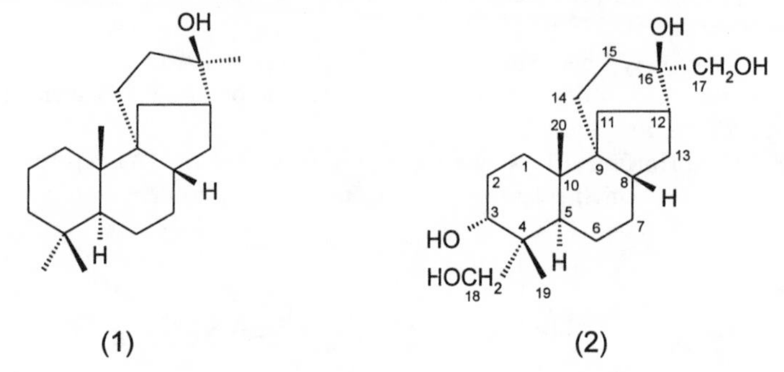

(1) (2)

We have utilized this organism to mediate the biotransformation of natural products with a number of objectives:- firstly to obtain biosynthetic information on enzyme: substrate structural relationships; secondly to explore the role of certain hydroxyl groups in directing a biosynthesis and having the role of 'control elements'; thirdly to prepare, biosynthetically, analogues of aphidicolin; and finally to examine the borderline between xenobiotic and biosynthetically patterned biotransformations. It is worth noting that rather than selectively protect the hydroxyl groups of aphidicolin, it is easier to make chemical modifications and then allow the organism to insert the hydroxyl groups characteristic of aphidicolin.

2 BIOSYNTHETIC INFORMATION FROM BIOSYNTHETICALLY PATTERNED BIOTRANSFORMATIONS

The stereochemical relationship between the substrate and the enzymatic oxidant is of interest in defining the geometry of the active site. The hydroxylations of the methyl groups, C-17 and C-18, are important steps in aphidicolin biosynthesis. Whilst a methyl group is a freely rotating entity, a $3\alpha\rightarrow 18$-oxetane ring (3) served to lock C-18 into a particular conformation. Some $3\alpha\rightarrow 18$-oxetane analogues of intermediates in the biosynthesis of aphidicolin were prepared and labelled stereoselectively at C-18 with deuterium (3). Biotransformation studies[8] using these substrates with *Cephalosporium aphidicola* showed that the 18-*pro*-R hydrogen atom was removed in the hydroxylation to give (4) suggesting that there was a *re* stereochemical relationship between the substrate and the oxidant for the normal hydroxylation (5).

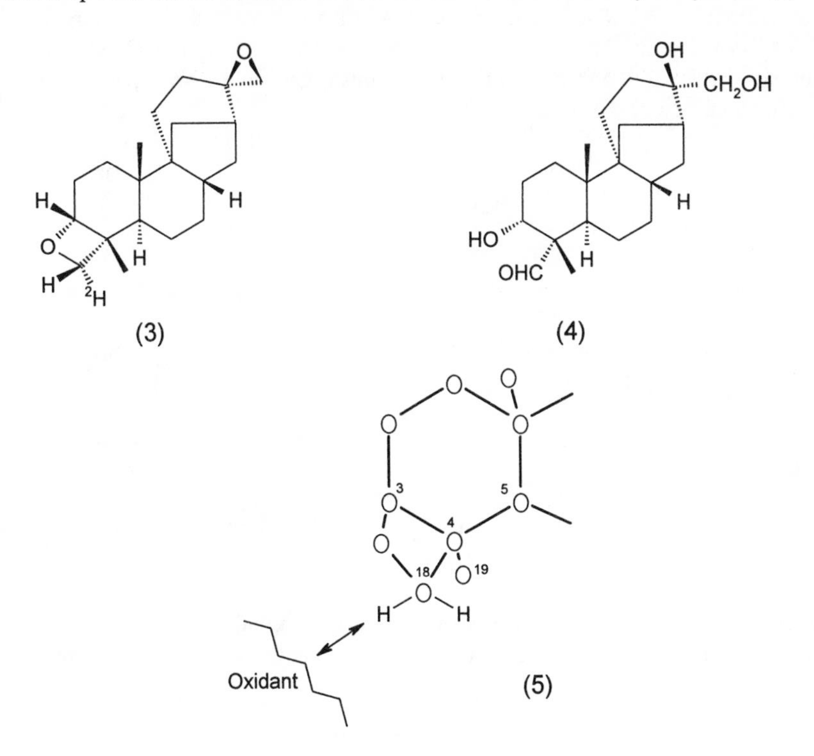

An alternative strategy was adopted to establish the stereochemical relationship between the substrate and oxidant at C-17. Hydroxylation of the 16-ethyl analogue proceeded preferentially to afford the 17(R)-homologue of aphidicolin suggesting that hydroxylation took place from the *re* face at this centre.[9]

3 SEQUENCE OF HYDROXYLATION

The biosynthesis of aphidicolin involves the sequential hydroxylation of aphidicolan-16β-ol (1) at C-18, C-3α and C-17 with further hydroxylation taking place at C-6β and C-11β. The factors governing a metabolic grid relationship between the variously hydroxylated aphidicolanes have been investigated.[3,10] Whilst 16β,17-dihydroxyaphidicolane (6) was converted into aphidicolin in 3.5% yield, 16β,18-dihydroxyaphidicolane (7) was converted into aphidicolin in 20.5% yield. 3α, 16β - Dihydroxyaphidicolane (10)[11] whilst possessing a 3α-hydroxyl group, lacks the 18-hydroxyl group which would have been inserted earlier in the biosynthesis. There were several possibilities for its biotransformation. Firstly the fungus might 'correct' the mistake by inserting the hydroxyl group (8) and then produce aphidicolin as normal. Secondly it might 'ignore' the mistake and produce the 3α, 16β, 17-triol (9) which is the 18-desoxy analogue of aphidicolin. Thirdly the fungus might reject the substrate or hydroxylate it using a different pathway. In practice the fungus converted the substrate to the 3α, 6β, 16β- (11) and 3α, 7β, 16β- (12) triols in rather low yield and, using labelled material, there was no detectable incorporation into aphidicolin. This suggested that 18-hydroxylation plays a directing role in the biosynthesis and that hydroxylation at C-18 cannot take place in the presence of a 3α hydroxyl group. A 3-hydroxyl group has a comparable effect on hydroxylation at C-19 of the *ent*-kaurene skeleton in *Gibberella fujikuroi.* In this organism a 19-oxygen function also appears to have a directing role.

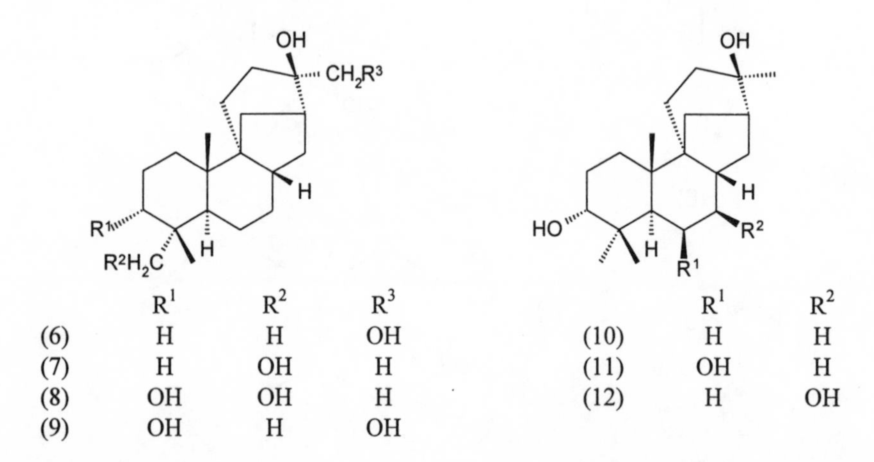

	R^1	R^2	R^3
(6)	H	H	OH
(7)	H	OH	H
(8)	OH	OH	H
(9)	OH	H	OH

	R^1	R^2
(10)	H	H
(11)	OH	H
(12)	H	OH

When an 18-hydroxyl group is present in un-natural aphidicolane substrates such as 16β,18-dihydroxyaphidicol-2-ene (13),[12] the fungus will produce aphidicolin analogues such as 2α, 3α -epoxy-16β, 17, 18-tri-hydroxyaphidicolane (14) and 6β , 16 β, 17, 18-tetrahydroxyaphidicol-2-ene (15).

4 SKELETAL VARIATIONS

The various families of tetracyclic diterpenoids possess a range of closely related carbon skeleta. The skeletal requirements for this biotransformation have been explored with a number of substrates. Whilst the A/B ring junction of the stemodane and aphidicolane diterpenoids is the same, the disposition of the bicyclo-3, 2, 1-octane ring system that constitutes rings C and D, is different. Incubation of stemodin (16) and stemodinone (22) from the plant *Stemodia maritima,* afforded a number of hydroxylated derivatives such as (17) - (21) from stemodin.[13]

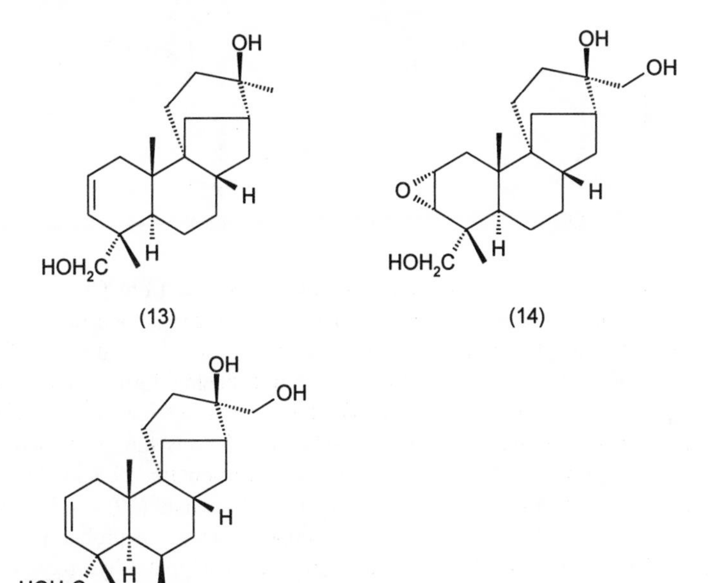

(13) (14)

(15)

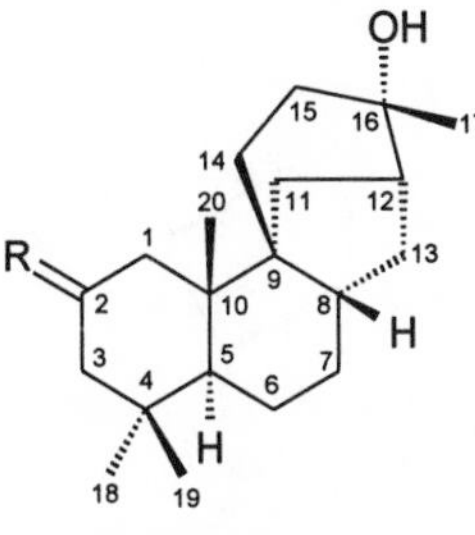

(16) R=α-OH, β-H
(22) R=O

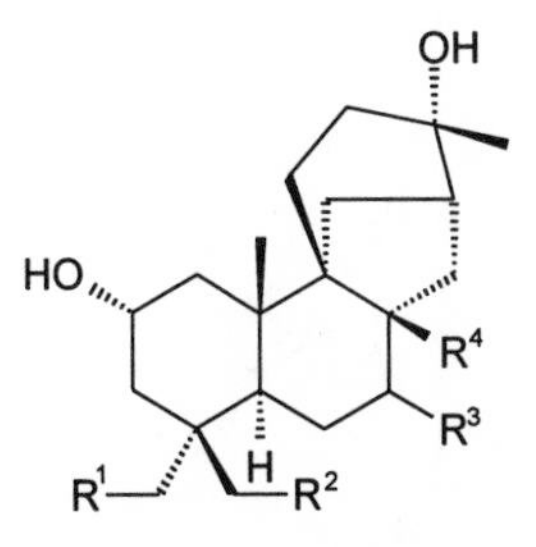

(17) R^1=OH, R^2=R^3=R^4=H
(18) R^1= R^3=R^4=H, R^2=OH
(19) R^1=R^2=R^4=H, R^3= α-OH, β-H
(20) R^1=R^2=R^4=H, R^3= α-H, β-OH
(21) R^1= R^2=R^3=H, R^4=OH

This biotransformation paralleled the xenobiotic microbiological hydroxylation of stemodin by other micro-organisms rather than the biosynthetically-patterned transformations associated with the aphidicolane metabolites. Although the hydroxylation at C-18 is common to both, there was no hydroxylation of the C-17 methyl group of stemodin or stemodinone.

ent-Kauranoid diterpenes are widespread amongst the Compositae and Labiatae. *ent*-16β, 19-Dihydroxykaurane (23) is an analogue of an intermediate in aphidicolin biosynthesis. Its biotransformation[14] by *C. aphidicola* was therefore of interest in the context of the borderline between analogue and xenobiotic biotransformation. The major product of the hydroxylation was the 11β-alcohol (24).

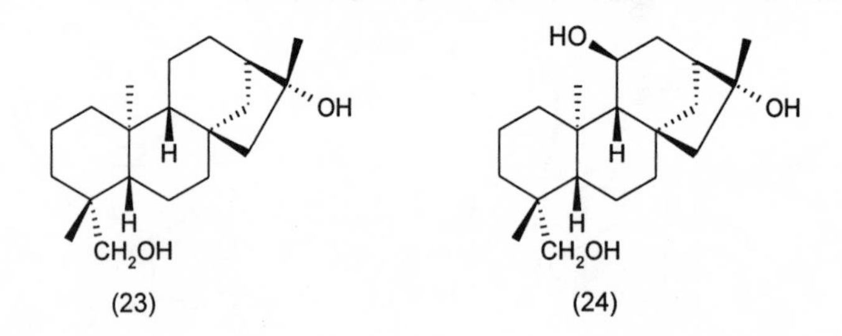

A number of highly hydroxylated *ent*-15-oxokaur-16-enes from Chinese medicinal plants possess tumour-inhibitory activity. The α-methylene ketone functionality can be introduced chemically into more readily available simpler kauranoid diterpenes such as xylopic acid (25). The possibility of introducing further functionality by microbiological means using *C. aphidicola,* has been investigated.[15, 16] The major metabolite from *ent*-19-hydroxykaur-16-en-15-one (26) was the 3α-alcohol, *ent*-3β, 16β, 19-trihydroxykauran-15-one (29) whilst the corresponding 19-acid (27) and ester (28) gave *ent*-11α , 16β -dihydroxy-15-oxokauran-19-oic acid (30) and its ester. The hydration of the exocyclic methylene was a disappointment. It is probably a reduction and hydroxylation since a nucleophilic addition would lead to a β-hydroxy-ketone.
The commercially available bicyclic diterpenoid, sclareol (31), from *Salvia sclarea,* was an interesting substrate [17] since it bears a similarity to the plausible bicyclic intermediates involved in the biosynthesis of aphidicolin.

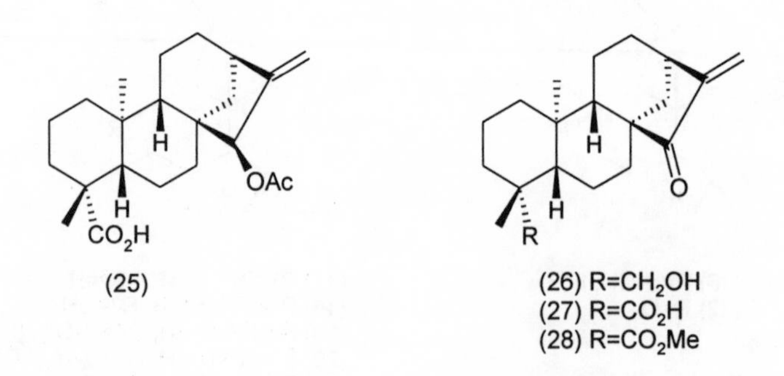

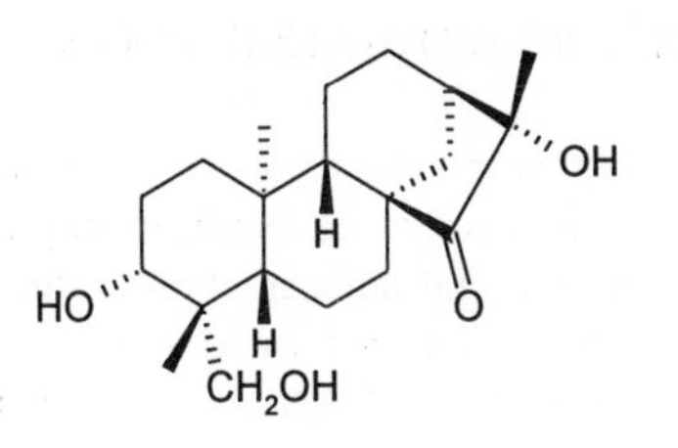

(29)

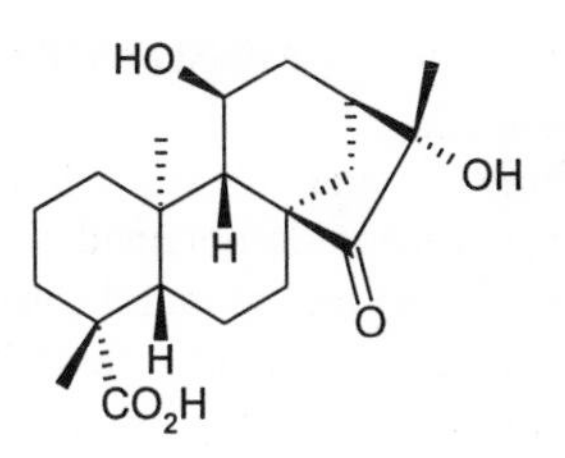

(30)

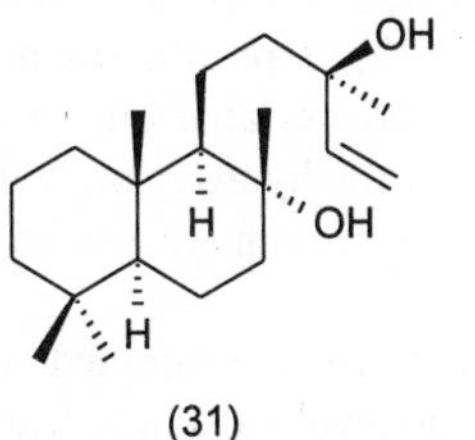

(31)

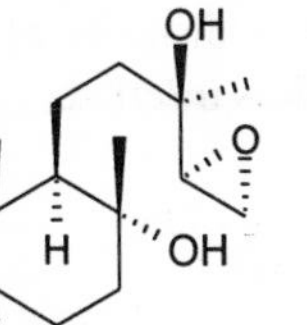

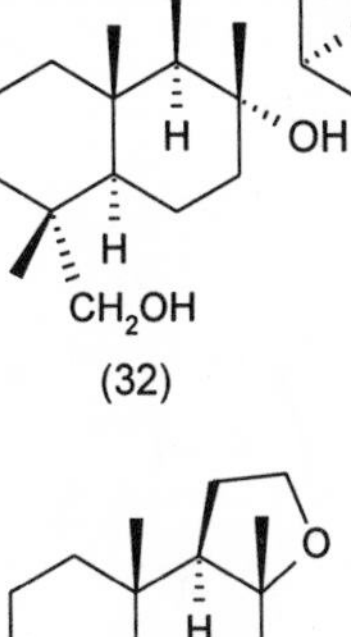

(32)

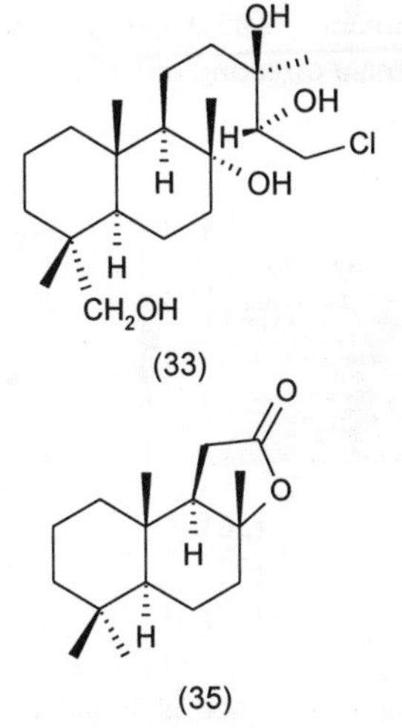

(33)

(34)

(35)

Furthermore the structure may be super-imposed on that of aphidicolin. Sclareol (31) was hydroxylated at C-3β, C-14 and C-18. The hydroxylation at C-18 is a typical aphidicolin biosynthetic step. One of the products (32, purified as the chlorohydrin 33) had a new oxygen function at C-14 with the generation of a new chiral centre. This centre has the same chirality as that formed in the hydroxylation of 17-methyl-3α, 16β, 18-trihydroxyaphidicolane. This suggested that the geometrical relationship of the substrate to the enzyme surface during the delivery of the oxygen was the same as that in aphidicolin biosynthesis. When the C-13 hydroxyl group of sclareol is superimposed on the C-16 hydroxyl group of aphidicolin, the C-14 of sclareol coincides with C-17 of aphidicolin.

The perfumery product Ambrox^{R} (34) and sclareolide (35) contain the six-membered rings A/B of aphidicolin together with a five-membered ring. These compounds were hydroxylated at C-3β, C-6β and C-12.

5 MONO- AND SESQUITERPENOID BIOTRANSFORMATIONS

Rings C and D of aphidicolin bear a formal relationship to a number of sesquiterpenoids obtained from essential oils. A number of these, exemplified by cedrol (36) and globulol (39), possess a tertiary alcohol adjacent to a methyl group reminiscent of aphidicolan-16β-ol (1). Initially their biotransformation was examined in the hope that we might observe the formation of a 1, 2-glycol[18, 19] - chemically quite a difficult transformation to carry out without rearrangements taking place. Although these expectations were not fulfilled, the products were of interest. At first sight the major products obtained from cedrol, 8-epicedrol and some relatives which involved hydroxylation at C-3 (37, 38), bore little resemblance to those from globulol (39) and epiglobulol which were hydroxylated on the methyl group adjacent to the cyclopropane ring (42). However if the hydroxy:methyl portion of the molecules are superimposed, the positions of hydroxylation are also co-incident.

This geometrical relationship has been examined with other substrates. Thus incubation of 4-*t*-amyl- and 4-*t*-butylcyclohexanols leads to hydroxylation on the 4-substituent. These can also be superimposed. There is a distance of about 7Å between a directing oxygen atom and the oxygen atom which is inserted.

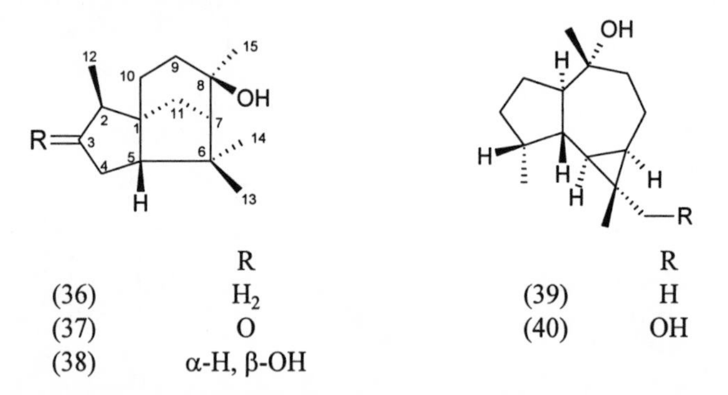

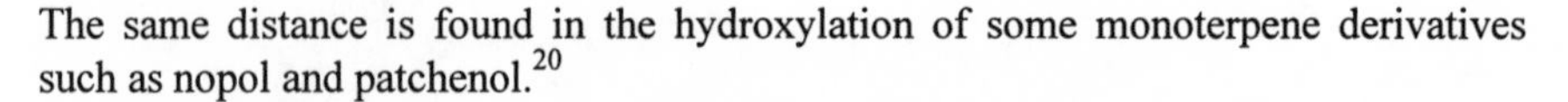

The same distance is found in the hydroxylation of some monoterpene derivatives such as nopol and patchenol.[20]

6 STEROID BIOTRANSFORMATIONS

In order to establish the pattern of xenobiotic transformations by this organism, the biotransformation of some typical steroids has been examined.[21,22] Progesterone (41) was hydroxylated at C-11α and then at C-6β to give the 6β, 11α -diol (43). There was a minor pathway which involved hydroxylation at C-17α (42) and then at C-12β (43). On the other hand testosterone and a number of its relatives were hydroxylated at C-6β with very little oxidation at C-11. Indeed the next most common site to be hydroxylated in the androgens was C-14α. The ease of hydroxylation at the allylic position of the unsaturated ketone was also found in smaller molecules such as the sesquiterpenoid from *Citrus* sp., nootkatone (45) which was hydroxylated at the corresponding allylic position.

Studies with a range of steroidal substrates have revealed that *Cephalosporium aphidicola* is capable of hydroxylating many different steroidal structural types. Whereas good transformations were obtained with mono-oxygenated sesquiterpenoids, the best steroidal transformations required two oxygen functions.

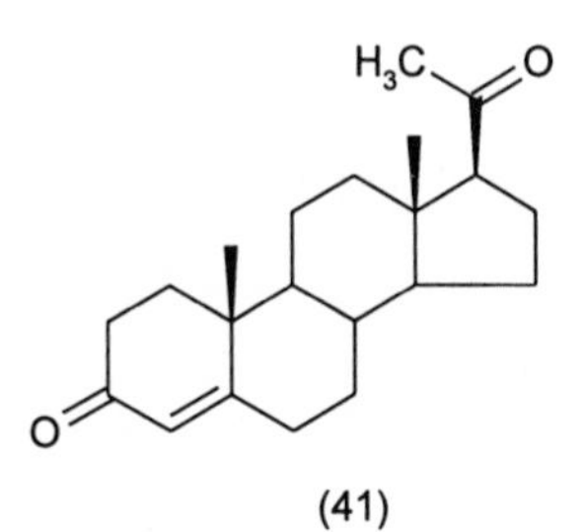

(41)

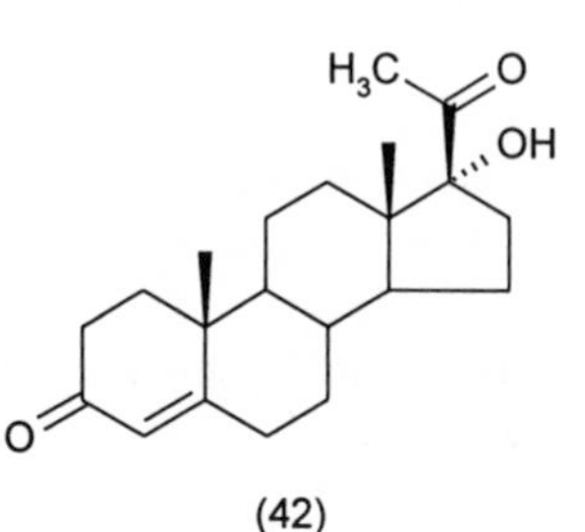

(42)

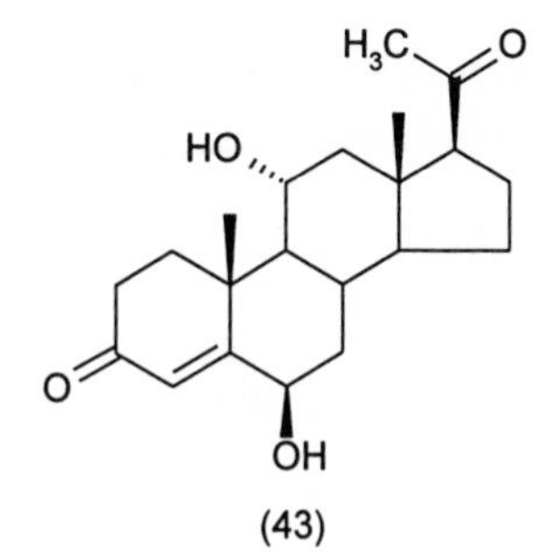

(43)

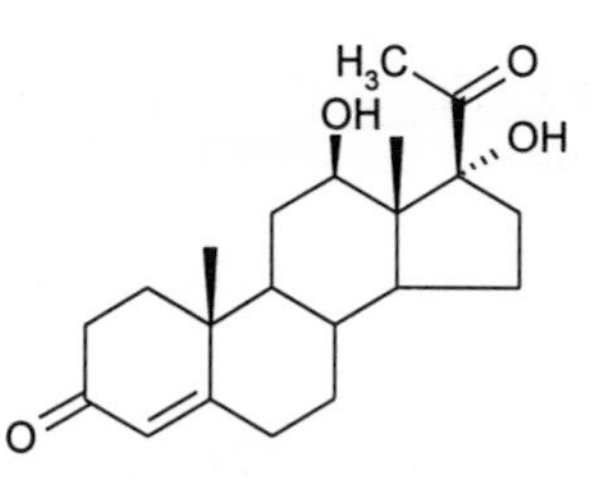

(44)

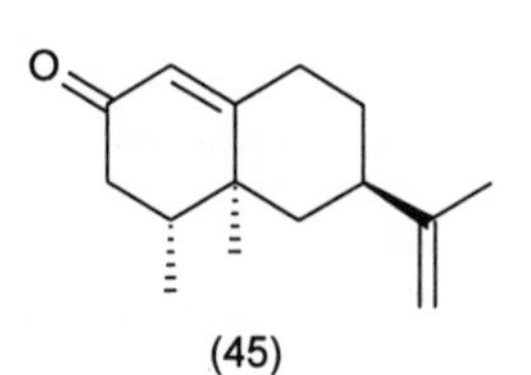

(45)

ACKNOWLEDGEMENTS

I am grateful to Drs A Farooq, R Guillermo, A Jarvis, H Nasir, A Parvez, J Takahashi and A Truneh, Miss S Arantes and C Bensasson and Mr C Hunter for their valuable experimental work and to Dr P B Hitchcock for the x-ray crystallographic studies.

REFERENCES

1. M J Ackland, J R Hanson and A H Ratcliffe, *J. Chem. Soc., Perkin Trans.* 1, 1984, 2751.

2. M J Ackland, J R Hanson, B L Yeoh and A H Ratcliffe, *J. Chem. Soc., Perkin Trans.* 1, 1985, 2705.

3. M J Ackland, J F Gordon, J R Hanson, B L Yeoh and A H Ratcliffe, *J. Chem. Soc., Perkin Trans.* 1, 1988, 1477.

4. M J Ackland, J F Gordon, J R Hanson and A H Ratcliffe, *J. Chem. Soc., Perkin Trans.* 1, 1988, 2009.

5. M J Ackland, J F Gordon, B L Yeoh and A H Ratcliffe, *Phytochemistry*, 1988, **27**, 1031.

6. J R Hanson, P B Hitchcock, A J Jarvis, E M Rodriguez-Perez and A H Ratcliffe, *Phytochemistry*, 1992, **31**, 799.

7. W Dalziel, B Hesp, K M Stevenson and J A J Jarvis, *J. Chem. Soc., Perkin Trans.* 1, 1973, 2841.

8. J R Hanson, P B Hitchcock, A G Jarvis and A H Ratcliffe, *J. Chem. Soc., Perkin Trans.* 1, 1992, 2079.

9. J F Gordon, J R Hanson and A H Ratcliffe, *J. Chem. Soc., Chem. Commun.*, 1988, 6.

10. J R Hanson, A G Jarvis and A H Ratcliffe, *Phytochemistry*, 1992, **31**, 3851.

11. J R Hanson, A G Jarvis, F Laboret and J A Takahashi, *Phytochemistry*, 1995, **38**, 73.

12. J R Hanson and A G Jarvis, *Phytochemistry*, 1994, **36**, 1395.

13. J R Hanson, P B Reese, J A Takahashi and M R Wilson, *Phytochemistry*, 1994, **36**, 1391.

14. J R Hanson, P B Hitchcock and J A Takahashi, *Phytochemistry*, 1995, **40**, 797.

15. M A D Boaventura, J R Hanson, P B Hitchcock and J A Takahashi, *Phytochemistry*, 1994, **37**, 387.

16. A B De Oliveira, J R Hanson and J A Takahashi, *Phytochemistry*, 1995, **40,** 439.

17. J R Hanson, P B Hitchcock, H Nasir and A Truneh, *Phytochemistry*, 1994, **36**, 903.

18. J R Hanson and H Nasir, *Phytochemistry*, 1993, **33**, 835.

19. J R Hanson, P B Hitchcock and R Manickavasagar, *Phytochemistry*, 1994, **37**, 1023.

20. A Farooq and J R Hanson, *Phytochemistry,* 1995, **40**, 815.

21. J R Hanson and H Nasir, *Phytochemistry*, 1993, **33**, 831.

22. A Farooq, J R Hanson and Z Iqbal, *Phytochemistry*, 1994, **37**, 723.

Higher Plants as a Clean Source of Semiochemicals and Genes for Their Biotechnological Production

A. J. Hick, J. A. Pickett*, D. W. M. Smiley, L. J. Wadhams and C. M. Woodcock

IACR-ROTHAMSTED, HARPENDEN, HERTFORDSHIRE AL5 2JQ, UK

ABSTRACT

Over the last ten years, much effort has been directed towards higher plants as a source of physiologically active compounds. The major activity sought has been toxicity against invertebrate pests and plant or animal pathogens, and direct physiological activity in dealing with human and animal disorders. However, higher plants can also contain compounds which act as signals (semiochemicals) modifying the behaviour and development of pest or pathogenic organisms and also in regulating hormone signalling in higher animals. Even where a semiochemical is biosynthesised within the pest or pathogenic organism itself, the biosynthetic pathways can be the same as, or closely related to, those expressed in higher plants. Thus, higher plants can be exploited as a source of semiochemicals, to provide genes for transferral into fermentation organisms to produce semiochemicals, or for the genetic modification of food and industrial crop plants so that they can deal more effectively with pests and pathogens directly.

1 INTRODUCTION

There is a tremendous demand from the public, and food retailers in particular, for semiochemicals to replace broad-spectrum toxicants in pest control. A great deal of research has been directed at this objective over the last 30 years but there has been a very limited penetration into agricultural systems. Even in certain countries where semiochemical use is promoted, such as Japan, only 10-15% replacement of pesticides has been made. However, two major developments are set to revolutionise use of semiochemicals. The first is the realisation that semiochemicals cannot be used alone but must be combined, as various types of semiochemicals, with population-reducing agents such as highly selective pesticides and biological control agents.[1] In some instances, particularly the control of human disease vectors in resource-poor areas, semiochemicals and perhaps associated use of biological agents appear to be the only way forward. In addition, agricultural practices in such regions could make use of semiochemical-based control measures if natural products were employed that could be obtained locally. The

* To whom correspondence should be addressed.

second important factor in achieving the promise of semiochemicals in pest control is the use of biotechnology, either for the production of semiochemicals that are already generated by higher plants as secondary metabolites, or for the production of insect pheromones and their precursors by closely related secondary metabolite pathways.[2,3] Furthermore, the modification of higher plant genetics to produce semiochemicals within crop plants for their own defence, or for defence of other crop plants and even animals, is now within sight and, in some cases, a reality. In spite of the intense research on semiochemicals for agricultural pest control and for the control of human or animal disease vectors, little work has been done on a sound scientific basis for pharmaceutical objectives. Nonetheless, semiochemicals of higher organisms could in turn allow regulation of certain hormonally-regulated processes where the pheromones, for example, act as primer pheromones causing subsequent hormonal changes. Thus, it may be possible to develop oestrus-regulating pheromones for use in rodent control and even for human birth control; such materials may also yield rationalised approaches to mood control and other psychological aspects in farm animals or human beings.

To exploit semiochemicals by biotechnological methods, it may be possible eventually to transfer to crop plants cassettes of genes for enzymes from a sequence yielding specific biologically active secondary metabolites. However, until this is technically feasible, opportunistic approaches must be developed, which involve either increased expression or reduced expression of endogenous genes for key enzymes within a secondary metabolism pathway already existing in the target plant.[4] The alternative is to transfer a single alien gene, preferably from another higher plant, so that substrate is diverted from secondary metabolites of low value through to a high value secondary metabolite. The latter has been clearly demonstrated by Hain's group,[5,6] who transferred stilbene synthase from groundnut, *Arachis hypogaea* (Fabaceae), to tobacco, *Nicotiana tabacum* (Solanaceae), thus allowing precursors, one molecule of *p*-coumaroyl-CoA and three molecules of malonyl-CoA, to be converted into the phytoalexin resveratrol, 3,4′,5-trihydroxystilbene. In subsequent work, it was demonstrated by the same group that such a transfer to *N. tabacum* plants protected against the pathogen.[7] In our own work, we are attempting to do this for production of semiochemicals rather than phytoalexins.

2 OILSEED RAPE SECONDARY METABOLISM FOR CROP PROTECTION

Oilseed rape, or canola as it is called in the U.S.A. and Canada, is an artificial species, *Brassica napus* (Brassicaceae). Like most other members of this family, its endogenous defence chemistry is based on the production of glucosinolates (4). During tissue damage by insect feeding or pathogen development, thioglucosidases, trivially termed myrosinases, convert the glucosinolates to the aglycones which, under certain conditions, can yield the organic isothiocyanates (5), known collectively as the mustard oils.[8,9] Such catabolites can be important in defence against poorly adapted or unadapted herbivores and fungal pathogens. However, certain of the catabolites, particularly the alkenyl isothiocyanates, can also be used by adaptive organisms to locate and develop on the crop.[10,11] The biosynthetic pathway towards these glucosinolates involves an oxidative decarboxylation of an amino acid (1) to an aldoxime (2) which then, via a thiohydroximate (3), is converted into a glucosinolate (4) (Scheme 1).[12] Hot and cold radiolabelled precursors of glucosinolates, typical of the alkenyl and aromatic types, have been synthesised and are contributing to enzymological and molecular genetic investigations. This work will provide the basis for creating crops that have the best

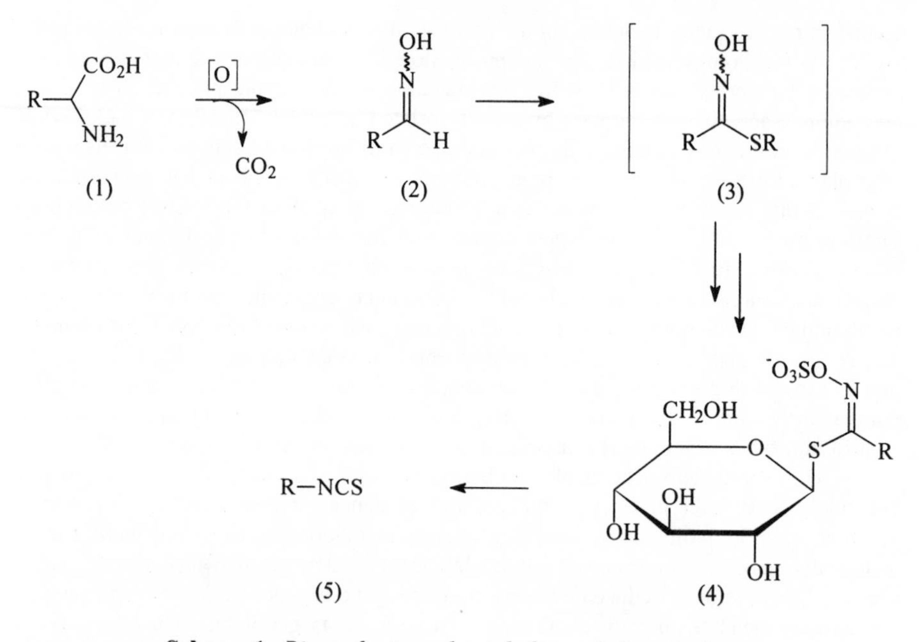

Scheme 1 *Biosynthesis and catabolism of glucosinolates*

defence based on glucosinolate catabolism, but which do not generate the main cues for pest or pathogen development and which do not contain large yields of glucosinolate in the seed.[3,12]

Although glucosinolates and their catabolites play important roles in insect and pathogen interactions with oilseed rape, semiochemicals produced via other pathways are also involved and need to be accommodated in developing strategies incorporating semiochemicals with biological control agents. By using electrophysiological recording techniques, either electroantennogram (EAG) or single cell recording, on the antennae of oilseed rape pests, volatiles entrained from intact *B. napus* plants can be analysed by high resolution gas chromatography and simultaneously investigated for neurophysiological activity.[11] Physiologically active compounds for the pod weevil, *Ceutorhynchus assimilis* (Coleoptera, Curculionidae), are presented in Table 1. Many of these components are common plant volatiles, e.g. the lipoxygenase products, while others give information on the family, e.g. the isothiocyanates which are typical of the Brassicaceae. Some compounds, detected by paired olfactory cells, may confer information on cultivar type or growth stage, since the insect could use this mechanism to determine relative proportions. For example, paired cells exist for compounds from different biosynthetic pathways (see Table 1) which probably relate to separate genetic aspects of the plant.[11] Similarly complex lists of active compounds exist for the other pests so far investigated. These include the flea beetle, *Psylliodes chrysocephala* (Coleoptera, Chrysomelidae),[3,13] and the pollen beetles, *Meligethes* spp. (Coleoptera, Nitidulidae).[10] It has now been found that, by incorporating non-isothiocyanate compounds into traps containing isothiocyanates, much lower release rates of the latter are required for attraction of oilseed rape pests (demonstrated principally for *Meligethes* spp.).[14] In the longer term, it is expected that, if the appropriate background of plant-derived compounds is provided, it will only be necessary to manipulate the isothiocyanate pathway, hence the major effort into the

Table 1 *Volatiles released by* Brassica napus *showing electrophysiological activity with* Ceutorhynchus assimilis

Type	*Compound*
Isoprenoids	sabinene
	1,8-cineole[a]
	(±)-linalool
	(*E*,*E*)-α-farnesene
Lipoxygenase pathway products	1-pentanol
	hexanal
	(*E*)-2-hexenal
	1-hexanol
	(*Z*)-3-hexen-1-ol[b]
	(*Z*)-3-hexenyl acetate[b]
	1-octen-3-ol
Isothiocyanates	allyl
	3-butenyl[c]
	4-pentenyl[c]
	2-phenylethyl[c]
Other amino acid derivatives	benzaldehyde
	benzyl alcohol
	phenylacetaldehyde
	2-methoxyphenol (guaiacol)
	2-phenylethanol[a]
	benzyl cyanide
	methyl salicylate[c]
	4-methoxybenzaldehyde (*p*-anisaldehyde)
	indole
	goitrin

[a,b,c] Compounds followed by the same letter are detected by cells frequently occurring in pairs.

biosynthesis and associated genetics of glucosinolate biogenesis.[12] Steps towards marketing plant semiochemical-based monitoring systems for some of the oilseed rape insect pests are being made and field trials on aspects of the integrated control strategies are in progress, particularly the push-pull or stimulo-deterrent diversionary strategy.[1]

3 APHID SEX PHEROMONES AND PARASITOID KAIROMONES

Aphids (Homoptera, Aphididae) are the main insect pests of northern European agriculture and it is likely that plant-sucking insects generally are set to become more important pests as world cropping patterns change with increased demands for food. The aphids originally appeared on the gymnosperms of the Jurassic and Cretaceous periods. Indeed, there are aphids, for example in the genus *Neophyllaphis*, still living on gymnosperms in the *Araucaria* genus (Pinaceae) (monkey-puzzle trees) in Chile that very closely resemble cretaceous fossils of aphids in the genus *Penaphis*.[15] A sub-family, the Aphidinae, evolved to exploit herbaceous angiosperms, which make very rapid growth in the summer and which the aphids colonise as asexually- or parthenogenetically-reproducing females. Nonetheless, to pass through the sexual stages, these insects must return to their primary host. An example is the black bean aphid, *Aphis fabae*, which feeds mainly on herbaceous members of the Fabaceae, i.e. legumes, in the summer and then returns to its primary host, the spindle tree, *Euonymus europaeus* (Celastraceae), to mate in the autumn. We have been successful in identifying the sex pheromones for many pest species in the Aphidinae, again, using pioneering work on aphid electrophysiology coupled with high resolution gas chromatography.[16-18] The compounds involved are cyclopentanoids (6) and (7), and these can be synthesised, although with considerable difficulty, particularly for the diastereoisomeric pair (8) and (9) (Figure 1).[17,19] These compounds, sometimes synergised by volatiles from the primary

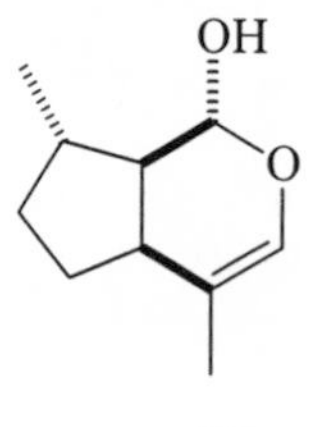

(6)

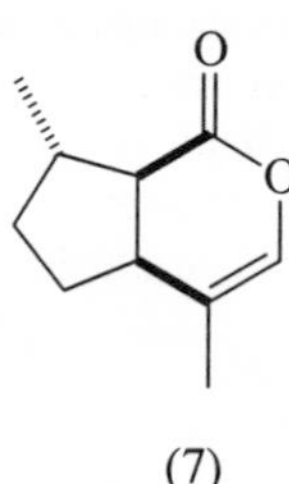

(7)

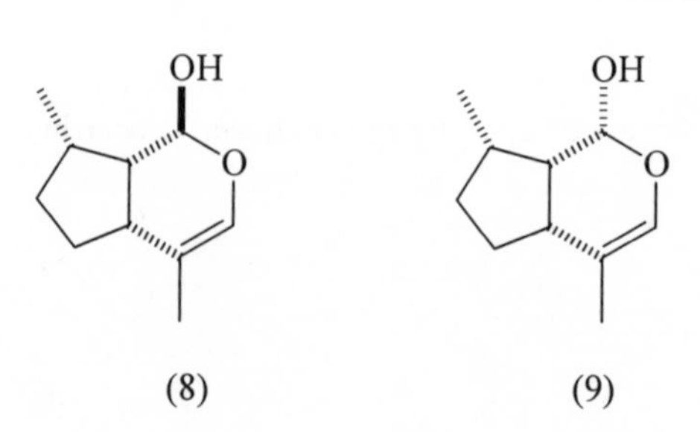

Figure 1 *Aphid sex pheromone components*

host, are potent attractants for male aphids.[19,20] In field trials with HRI, East Malling, thousands of male hop aphids, *Phorodon humuli*, were caught in Petri-dish water traps releasing the cyclopentanoids (8) and (9), compared to a few hundred caught over the same period in a suction trap sampling 500 m^3 of air per hour.[19] In Korea, collaborators working with a complex of orchard aphids have found it necessary to "detune" the water trap by using a white background, as so many thousands of aphids were being caught (K. S. Boo, personal communication).

An exciting development from this study is the demonstration that aphid parasitoids, as suggested earlier,[21] are attracted by aphid sex pheromone components[22] and may use the pheromone in the autumn to locate hosts. By electrophysiological studies on the antennae of these insects, it has been possible to show that they retain the ability to detect aphid sex pheromones throughout the summer,[23] and we have already demonstrated that plant-derived synthetic pheromone can be used to increase parasitisation of aphids on potted plants, including cereals and beans, standing in the field.[24]

Although catching males is not adequate as a method of aphid control, we are at present investigating the incorporation of fungal pathogens, particularly *Verticillium lecanii*, into a trap that attracts and holds male insects while they pick up the pathogen and then releases them to transfer inoculum to the mating population on the primary host (C. A. M. Campbell, personal communication). However, we can already see the practical value of the aphid sex pheromone for its kairomonal activity in attracting parasitoids and have begun a project, funded by the appropriate agencies, for commercially-oriented development.

The initial successes in field trials have created a considerable demand for the aphid sex pheromone cyclopentanoids. Although we have devised new synthetic routes to these compounds,[17] and further work is in progress, the ultimate objective is to develop a biotechnological production. Use of genetically modified *Agrobacterium tumefasciens* at the Institute of Food Research, Norwich, has produced callous tissue from certain members of the Lamiaceae, e.g. *Nepeta racemosa*, which produce cyclopentanoids either the same as, or closely related to, the aphid sex pheromone components (M. J. C. Rhodes, personal communication). However, successful tissue culture has not yet been developed for commercial production. With the Biochemistry and Physiology Department at IACR-Rothamsted, we have also been exploring the putative biosynthetic pathway (Scheme 2) to these compounds.[25,26] Already, some transformed plants have been created and production of cyclopentanoids and precursors is under investigation (D. L. Hallahan, personal communication).

4 PLANT-DERIVED ANTIFEEDANTS

Because of the success with the cyclopentanoid aphid sex pheromone components, we have returned to sesquiterpenoid dialdehydic antifeedants in the drimane class,[27] on which we have worked in the past and which showed activity in model field trials against barley yellow dwarf virus transmission by the bird-cherry-oat aphid, *Rhopalosiphum padi*.[28] We originally produced material by liquefied carbon dioxide extraction of field-grown *Polygonum hydropiper* (Polygonaceae). However, it has proved difficult to study the enzymology of the process in this plant. The objective was to isolate the cyclase, which might then be transferred to the crop plant to act on endogenous farnesyl pyrophosphate, the precursor necessary for drimane production.[2,29] Recent collaborative studies on the mode of action, with Imperial College at Silwood Park,[30-33] promise to provide better use

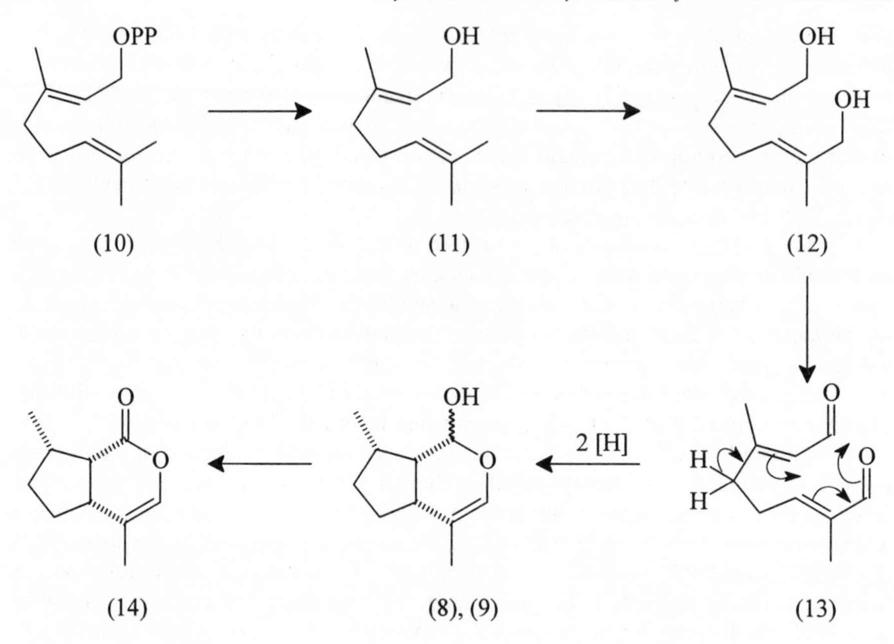

Scheme 2 *Proposed biosynthetic pathway towards cyclopentanoids in* N. racemosa. *The cyclisation is shown as a concerted step for convenience*

of antifeedants and there is a strong demand for such compounds in a European Union-funded project in southern Europe. Added to this is the success of other groups in taking enzymological studies on cyclases for other monoterpenoids[34] sesquiterpenoids[35] and higher terpenes[36] towards the isolation of associated plant genes. We are therefore devising a strategy for re-investigating the drimane cyclase and are considering transferring the study from higher plants to basidiomycetes, for example *Gloeophyllum odoratum* (C. P. Brookes, personal communication), known to produce drimenol.[18] Production directly from cultivated basidiomycete could also be used since fungiculture is practised widely in the Far East. Drimenol thus produced would then be converted[27] to the required dialdehydes by methods already available for such processing.

Earlier work on the clerodane antifeedants (Figure 2) of *Ajuga* spp. (Lamiaceae) has shown that ajugarin I (15) from *A. remota* can be used at very low levels against Coleoptera, particularly Chrysomelids such as the mustard beetle, *Phaedon cochleariae*,[37] and also the major world pest, the Colorado potato beetle, *Leptinotarsa decemlineata*. In simulated field work, we have electrostatically sprayed this antifeedant to protect the top of the plant whilst slow-acting but selective insect growth regulant pesticides, such as teflubenzuron, destroy the population feeding on the lower leaves before new growth can be damaged.[38] A more active compound, 14-hydro-16-hydroxyajugapitin (16), would be of immense value but is only available in very low yield from, for example, the ground-pine, *Ajuga chamaepitys* which, with some plant breeding, could provide a commercial source of this antifeedant as a new industrial crop.[39]

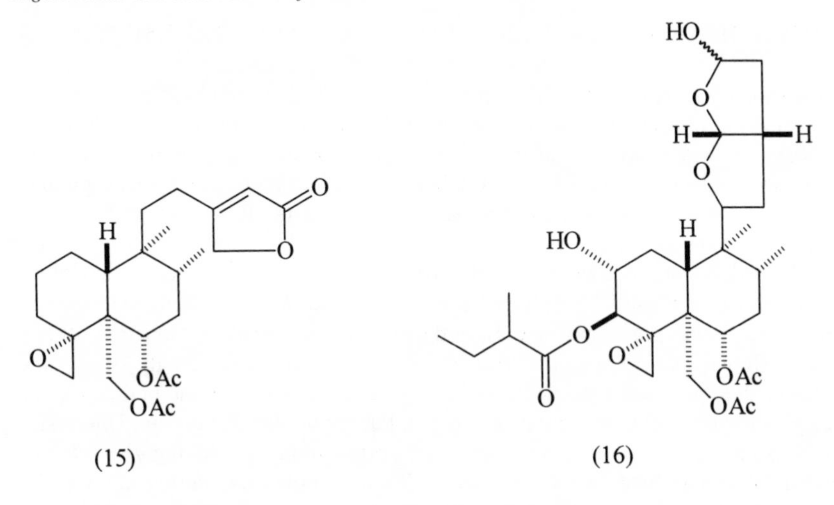

Figure 2 *Clerodane antifeedants from* Ajuga *spp.*

Antifeedants have also been shown to have high activity against the two-spotted spider mite, *Tetranychus urticae* (Acari: Tetranychidae). Although this mite is a major pest of hops, *Humulus lupulus* (Cannabiaceae), certain resin components, the β-acids or lupulones (17)-(19) (Figure 3), which are waste products from hop extraction processing, significantly reduce feeding when sprayed electrostatically on to hop plants.[40] We have now returned to this subject and are looking at more active components of *H. lupulus*, but the plant itself could be seen as a ready source of these materials.[41]

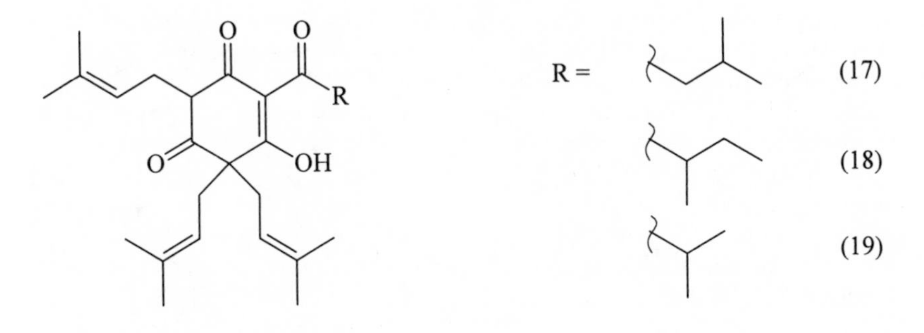

Figure 3 β-*acids from* H. lupulus

5 MOLECULAR GENETIC TARGETS FOR PLANT-SUCKING INSECT PESTS

As the Biochemistry and Physiology Department at IACR-Rothamsted has a major funded unit for developing robust cereal transformation (P. A. Lazzeri, personal communication), it is necessary to locate targets that are immediately available for creating transgenic cereal crops with enhanced semiochemical defences built in. To this end, investigations have started on host plant location and avoidance. We were the first to show that many insect pests, including the plant-sucking insects, have olfactory nerve cells, originally thought to be redundant, specifically for the identification and subsequent behavioural avoidance of non-host plants.[42-44] The bird-cherry-oat aphid, *R. padi*, needs to locate the bird-cherry tree, *Prunus padus* (Rosaceae), in the autumn for sexual reproduction. It has already been shown that compounds from *P. padus* synergise the activity of the sex pheromone for this species.[20] One electrophysiologically active compound in the primary host, methyl salicylate (23), was predicted by a collaborator, Jan Pettersson, University of Uppsala, Sweden, to be a dispersal agent for spring migrants leaving *P. padus* to colonise their secondary host (cereals, Poaceae). We have now used methyl salicylate in slow-release formulations over three field seasons to reduce aphid feeding by up to 50%.[1,45] Once this level of reduction has been achieved, increasing the dose will not increase the effect, but as a component of a "push-pull" strategy, for example incorporating the attractants for parasitoids mentioned earlier, we believe we could obtain a robust level of control and these aspects are currently under investigation. We have also found specialised olfactory cells for methyl salicylate (23) in the antennae of other cereal aphids. Although these species do not necessarily need to disperse from a host containing methyl salicylate, they may use this compound, because of its relationship to salicylate signalling, to avoid plants with induced phenylalanine ammonia lyase-based defence. The likely pathway to methyl salicylate involves methylation of salicylic acid (22), which we are currently investigating by cold isotopic labelling. Genes from higher plants associated with salicylate metabolism are also available (Scheme 3) (J. Draper, personal communication).

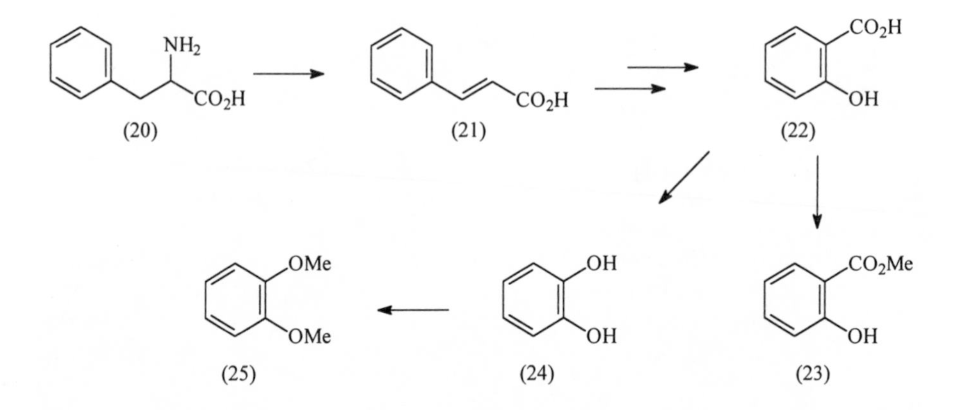

Scheme 3 *Biosynthesis of methyl salicylate and veratrole*

When aphids or other herbivores attack plants, a range of volatiles is generated. We have shown that on the same plant species and cultivar, the broad bean *Vicia faba* (Fabaceae), different aphids, for example *Acyrthosiphon pisum*, *Aphis fabae* and *Megoura viciae*, produce a different range of compounds. These compounds relate to inducible phenylalanine ammonia lyase and lipoxygenase pathways and also to the oxidation of isoprenoids. One of the latter found typically is the homoterpenoid (27), the biosynthesis of which from nerolidol (26) has been elucidated by elegant work from Boland's group (Scheme 4).[46] Other compounds include more obvious oxidation products from isoprenoids, for example 6-methyl-5-hepten-2-one (28). Much insight into oxidative stress pathways has been provided by Klessig's work in his discovery that salicylate causes blocking activity of the enzyme catylase on hydrogen peroxide,[47] leaving this to fuel oxidative stress-related biosynthesis. For our part, we are devising an approach to biotechnological development of these isoprenoid oxidation products and are working on strategies for exploiting the pathways in repelling herbivorous insects such as the aphids, but more particularly in attracting parasitoids by means of the associated learned responses that they are known to exploit in locating a suitable host ecosystem.[48]

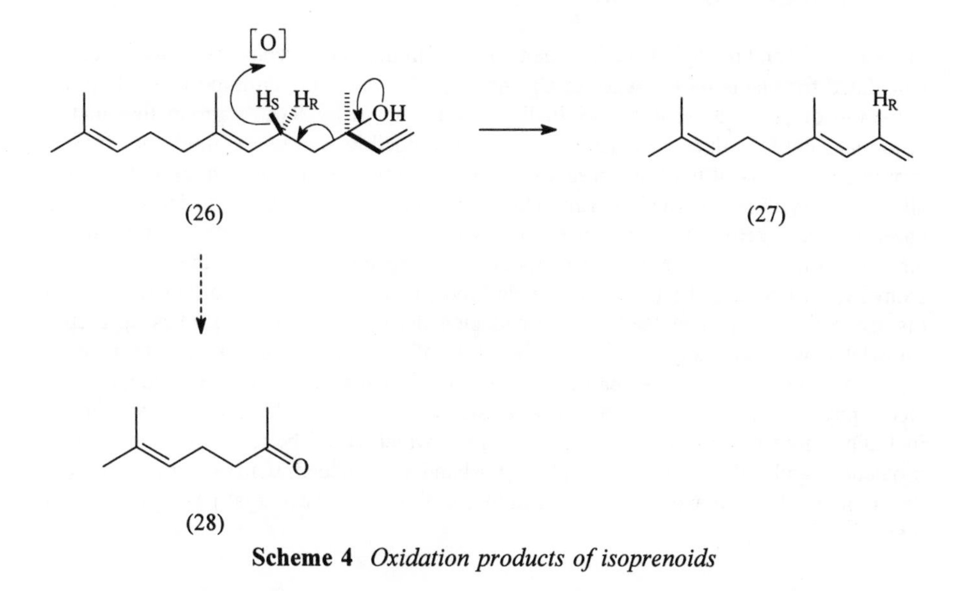

Scheme 4 *Oxidation products of isoprenoids*

The brown planthopper, *Nilaparvata lugens* (Homoptera, Araeopidae), a major pest of rice, *Oryza sativa* (Poaceae), has also been investigated by electrophysiology coupled-high resolution gas chromatography. A number of highly active compounds have been found in rice plants colonised by *N. lugens*, including 1,2-dimethoxybenzene or veratrole (25). The biosynthesis for this compound involves oxidative decarboxylation of salicylate (22). The *Nah*G gene employed to demonstrate an aspect of signalling by salicylate is available,[49] as are similar genes for enzymes from higher plants (J. Draper, personal communication). Genetic manipulation based on an anti-sense strategy could be employed to prevent rice plants generating the 1,2-dihydroxybenzene, catechol (24), precursor for this highly active signal molecule.

The appropriate expression of such genes in transgenic crop plants could allow the "push-pull" approach to be developed, exemplified hypothetically in Figure 4.

"PUSH" (away from the crop)	"PULL" (towards the crop)
Harvestable Crop:-	*Trap Crop:-*
signal pathway suppressed	signal pathway expressed
	+ biological control agent, e.g. fungal pathogen

Figure 4 *Stimulo-Deterrent Diversionary (Push-Pull) Strategy for transgenic crops modified for pest colonization signals*

6 PLANT SIGNALLING

Jasmonic acid and methyl jasmonate, generated from the lipoxygenase pathway, elegantly elucidated for plants by the work of Crombie *et al.*,[50] can have an important role in the induction of plant defence.[51] Originally, it was shown by Ryan's group that methyl jasmonate could induce production of wound-inducible proteinase inhibitors by, for example, members of the Solanaceae.[52,53] Zenk's group has also shown the induction of alkaloid biosynthesis by methyl jasmonate in pioneering studies on secondary metabolite-based defence chemistry.[54] We have also shown that methyl jasmonate, permeated into air above oilseed rape plants, causes induction of specific aspects of the glucosinolate pathway, particularly the production of indolyl glucosinolates,[55] important in reducing disease development and feeding by unadapted herbivores. In our studies on methyl salicylate, we have suggested that the role of this compound is as an external representative of salicylate signalling for defence.[45] Thus, insects may provide, through electrophysiology coupled-chemical analysis, a means of identifying new signals, including those released on pathogen damage,[56] which could be capitalised on for crop protection and other aspects of plant husbandry. Indeed, Ciba-Geigy is already developing salicylate analogues which induce defence against fungal pathogens in crop plants.[57]

7 HUMAN DISEASE VECTORS

In the early 1980s, we identified and patented the first mosquito pheromone (31).[58-60] Since then, trials based on collaborations in nine countries throughout the world have shown the efficacy of this pheromone in attracting mosquitoes to oviposit. In Kenya, we combined use of the pheromone, formulated in an effervescent tablet, with a juvenile hormone-based insecticide, pyriproxyfen, to destroy the larvae.[61] However, to provide cheap material for resource-poor regions where the mosquito targets, principally in the *Culex* genus, cause major disease problems, we have turned to plants for production of the pheromone. The phytochemical precursor required is 5-hexadecenoic acid (29) (Scheme 5). We developed, in collaboration with scientists at the Abubakar Tafawa Balewa University, Nigeria, GC-MS analysis for this compound, involving epoxidation and detection of the α-cleavage fragments which, in the case of the epoxide in the 5 position (Scheme 5), are substantially different from the main plant components such as 9-hexadecenoic (palmitoleic) acid. *Kochia scoparia* (Chenopodiaceae), the summer cypress, already known to produce this compound, has proved to be an appropriate plant source (T. O. Olagbemiro, personal communication). Crude acids are obtained by saponification of an extract of *K. scoparia* and are then converted, through a catalytic and recyclable reagent, osmium tetroxide, to the *erythro*-5,6-dihydroxyhexadecenoic acid (30). This is then converted into the pheromone in one step by treatment with acetic anhydride in pyridine, both of which are cheap and readily available reagents (Scheme 5).

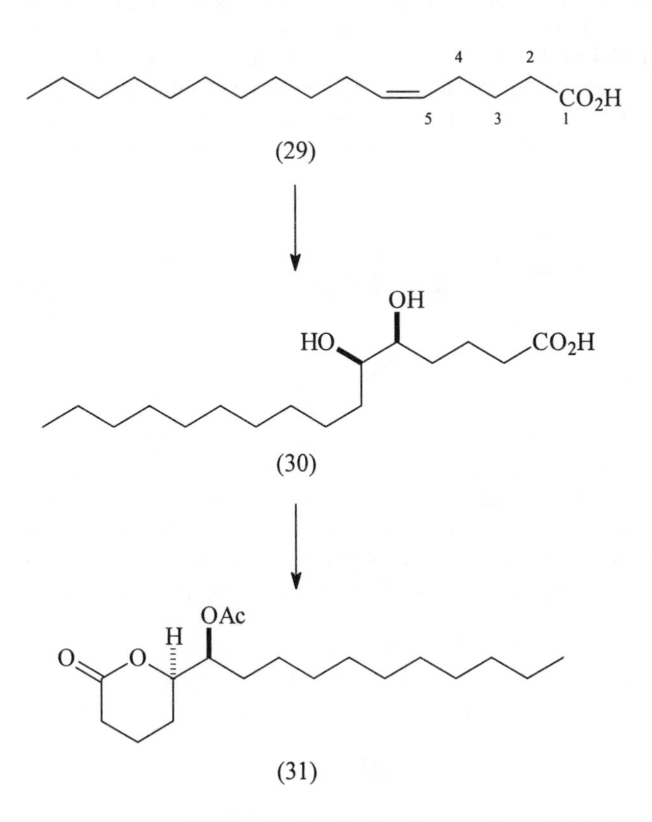

Scheme 5 *Synthesis of mosquito oviposition pheromone*

Contamination by the by-products of this process doesnot appear to present a problem. Although the *erythro*-dihydroxy acid yields the natural plus the inactive enantiomer of the pheromone, the overall yield is rendered satisfactory by the absence of the two inactive enantiomers that would arise from the *threo* stereochemistry. Replacement of the juvenile hormone insecticide by a cheap recyclable biological agent is the next objective, and in collaboration with the Institut Pasteur, the oomycete *Lagenidium giganteum* has been chosen (B. Papierok, personal communication). Recently, with the University of Aberdeen,[62] a breakthrough has been made in completing the oviposition pheromone cue (A. J. Mordue, personal communication), enabling use of relatively clean water, necessary for the continued development of a strategy utilising *L. giganteum*.

Working with the University of Keele and, in turn, scientists in Brazil, a number of races of sandfly, *Lutzomyia longipalpis*, which transmit various forms of the debilitating disease leishmaniasis, have been investigated for their sex pheromone composition. The sex pheromone is produced by the male and attracts females, the disease-vectoring gender of this insect; this situation is quite unusual for insects, where the sex pheromone is usually generated by the female to attract males. Our tentative identification of the pheromone for two strains as being 9-methylgermacrene-B and 3-methyl-α-himachalene provides a new development in understanding the chemical ecology of human disease vectors.[63,64] Although we are currently working on *de novo* synthesis of these compounds, we are at the same time developing a biotechnological route based on germacrone from *Geranium macrorrhizum* (Geraniaceae) and α-himachalene from the Himalayan deodar, *Cedrus deodara* (Pinaceae). Already, large amounts of germacrone are produced in Bulgaria by commercial cultivation of *G. macrorrhizum*, stimulated by colleagues in the Netherlands interested in bioactive compounds from chicory, *Cichorium intybus* (Asteraceae),[65] based on the germacrene skeleton (Figure 5).

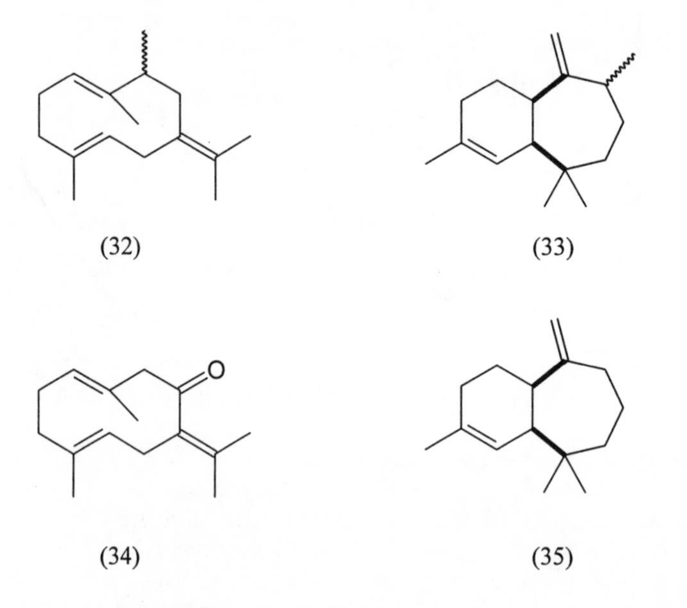

Figure 5 *Sandfly pheromone components and chemical precursors*

Finally, our attention is turning towards the mosquito vectors of malaria, particularly *Anopheles gambiae* s.s., and using our electrophysiological methods coupled with high resolution gas chromatography, we are attempting, with others, to identify compounds involved in mosquito/host interactions.[66] It may also be possible in this case, and in others in the future, to use a combinatorial chemistry approach to find active semiochemical analogues. The approach being adopted involves replacing the gas chromatograph, as a source of separate volatile chemical samples, with a combinatorial chemical library which is sequentially tested using an autosampler and gas chromatographic injector linked directly to the electrophysiological preparation. This would allow the site of action in the receptor protein, and the other odorant-binding proteins, to be probed using a "chemical genetics approach". The approach would also provide compounds that could incorporate a photo-affinity label which, with further radioisotope labelling, could enable isolation of the odorant binding proteins from insect antennae. This in turn would facilitate genetic approaches to blocking essential olfactory processes so that, for example, mosquitoes could not locate human hosts.[66]

8 CONCLUSIONS

Thus, a number of semiochemicals have been identified, with potential or already demonstrated value in pest control, that can be readily obtained from higher plants and which, in the longer term, will be available by biotechnological means employing recombinant DNA techniques. Further into the future, we can see crop plants being modified by molecular genetic techniques to produce a more useful range of semiochemicals for protection by the "push-pull" strategy, incorporating biological control agents. By exploiting higher plants as a source of semiochemicals and the genes for their biotechnological production, it would be possible to find alternatives where necessary, or desired, for broad-spectrum eradicant pesticides.

Acknowledgements

This work was in part supported by the United Kingdom Ministry of Agriculture, Fisheries and Food. IACR receives grant-aided support from the Biotechnology and Biological Sciences Research Council of the United Kingdom.

References

1. J. A. Pickett, L. J. Wadhams and C. M. Woodcock, *Agriculture, Ecosystems and the Environment*, submitted.
2. G. W. Dawson, D. L. Hallahan, A. Mudd, M. M. Patel, J. A. Pickett, L. J. Wadhams and R. M. Wallsgrove, *Pestic. Sci.*, 1989, **27**, 191.
3. J. A. Pickett, T. M. Butt, K. J. Doughty, R. M. Wallsgrove and I. H. Williams, *Proc. Ninth Int. Rapeseed Cong., Rapeseed Today and Tomorrow, 4-7 July, 1995, Cambridge, U.K.*, Volume 2, F1, Dorset Press, Dorchester, p. 565.
4. J. A. Pickett, L. J. Wadhams and C. M. Woodcock, 'Phytochemistry and Agriculture', T. A. van Beek and H. Breteler (eds.), Clarendon Press, Oxford, 1993, p. 62.
5. R. Hain, B. Bieseler, H. Kindl, G. Schröder and R. Stöcker, *Plant Mol. Biol.*, 1990, 325.
6. R. Hain, H. J. Reif, R. Langebartels, P. H. Schreier, R. H. Stöcker, J. E. Thomzik, K. Stenzel, H. Kindl and E. Schmelzer, *Brighton Crop Prot. Conf. - Pests and Diseases - 1992*, 757.
7. R. Hain, H. J. Reif, E. Krause, R. Langebartels, H. Kindl, B. Vornam, W. Wiese, E. Schmelzer, P. H. Schreier, R. H. Stöcker and K. Stenzel, *Nature.*, 1993, **361**, 153.
8. G. R. Fenwick, R. K. Heaney and W. K. Mullin, *CRC Crit. Rev. Food Sci. Nutr.*,

1983, **18**, 123.
9. G. R. Fenwick, R. K. Heaney and R. Mawson, 'Toxicants of Plant Origin', P.R. Cheeke (ed.), CRC, Boca Raton, 1989, Vol. 2, p.1.
10. M. M. Blight, J. A. Pickett, J. Ryan, L. J. Wadhams and C. M. Woodcock, *Proc. Ninth Int. Rapeseed Congress, Rapeseed Today and Tomorrow, 4-7 July, 1995, Cambridge, U.K.*, Volume 3, I25, Dorset Press, Dorchester, p. 1043.
11. M. M. Blight, J. A. Pickett, L. J. Wadhams and C. M. Woodcock, *J. Chem. Ecol.*, 1995, **21**, 1649.
12. G. W. Dawson, A. J. Hick, R. N. Bennett, A. Donald, J. A. Pickett and R. M. Wallsgrove, *J. Biol. Chem.*, 1993, **268**, 27154.
13. M. M. Blight, J. A. Pickett, L. J. Wadhams and C. M. Woodcock, *Asp. Appl. Biol.*, **23**, 329.
14. L. E. Smart, M. M. Blight and A. J. Hick, *J. Chem. Ecol.*, submitted.
15. E. A. Jarzembowski, *Cretaceous Research*, 1989, **10**, 239.
16. J. A. Pickett, L. J. Wadhams, C. M. Woodcock and J. Hardie, *Annu. Rev. Entomol.*, 1992, **37**, 67.
17. G. W. Dawson, J. A. Pickett and D. W. M. Smiley, *Bioorg. Med. Chem.*, 1996, **4**, 351.
18. J. A. Pickett, L. J. Wadhams and C. M. Woodcock. First steps in the use of aphid sex pheromones. In: *Pheromone Research: New Directions*, R.T. Cardé and A.K. Minks (eds.), Chapman and Hall, New York, in press.
19. C. A. M. Campbell, G. W. Dawson, D. C. Griffiths, J. Pettersson, J. A. Pickett, L. J. Wadhams and C. M. Woodcock, *J. Chem. Ecol.*, 1990, **16**, 3455.
20. J. Hardie, J. R. Storer, S. Nottingham, J. Harrington, L. A. Merritt, L. J. Wadhams and D. Wood, *Brit. Crop Prot. Conf. - Pests and Diseases*, 1994, 1223.
21. G. W. Dawson, D. C. Griffiths, N. F. Janes, A. Mudd, J. A. Pickett, L. J. Wadhams and C. M. Woodcock, *Nature*, 1987, **325**, 614.
22. J. Hardie, A. J. Hick, C. Höller, L. A. Merritt, S. F. Nottingham, W. Powell, L. J. Wadhams, J. Witthinrich and A. F. Wright, *Ent. Exp. Appl.*, 1994, **71**, 95.
23. J. Hardie, R. Isaacs, F. Nazzi, W. Powell, L. J. Wadhams and C. M. Woodcock, *Proc. 5th Int. Symp. Global IOBC Working Group: Ecology of Aphidophaga*, 1993, 29.
24. J. A. Pickett, L. J. Wadhams and C. M. Woodcock, *Brighton Crop Prot. Conf. – Pests and Diseases, 1994*, 1239.
25. D. L. Hallahan, S-M. C. Lau, P. A. Harder, D. W. M. Smiley, G. W. Dawson, J. A. Pickett, R. E. Christoffersen and D. P. O'Keefe, *Biochimica et Biophysica Acta*, 1994, **1201**, 94.
26. D. L. Hallahan, J. M. West, R. M. Wallsgrove, D. W. M. Smiley, G. W. Dawson, J. A. Pickett and J. G. C. Hamilton, *Arch. Biochem. Biophys.*, 1995, **318**, 105.
27. Y. Asakawa, G. W. Dawson, D. C. Griffiths, J–Y. Lallemand, S. V. Ley, K. Mori, A. Mudd, M. Pezechk–Leclaire, J. A. Pickett, H. Watanabe, C. M. Woodcock and Z-n. Zhang, *J. Chem. Ecol.*, **14**, 1845.
28. J. A. Pickett, G. W. Dawson, D. C. Griffiths, A. Hassanali, L. A. Merritt, A. Mudd, M. C. Smith, L. J. Wadhams, C. M. Woodcock and Z-n. Zhang, 'Pesticide Science and Biotechnology', R. Greenhalgh and T. R. Roberts (eds.), Blackwell Scientific Publications, 1987, p. 125.
29. D. L. Hallahan, J. A. Pickett, L. J. Wadhams, R. M. Wallsgrove and C. M. Woodcock, 'Plant Genetic Manipulation for Crop Protection', A. M. R.

Gatehouse, V. A. Hilder and D. Boulter (eds.), C.A.B. International, Wallingford, 1992, p. 215.

30. G. Powell, J. Hardie and J. A. Pickett, *Ent. exp. appl.*, 1993, **68**, 193.
31. G. Powell, J. Hardie and J. A. Pickett, *Ent. Exp. Appl.*, 1995, **74**, 91.
32. G. Powell, J. Hardie and J. A. Pickett, *Physiological Entomology*, 1995, **20**, 141.
33. G. Powell, J. Hardie and J. A. Pickett, *J. Appl. Entomol.*, 1996, **120**, 241.
34. T. W. Hallahan and R. Croteau, *Arch. Biochem. Biophys.*, 1989, **269**, 313.
35. D. E. Cane, D. B. McIlwaine and P. H. M. Harrison, *J. Am. Schem. Soc.*, 1989, **111**, 1152.
36. E. J. Corey, S. P. T. Matsuda and B. Bartel, *Proc. Natl. Acad. Sci. USA*, 1993, **90**, 11628.
37. D. C. Griffiths, A. Hassanali, L. A. Merritt, A. Mudd, J. A. Pickett, S. J. Shah, L. E. Smart, L. J. Wadhams and C. M. Woodcock, *Proc. Brighton Crop Prot. Conf. – Pests and Diseases*, 1988, 1041.
38. D. C. Griffiths, S. P. Maniar, L. A. Merritt, A. Mudd, J. A. Pickett, B. J. Pye, L. E. Smart and L. J. Wadhams, *Crop Prot.*, 1991, **10**, 145.
39. F. Camps, J. Coll, O. Dargallo, J. Rius and C. Miravitlles, *Phytochemistry*, 1987, **26**, 1475.
40. P. I. Sopp, A. Palmer and J. A. Pickett, *SROP/WPRS Bulletin XIII/5*, 1990, 198.
41. G. Jones, C. A. M. Campbell, B. J. Pye, S. P. Maniar and A. Mudd, *Pestic Sci.*, 1996, **47**, 165.
42. S. F. Nottingham, J. Hardie, G. W. Dawson, A. J. Hick, J. A. Pickett, L. J. Wadhams and C. M. Woodcock, *J. Chem. Ecol.*, 1991, **17**, 1231.
43. J. Hardie, R. Isaacs, J. A. Pickett, L. J. Wadhams and C. M. Woodcock, *J. Chem. Ecol.*, 1994, **20**, 2847.
44. J. A. Pickett, L. J. Wadhams and C. M. Woodcock, *Proc. 1st Int. Conf. on Insects: Chemical, Physiological and Environmental Aspects, September 26-29, 1994, Ladek Zdroj, Poland*, D. Konopińska, G. Goldsworthy, R. J. Nachman, J. Nawrot, I. Orchard, G. Rosiński and W. Sobótka (eds.), University of Wrocław, 1995, p. 126.
45. J. Pettersson, J. A. Pickett, B. J. Pye, A. Quiroz, L. E. Smart, L. J. Wadhams and C. M. Woodcock, *J. Chem. Ecol.*, 1994, **20**, 2565.
46. W. Boland, Z. Feng and J. Donath, 'Flavor Precursors', P. Schreier (ed.), Allured Publ. Comp., Wheaton, IL, U.S.A., 1993, p. 123.
47. D. F. Klessig and J. Malamy, *Plant Mol. Biol.*, 1994, **26**, 1439.
48. J. A. Pickett, W. Powell, L. J. Wadhams, C. M. Woodcock and A. F. Wright *Proc. 4th European Workshop on Insect Parasitoids, April 3-5 1991, Perugia, Italy*, F. Bin (ed.), REDIA, Vol. LXXIV, n.3, Appendice, p. 1.
49. T. Gaffney, L. Friedrich, B. Vernooij, D. Negrotto, G. Nye, S. Uknes, E. Ward, H. Kessmann and J. Ryals, *Science*, 1993, **261**, 754.
50. L. Crombie and D. O. Morgan, *J. Chem. Soc. Perkin Trans. 1*, 1991, 581.
51. R. A. Creelman and J. E. Mullet, *Proc. Natl. Acad. Sci. USA*, 1995, **92**, 4114.
52. E. E. Farmer and C. A. Ryan, *Proc. Natl. Acad. Sci. USA*, 1990, **87**, 7713.
53. E. E. Farmer and C. A. Ryan, *The Plant Cell*, 1992, **4**, 129.
54. H. Gundlach, M. J. Müller, T. M. Kutchan and M. H. Zenk, *Proc. Natl. Acad. Sci. USA*, 1992, **89**, 2389.
55. K. J. Doughty, G. A. Kiddle, B. J. Pye, R. M. Wallsgrove and J. A. Pickett, *Phytochemistry*, 1995, **38**, 347.

56. K. J. Doughty, M. M. Blight, C. H. Bock, J. K. Fieldsend and J. A. Pickett, *Phytochemistry*, in press.
57. H. Kessmann, T. Staub, C. Hofmann, P. A. Goy, E. Ward, S. Uknes and J. Ryals, *10th Symp. Research Committee for Bioactivity of Pesticides, Nagano, Japan*, 1993, 29.
58. B. R. Laurence and J. A. Pickett, *J. Chem. Soc., Chem. Commun.*, 1982, 59.
59. B. R. Laurence, K. Mori, T. Otsuka, J. A. Pickett and L. J. Wadhams, *J. Chem. Ecol.*, 1985, **11**, 643.
60. G. W. Dawson, A. Mudd, J. A. Pickett, M. M. Pile and L. J. Wadhams, *J. Chem. Ecol.*, 1990, **16**, 1779.
61. W. A. Otieno, T. O. Onyango, M. M. Pile, B. R. Laurence, G. W. Dawson, L. J. Wadhams and J. A. Pickett, *Bull. Ent. Res.*, 1988, **78**, 463.
62. A. J. Mordue (Luntz), A. Blackwell, B. S. Hansson, L. J. Wadhams and J. A. Pickett, *IOBC/WPRS Bulletin*, 1993, **16**, 335.
63. J. G. C. Hamilton, G. W. Dawson and J. A. Pickett, *J. Chem. Ecol.*, accepted.
64. J. G. C. Hamilton, G. W. Dawson and J. A. Pickett, *J. Chem. Ecol.*, accepted.
65. D. P. Piet, R. Schrijvers, M. C. R. Franssen and A. de Groot, *Tetrahedron*, 1995, **51**, 6303.
66. J. A. Pickett and C. M. Woodcock, 'Olfaction in Mosquito-Host Interactions', G. Cardew (ed.), CIBA Foundation Symposium No. 200, John Wiley, Chichester, in press.

Industrial Production of Amazonian Natural Products

Benjamin Gilbert

FUNDAÇÃO OSWALDO CRUZ – FAR-MANGUINHOS, RUA SIZENANDO NABUCO, 100, 21041-250 RIO DE JANEIRO, BRAZIL

1 INTRODUCTION

Although the Amazonian forest is known to be a rich source of many natural products useful for medicinal, cosmetic, or nutritional purposes, in general these have not become available on the international market in quantities sufficient to sustain an industrial line of production. This may be attributed to a number of reasons:

- The lack of an adequate commercial transport network within the region
- Human diseases such as malaria that have driven out producers from the remoter areas
- Plant diseases or insect pests that have destroyed plantations
- Disinterest of the Amazonian inhabitant for money and interest rather for improved living, educational and sanitary conditions not available to the individual by means of money
- Failure of buyers to realise that the Amazonian inhabitant wants to exchange his products for these benefits, without which he will not collaborate
- The failure of international companies to set up "finished product" production in the region so that a major part of the benefit remains there.

A number of new factors now permit the industrial scale production of several natural oils, some high vitamin sources, some aromas or flavours, one or two pharmaceuticals and some insecticides or insect repellents in multi-ton lots almost immediately if the relevant principles above are respected. Many other products can be produced by mixed plantation techniques and would be available in less than a decade on the industrial scale. Among these factors are:

- International finance, often non-returnable, is available for sustainable development projects
- Research and development work over the past 30 years, much of it hidden in non-English literature, defines many products chemically or provides agro-forestry

methodology. Much field work has been conducted by NGO's with multi-ton scale production
- Accessibility is better with daily air services to many locations
- Brazilian quality control and packing technology has acquired world standards as a result of the extensive export of manufactured products
- Amazonian mining companies are willing to give logistic support.

This paper discusses only those products whose technical and economic exploration can be considered as a short to medium term operation. Natural pharmaceuticals whose isolation and identification are still pending and whose utilisation will depend upon several years of assessment and follow-up have not been considered.

2 MARKETS

Only three main markets have been contemplated: cosmetics (including over-the-counter pharmaceuticals); pesticides and similar products; and food products, markets for all of which a world demand often already exists but which is unfulfilled due to the non-existence of suppliers. Other markets, such as those for fibres and rubber are not dealt with in this review although they merit attention at a second stage.

2.1 Cosmetics and Over-the-counter Pharmaceuticals Market

The American retail market for cosmetics has been estimated at approximately $18 billion; the world market is perhaps in excess of $ 30 billion, the market for raw materials being about one-tenth of this size[1]. In order to make the export of Amazonian products feasible as a benefit to the region it is desirable that they be marketed in their final retail form, because, taking into account the high costs of Amazonian raw materials, which are often hauled over long distances by precarious transportation means, the marketing of crude products does not bring a sufficient benefit to the producer or collector to make his work worth while. Processing to a higher added value and packaging for the retail market in the region are entirely feasible and make the *operation profitable to the local community.*

To exemplify a product of this class one can consider a formula of *Brazil nut oil* (base) containing either *andiroba* (*ca.* 20%) or *copaíba* (*ca.* 10%) oil or both, for skin treatment and healing of wounds and skin rash. Such formulae have been commercialised both locally and internationally. The three components can be marketed individually and in varied forms such as creams and lotions. These can conveniently also contain natural pigments and perfumes. In order to judge whether such a formula can reach the world market one must examine the availability of these components.

2.1.1 *Brazil Nut Oil.* The states of Para, Amazonas and Acre offer industrial-scale production; and the raw material is also readily available in the states of Amapa, Rondonia and Mato Grosso. Production requires scalding, shelling, pressing, filtering and packaging. It is adaptable to 50 litre/hour for small communities or to a larger scale for private companies. Harvest of the nuts ranges over as much as 6 months.

2.1.2 *Andiroba Oil.* Industrial production of an oil with uniform quality is just beginning in the state of Amapa, non-industrial production is available from various sources and a few tons per year of this "home-made" product are available with variable quality. Plantation of the tree in the state of Para has been undertaken by the Ministry of Agriculture and the propagation technology for this is available from the Eastern Amazonian Agroforestry Research Centre (CPATU-EMBRAPA) in Belém. The nuts are harvested in April and May in most states. Production of both this oil and that of the Brazil nut could reach an industrial scale to supply any foreseeable market.

2.1.3 *Copaíba Oil.* This is manually extracted by boring the trunk of the tree without causing it apparent damage. Availability is perhaps a few hundred tons per year; supply dwindles apparently due to low prices. It is not a seasonal product. It is available in all states of Northern Brazil with botanical and chemical variations. It is sustainable on the basis of natural occurrence; enrichment by planting is however highly recommended to offset major losses incurred due to the timber industry. There are a number of alternative trunk exudate balsams also available which include copal *(Hymenaea courbaril)* and "breu branco" *(Protium spp.)* but these will remain products of relatively limited availability until real sustainable agro-forestry replaces the present predatory timber industry. Other products are now considered.

2.1.4 *Buriti, tucuman and piquiá pulp oils.* These oils are characterised by a very high content of β-carotene and probably also other skin nutrient and protective elements. Natural sources of buriti and tucuman are virtually inexhaustible - a single existing stand of buriti palms at the mouth of the Jari River could supply about 40,000 tons of oil annually. Although recent studies deny the applicability of synthetic β-carotene for cancer prevention, there is ample evidence of the cancer preventative effect of foodstuffs containing natural β-carotene, which in these oils is accompanied by other free-radical absorbers or oxidation inhibitors, not always identified but certainly including tocopherols or tocotrienols. The harvest of ripe buriti, tucuman and piquiá nuts by community pickers at their respective regions of occurrence would be capable of supporting an industrial production of hundreds of tons per month of the first two products and a little less of the latter. As in the cases of Brazil nut and andiroba the central process is oil-pressing. In the case of palm fruits there are usually preceding stages of cooking (sterilisation), separation of pulp from nut and drying of the pulp (optional), followed by filtration and separation of water, when present.

2.1.5 *Fine toilet soap.* This requires an oleic-type plant oil available from several oil-bearing fruits (mainly palm fruits, including, besides those mentioned above, *pupunha* and *African palm*) and a lauric oil extracted from palm species such as *babaçu* and *muru-muru,* or fat bearing species such as *ucuúba* (*Virola surinamensis*). Fine toilet soap also requires scents which can, for example, be obtained by steam distillation of some herbaceous species (*erva do Marajó, pataquera etc.* belonging often to the botanical families Lamiaceae or Verbenaceae), and some pigment (see below). Associated with the production of the two plant oils - oleic and lauric, which also have dietary applications, soap can be produced entirely from natural Amazonian components (except sodium hydroxide). A typical local operation might employ a 10 ton/day plant of a design which is commercially available at approximately $ 250,000. The local market absorbs a significant part of production, which will offset eventual foreign market fluctuations. Fine toilet soap must be adequately packaged and be accompanied by a printed leaflet describing its origin and properties. The incorporation of certain scents, such as that of *Hyptis* spp will afford bacterial and fungicide activity and the same is true for certain dyes.

2.1.6 *Essential oils - scents.* An important product of the past was *rosewood oil,* which was exploited until the virtual extinction of the species. The main component of this oil is *linalool,* which remains the leader in terms of the number of products which contain it, be they either fine or cheap perfumes. Most linalool marketed today is synthetically made. The revitalisation of the rosewood industry can be envisaged in two ways - both of them dependent on growing the species, using a mixed plantation method to prevent pest infestation. The preferred option is the working of the leaves, just as *Eucalyptus citriodora* is worked in the state of São Paulo. The other method, which would be to distil the trunk wood, would require a much longer time span and could not be regarded as a short to medium term sustainable operation.

Other sources of linalool exist and the one which seems to show best promise is *sacaca, Croton cajucara.* With adequate selection of a linalool-rich variety, this industry could succeed from its first year of activities. Apart from an appropriate plant source the producer would need the collaboration of an analytical laboratory for quality control. Through the Natural Products Division of the National Amazon Research Institute (INPA-CPPN) in Manaus, or the Museu Goeldi or CPATU (mentioned above) in Belém, or of one of several regional Universities, this collaboration could be obtained without any major problem. Rosewood linalool commanded a price in excess of $ 30 in the world market. An equivalent oil from a natural source such as sacaca would be expected to reach similar levels. Pataquera and erva-do-Marajó oils were mentioned above in the item Toilet Soap, and there are naturally many others, among them *puxuri-maior*, *Licaria puchury-major*, a well-known cultivable tree, whose leaves yield a pleasantly scented oil.

2.1.7 *Pilocarpine.* This is not in itself an over-the-counter pharmaceutical product although crude extracts of the plant which contain it are. It is isolated from *Pilocarpus jaborandi* or *P. microphyllus* as the natural base and transformed into its dihydrochloride at an internationally accepted standard of purity *in Northern Brazil*, and is perhaps the only allopathic therapeutic drug in world-wide use provided by the eastern Amazon today. It is produced by Merck (Darmstadt), out of their Maranhão facility and also by several domestic manufacturers, among them PVP - Produtos Vegetais Piauí, in Parnaíba - state of Piauí. According to data from Banco do Brasil, an annual average of 9 tons of pilocarpine were exported in the past 5 years, at an average price of $1,300/kg fob. Considering an alkaloid recovery rate of 1% on dry leaves, this exportation volume would correspond to a consumption of 900 tons of dry leaves, an amount which can only be supported through re-planting, which is now a major Merck operation

2.2 Market for Insecticides, Pesticides and the like.

In 1987, the world pesticide market was estimated at $ 7 billion[2]. About 30% of this market refers to applications under pressure for substitution by natural agents (especially for market garden produce and domestic applications). Thus, this natural insecticide market, which existed in the past before the appearance of DDT and BHC, is growing once again. Today, the major suppliers are Peru and Costa Rica. Four individual products may be mentioned.

2.2.1 *Derris.* This insecticide, which is extracted from timbó root, *Lonchocarpus (Derris) urucu*; could be produced immediately in commercial quantities based on dense occurrences of the plant along the northern bank of the Eastern Amazon as well as in several other localities.

2.2.2 *Quassia.* The aqueous extract of the wood of "quina" (*Quassia amara L.* or *Picrasma excelsa:* similar products in some other Simaroubaceous species such as *Simaba cedron*) is a long known insecticide and insect antifeedant as well as having use as a bitter principle in drinks including beer and as an antiprotozoal remedy in herbal medicine[3]. As an insecticide/insect repellent it is particularly effective against Aphidae. Quassia planting technique is under investigation at CPATU in Belém and at the Agricultural Technology Research and Educational Centre (CATIE) in Costa Rica.

2.2.3 *Andiroba.* Mentioned under 2.1.2 above, the oil and other part of the plant (*Carapa guianensis* Aubl.) have a traditional use as insect repellents, although they are probably essentially anti-feedants[3]. This is a viable industrialisable product and as mentioned above, manufacture in the region has begun.

2.2.4 *Safrole.* This is the raw material for the production of piperonyl butoxide, the main insecticide synergist presently in use. It is distilled from the leaves of *Piper hispidinervium* using a process developed by the Goeldi Museum in Belém with help from the British Natural Resources Institute. Cultivation of the source plant is under way at CPATU in Belém and proves to be feasible in relatively poor soils. There is a big market for safrole which commands about $4 per kilogram on the world market

2.3 The Foodstuffs Market

This market comprises a wide variety of classes, among which one may list the plant oils mentioned above, as well as scents and flavours, dyes, chiclés, antioxidants and food preservation agents, thickeners or gums, emulsifiers or wetting agents, sweeteners or bitter principles, vitamins, pro-vitamins or special nutrients of some type or other. Briefly discussing these classes - most are not exploited outside the region and therefore have not been industrially developed - we can earmark some products carrying an immediate potential for both the domestic and the foreign markets:

2.3.1 *Plant Oils. Buriti, Tucuman and Pupunha.* These palms have been mentioned above. Pupunha has the best agricultural technology data, and its cultivation is being made in several localities, notably at the National Amazon Research Institute (INPA), Manaus and in Costa Rica. Among the many products yielded by this palm tree, oil and flour made from the fruit rank second and third behind the *palmito*, or palm-heart, rated the best among palmitos of Amazonian origin. *Patauá,* another palm tree, abundant in certain regions, yields an oil from its fruit pulp which is comparable to olive oil[4]. No native oil can compete with the African palm in the commodities market - therefore, if plant oils are to be economically feasible, they must contain some special property - carotene, flavour or medicinal characteristics which set them apart.

2.3.2 *Aromas and Flavours.* This is a huge and largely untapped market. The main investments are being made in fruits which are native to or can be grown in the Amazon, among which feature cashew, camu-camu, cupuaçu, guava, açaí and passion fruit are the most important. Acerola is not native to the region; however, it is being widely planted. Many other fruits worthy of attention also exist; they have been collected and stored in genetic banks (and unfortunately often lost afterwards) especially by the INPA in Manaus. This is an area where the investor will have a guaranteed return, because foreign markets are eager for new flavours which command high prices and guaranteed sales. The production of nectars and frozen juices has permitted the widespread commercialisation of fruit drinks and ice-creams in Southern Brazil in recent years.

2.3.3 *Food Colours.* This is a smaller market (Brazil, as an illustration, imports $ 1 million per year of natural dyes); however stable or growing due to the banning of most synthetic dyes from food products. Four immediately exploitable pigments, whose production technology is under investigation by POEMA, a natural product technology group, at the Federal University of Pará, are listed below. There are other products less technically developed, such as, for instance, a native indigo. *Bixin / urucu / achiote,* a natural carotenoid acid ester, is used as a skin dye by the native indians for centuries and generally regarded as safe. It is under heavy competition abroad, and the production of the bright red-orange pigment *without the utilisation of organic solvents or alkali* is necessary to gain a market share. This process is already available from the Engineering Research and Postgraduation Department (COPPE) of the Federal University of Rio de Janeiro. The dissemination of an additive-free "achiote," without the possibility of synthetic contaminants, can create a market for this product. *Carajuru,* a flavonoid quinone-type red pigment derived from the fermented leaves of *Arrabidaea chica,* a climbing plant easily cultivated, was marketed in the past under the name "chica red." Since the source plant is one of the most often used in the Amazon (especially for skin rash but also for several conditions which require antibiotic and anti-inflammatory action), the pigment could have a special value for cosmetics (lipstick) and disinfecting soaps. *Genipapo* is a fruit which contains the iridoid monocyclic terpene genipin. This compound reacts with natural mono-amines such as amino acids to generate a stable blue dye (a bisaza-azulene). On the indians' skin, the pigment turns black probably due to some other, concomitant reaction. Its utilisation by the natives indicates its lack of toxicity and the development of this product for the cosmetics industry is feasible. Genipapo is a relatively abundant, easy-to-grow fruit tree. *Turmeric* (*Curcuma longa* L.). This yellow pigment, which owes its colour to diferuloylmethane or curcumin, is not native; the plant is Asiatic. The production of pigment from the rhizome therefore would face competition from Asiatic producers, and would lack the Amazonian character. However, the healing and insecticidal properties of curcumin justify investigating this pigment as an ingredient of medicinal soaps or other cosmetics. Its long-standing utilisation in curry and several other foodstuffs is strong evidence of its safety.

2.3.4 *Chicle.* In the past, before the growth of the synthetic rubber industry which presently meets the demands of the chewing gum market, the Adams Company, at least in Brazil, harvested its chicle in the north of the State of Pará by bleeding the trunk of the tree *Manilkara bidentata*. There is perhaps a possibility of restoring this market for natural chicle; at any rate, there is an embryonic market for the related *sorva* (*Couma utilis*) and edible latexes from a number of latex yielding trees, exploited by the NGO Vitória Regia, Manaus.

2.3.5 *Vitamins and Provitamins.* Three products are outstanding in this category: *provitamin A* or *β-carotene* from the already-mentioned buriti, tucuman and piquiá oils[5]; *vitamin C*, of which the Amazon has the richest natural source - camu-camu, and *vitamin E - tocopherol* and *tocotrienols*, present in plant oils and evidently responsible for the conservation of unsaturated fatty acids and carotenes which make up the main components of fruit pulp oils. The three sources of provitamin A are sufficiently abundant to preclude an immediate cultivation requirement. Harvesting can be organised among communities and processing does not fundamentally differ from that of African palm oil, well known in the region. *Camu-camu* as a source of vitamin C was the object of an introduction attempt in Florida, USA in the past; however, it could be produced with relative ease in western Amazon, in areas under seasonal flooding which are useless for any other type of cultivation[6]. Its production could possibly be associated with fish culture -- another food product where demand outstrips supply. The sources of vitamin E will have to be scientifically assessed as a first step. The production of natural vitamin E from plant oils is a known industrial process; however, a plant oil can be directly used in some applications, thus preserving a natural formula which may contain synergistic elements in its composition.

2.3.6 *Special nutrients.* The Amazon people is a great consumer of açaí juice, which is said to possess valuable nutrient properties, for example, to combat anaemia. The juice is extracted from the pulp of the fruit which has a 6 months season. The açaí palm tree has suffered predatory exploitation for its palmito, but since it sprouts back very easily, recovery is not difficult. Another group of palm trees with nutritional properties is the *Astrocaryum* and *Jessenia* spp. group, among feature which *bacaba* and *pataná.* Patauá juice was shown to be practically equivalent to human milk, both in terms of fat and protein content and in terms of the amino-acid composition of the latter. Production requires crushing and some type of spray-drying system to give a powdered "milk."

2.3.7 *Other Food Additives.* There seems to be scant industrial or technological initiative to develop markets for antioxidants, food preservation agents, thickeners or gums, emulsifiers or wetting agents, sweeteners or bitters available from Amazonian plants. Thickeners and gums are perhaps the easiest to develop in the short run. One bitter agent, quassia (see above, insecticides) was marketed in the past to the beverage industry.

2.3.8 *Fish.* The Amazon basin is capable of producing fish by replenishing the harvested species. Replenishment techniques are known and were developed in the Amazon for some economically feasible species by INPA, Manaus. There are several areas which are adequate for this activity within private properties, for example in the Tefé River, but the repopulating of the main rivers deserves attention for the benefit of riverine populations. The foreign market for Amazonian fish is very large and there are established means for harvesting, transport and distribution.

3 CONCLUSION

The industrial development of Amazonian natural products is perfectly feasible, would bring economic betterment to a large and increasing rural population, would satisfy a repressed demand in the world market and would at the same time provide a means of preserving and restoring the native forest of the Amazon basin.

REFERENCES

1. B. Gilbert in P. R. Seidl, O. R. Gottlieb and M. A. C.Kaplan, editors 'Chemistry of the Amazon', ACS Symp. Ser., **588**, Amer. Chem. Soc., Washington, 1995, p. 20 and citations therein.

2. Ch. von Szczepanski in L. Crombie, editor 'Recent Advances in the Chemistry of Insect Control', II, Roy. Soc. Chem., Cambridge, UK, 1990, p. 1.

3. P. Grenand, C. Moretti and H. Jacquemin, 'Pharmacopées Traditionelles en Guayane', ORSTOM, Paris, 1987, pp.289 and 399.

4. M. J. Balick, Econ. Bot., 1981, **35**, 261.

5. W. B. Mors and C. T. Rizzini, 'Useful Plants of Brazil' Holden Day, S. Francisco, 1966, p. 28.

6. C. M. Peters and A. Vasquez, Acta Amazonica, 1986/87, **16/17**, 164.

Subject Index